Designing Faultless Mechanical Products Based on Advanced Reliability Analysis

Serving as a foundational guide to the field, this book is an essential resource for professionals involved in the development of mechanical systems and related industries. It provides a comprehensive introduction to mechanical systems design and offers a detailed overview of the current advancements in mechanical systems design, as well as insights into future trends and applications.

Designing Faultless Mechanical Products Based on Advanced Reliability Analysis focuses on the mechanical system based on mechanism and parametric accelerated life testing (ALT). It describes basic reliability concepts, parametric ALT plans, failure mechanism and design, and reliability testing with acceleration factor and sample size equation. A generalized life-stress failure model with a new effort concept has been derived and recommended to calculate the acceleration factor of the mechanical system, which is discussed in the book and should help companies to improve product reliability and avoid recalls due to product/structure failures in the field. As the faulty designs of mechanical products are experimentally identified by this new reliability design method—parametric ALT—the mechanical/civil engineering system might improve in reliability by the increase in lifetime and the reduction in failure rate.

This book offers a thorough exploration of design concepts for professional engineers, college- and university-level lecturers, researchers, and design managers specializing in reliability. It is also beneficial for senior-level undergraduates and graduates studying mechanical, industrial, and civil engineering. Companies can utilize this resource to enhance product reliability and prevent recalls resulting from product or structural failures.

Advanced Research in Reliability and System Assurance Engineering

Series Editor: Mangey Ram, Professor, Graphic Era University, Uttarakhand, India

Social Networks: Modeling and Analysis
Niyati Aggrawal and Adarsh Anand

Operations Research: Methods, Techniques, and Advancements
Edited by Amit Kumar and Mangey Ram

Statistical Modeling of Reliability Structures and Industrial Processes
Edited by Ioannis S. Triantafyllou and Mangey Ram

Industrial Reliability and Safety Engineering: Applications and Practices
Edited by Dilbagh Panchal, Mangey Ram, Prasenjit Chatterjee, and Anish Kumar Sachdeva

Reliability and Maintenance Modeling with Optimization: Advances and Applications
Edited by Mitsutaka Kimura, Satoshi Mizutani, Mitsuhiro Imaizumi, and Kodo Ito

Intelligent Prognostics for Engineering Systems with Machine Learning Techniques
Edited by Gunjan Soni, Om Prakash Yadav, Gaurav Kumar Badhotiya, and Mangey Ram

Statistical Methods and Applications in Systems Assurance and Quality
Edited by Sotiris Bersimis, Polychronis Economou, and Athanasios Rakitzis

Designing Faultless Mechanical Products Based on Advanced Reliability Analysis
Seongwoo Woo

For more information about this series, please visit: https://www.routledge.com/Advanced-Research-in-Reliability-and-System-Assurance-Engineering/book-series/CRCARRSAE

Designing Faultless Mechanical Products Based on Advanced Reliability Analysis

Seongwoo Woo

CRC Press
Taylor & Francis Group
Boca Raton London New York

CRC Press is an imprint of the
Taylor & Francis Group, an **informa** business

Designed cover image: Seongwoo Woo

First edition published 2026
by CRC Press
2385 NW Executive Center Drive, Suite 320, Boca Raton FL 33431

and by CRC Press
4 Park Square, Milton Park, Abingdon, Oxon, OX14 4RN

CRC Press is an imprint of Taylor & Francis Group, LLC

ISBN: 978-1-032-58657-1 (hbk)
ISBN: 978-1-032-58785-1 (pbk)
ISBN: 978-1-003-45147-1 (ebk)

DOI: 10.1201/9781003451471

Typeset in Palatino
by Apex CoVantage, LLC

Contents

Preface

To live a comfortable life, customers seek to use reliable products that perform well and are easy to use. Over the course of the 20th century, new sophisticated mechanical products or structures such as automobiles, airplanes, refrigerators, space shuttles, rockets, and bridges were designed and constructed through product design processes. The design process may be normally broken down as follows: 1) define the engineering problems, 2) evolve the product—prototype, design, and testing, 3) manufacturing. In a mechanical product, power transmission is utilized to create mechanical advantages via forces and movement by adopting appropriate mechanisms.

Due to competitive situations in the global marketplace, manufacturers must yearly design the newly developed features. After being integrated into the product's specifications, these newly discovered attributes are rapidly developed, combined into the system, and shipped to the marketplace. Under these circumstances, there may be an inadequate period for engineers to test new features in the limited period provided for scientific tests. If a significant number of failures occur in the marketplace due to design problems, the manufacturer's brand and business may be influenced negatively. To stop these market failures, reliability testing such as parametric ALT should be carried out before the product releases. Though the research of reliability engineering—a considerable quantity of ideas, hypothesis, and written work on reliability—has been intensified to stop the disasters of the past century, the reliability design of product has become more complicated because mechanical products demand higher performance and yearly cost reduction. An up-to-date book for bringing attention to the new structured method of reliability design is still in demand.

Mechanical fatigue explains over 70–80% of constructional failures in machine components and load-bearing parts. These failures frequently begin as cracks from stress concentrations such as notches, sharp edges, thin surfaces, grooves, etc. and grow when exposed to repeated (impact) loads. For example, random vibration and internal design defects in an engine was suspected to have caused the crashes of Boeing 737 MAX airplanes, which used CFM International LEAP-1B engines with 12% less power consumption and 7% lighter weight, compared with the sizes/specs of other models. Notable work focuses on low-cycle fatigue in the turbine engine. Fatigue may be governed by the mean stress, cyclic stress amplitude, or stress proportion, R ($=\sigma_{min}/\sigma_{max}$), and indicated as the portion of the minimum cyclic stress to the maximum cyclic stress as a reverse of the accelerated factor (AF) in the parametric ALT. Recent diffusion and transport studies in the area of quantum mechanics and statistical physics give engineers a new perspective and help them understand how fatigue due to loads materializes from nano-size imperfections that are not outlined by the strength of materials based on stress–strain concepts and current methodologies, including typical thermodynamics and physical chemistry.

Due to the repeated appearance of disasters for newly designed products, reliability has become a steadily more critical element to contemplate due to price, competitiveness, public demand, and adaptation of new technology. The most successful way to prevent a reliability disaster is to employ the reliability-embedded design process involving its structured method in parallel with the established design process. Moreover, to reduce such failures due to any constructional defects in future introductions of the system, novel systematic methods, such as parametric ALT, may be required to improve reliability, which evaluates the product's anticipated lifetime and conducts multiple redesigns utilizing a logical approach.

From the viewpoint of economics, manufacturers must watch their quality cost, even given the possibility of a catastrophic failure in the product's anticipated life due to the presence of Product Liability Law in the marketplace. Based on system engineering, all mechanical products, from tires to electric parts, are systemically created from the construction and may be inclined to deteriorate or fail suddenly by periodic (random) loads exerted on the design defect (weakest link). Namely, the mechanical product can eventually be cracked and fractured due to fatigue that may be caused by stresses (or loads) applied at regular intervals in its life. When products are exposed to repeated (random) loads, they start to crack due to design flaws, spread, and rupture. If failure occurs, the product may no longer satisfy the established workability specifications. To prevent product failure over its lifetime, a product should be designed by means of reliability testing to robustly endure a number of different loads for its life span.

The major motivations for writing this book are to 1) describe the evoluting necessity for the reliability-embedded design process and its organized design method—parametric ALT and 2) stress design based on quantum mechanics. These structured reliability procedures incorporate 1) an ALT scheme on a product BX lifetime, 2) a vibrational load modeling to reproduce the field phenomena, (3) an ALT with alterations, and 4) a judgment of whether the designs attain the identified BX lifetime. It creates the reliability quantitative test specifications (RQ) in a manner conforming with the reliability target. A parametric ALT approach thus evaluates the reliability of the product when exposed to repeated stresses. This organized series of steps will be more widely relevant to mechanical systems, such as automobiles, airplanes, appliances, and construction machinery, that require power transmission. A quantum-conveyed failure model and sample size expression built to RQ specifications are supplied to replicate such a failure in real time. Numerous examples of applying this series of steps are also presented in Chapter 8.

This book consists of nine chapters. Chapter 1 provides the current features and the necessity for reliability engineering given the progress of current technology. It discusses historical reliability disasters and their basic causes. Chapter 2 describes the most critical basic definitions of statistics and probability theory, the mathematical prerequisites of reliability engineering, and the most notable outlooks of reliability engineering as evolved in the past century. It will help to clarify the fundamental concepts of reliability methodologies that are examined in Chapter 8. Chapter 3 through Chapter 6 supply mechanism design, load analysis, mechanical system design (strength and stiffness), and mechanical product failures and quantum mechanics. Chapter 7 explains the basic concepts of parametric ALT, which is the main part of this book. Chapter 8 also supplies the case investigations that are practical in a number of different engineering areas. Chapter 9 presents the future features of parametric ALT, which will evolve as system engineering in a mechanical engineering context.

Author

Seongwoo Woo earned BS and MS degrees in mechanical engineering and earned a PhD in mechanical engineering at Texas A&M University, USA. His major interests include energy systems, such as HVAC and heat transfer, the optimal design and control of refrigerators, the reliability design of mechanical components, and failure analysis of thermal components in the marketplace using nondestructive methods, such as scanning electron microscopy and X-rays. In particular, he has developed parametric accelerated life testing (ALT) as a new reliability methodology, which determines whether there is a design fault in a mechanical system that is subjected to repetitive stress. Using parametric ALT, engineers can find design faults before the product is launched and thus avoid failures.

From 1992 to 1997, Dr. Woo worked in the Agency for Defense Development in Chinhae, South Korea, where he was the researcher in charge of the development of naval weapons systems. From 2000 to 2010, he worked as a senior reliability engineer in the Side-by-Side Refrigerator Division, Digital Appliances, Samsung Electronics, where he focused on enhancing the lifetimes of refrigerators using parametric ALT. He is currently a professor in the Mechanical Department, Ethiopian Technical University, Addis Ababa.

1

Introduction to the Necessity of a Structured Design Method

1.1 Introduction

People want to live wealthy and diversified lives. New capabilities and technologies not available before the 20th century—mechanical systems such as automobiles, airplanes, refrigerators, etc.—have evolved in modern society. To sustain a competitive business position in the marketplace, a product must be designed to serve numerous (intended) functions and satisfy a variety of customer needs by means of superior performance, ease of use, and reliability: 1) quick response, 2) high-performance control, 3) no noise, 4) precise control over a wide frequency band, 5) long life, 6) compactness and transportable weight, 7) good reliability, 8) low price, 9) sophisticated hardware design, 10) contamination resistance, and 11) high energy efficiency. Mechanical products with the well-devised designs can provide enormous advantages to customers in the field (Figure 1.1) [1].

However, after inclusion in the product's design specifications, these newly developed features are swiftly developed, combined into the system, and shipped to the global market. Under the severe circumstances to which a product may be subjected and situated by customers, there may be insufficient time to test those features in the limited time provided for scientific tests. Thus, after some period, the product can reveal potential field issues—and recalls—under customer usage or environmental conditions due to structural defects. Consequently, the manufacturer's brand and market share can be influenced negatively in the field. As stated by an investigation on components failure performed by the U.S. Navy, 43% of the failures in product were ascribed to design, 30% to functioning and maintenance, 20% to manufacturing, and 7% to numerous other factors. Although the design cost bears only 5% of the whole product price, it is known that the cost effect in a recall impacts 70% of the real quality cost (Figure 1.2). In the early design stage, just $1 is required to identify and modify a design flaw. On the other hand, the cost might grow to $10 after the design phase, $100 at the beginning of manufacturing, and $1,000 during mass production.

DOI: 10.1201/9781003451471-1

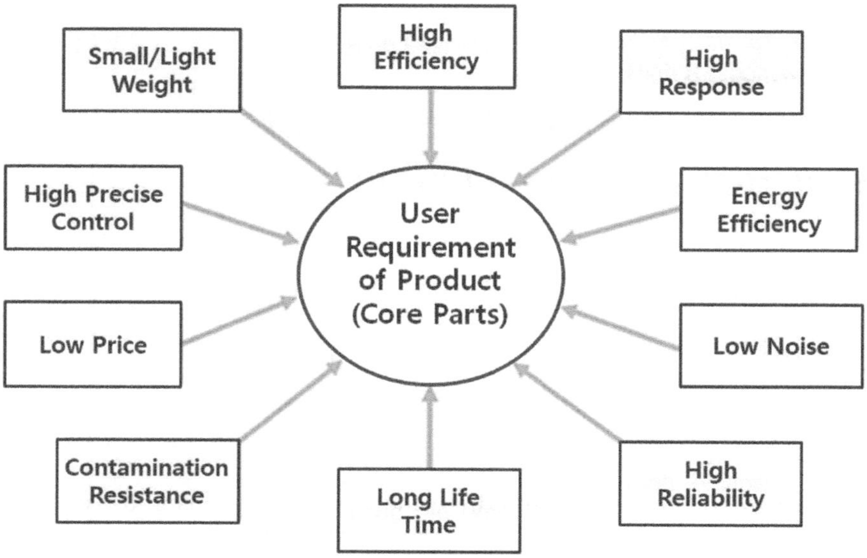

FIGURE 1.1
Customer needs of a product.

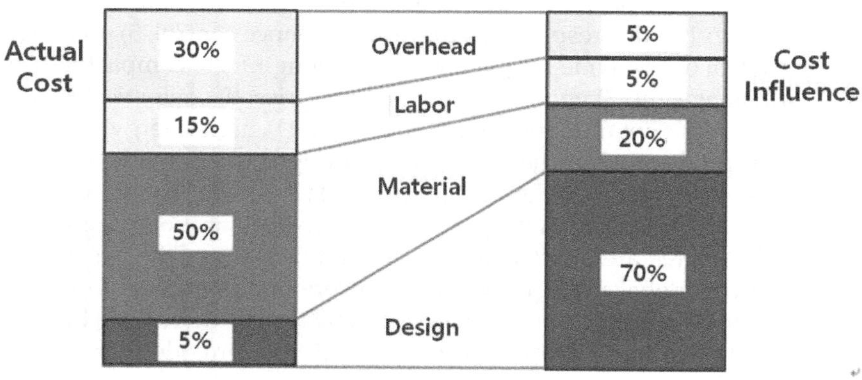

FIGURE 1.2
Leverage in product design.

In the field, machines, operating structures, and their elements are exposed to countless various cyclic, static, or dynamic loads. After the accumulation of damage in the structure, mechanical failure—fatigue—is found and develops as a type of crack. Depending on the type of material, the phenomena of a fatigue fracture will be differently propagated on its surface and observed from the field samples. Owing to recent quantum mechanics, fatigue, fracture, creep, or environmentally assisted cracking, among other critical and subcritical processes, are still progressing and manifesting as open problems when using something like the quantum model to explain basic phenomena. To properly secure the safety and optimal durability of components in the structure, simulations and testing of a newly designed products are also required to resolve the outbreak of some factors—notches, the crack growth rate, voids, the impact of microstructure, the environment, etc. Attention is paid to the primary outlooks, which describe the material response at the atomic and microstructural levels and that can explain the integrity of a structural component in the developmental engineering steps.

The implementation of all these new features and technologies in product development influences a broad span of zones where structural safety is a crucial issue such as aircraft, nuclear power plants, automobiles, civil engineering constructions, oil and gas, naval constructions, pressurized appliances, etc. Structural failures in any of these types of product can clearly have significant results with regard to human lives, environmental disasters, or economic losses. To circumvent those structural failures, it is required to perceive the dissimilar failure mechanisms producing subcritical and critical processes in the constructional materials and to evolve evaluation skills and management steps for equivalent constructions. The product development period can vary from a few months to several years. The process depends on several factors, including the complexity of the product, available resources, methodologies used, market demand, technology trends, business strategy, and user feedback. As a multistep process in system engineering, cycles of product development include marketing, the formation of ideas, validation, prototyping, design development, launch, and improvement through market feedback. The development time of products like automobiles has been shortened from 65 to 24 months, while reliability has grown from 0.9 to 0.99 (Figure 1.3).

Due to such swift delivery, products with new features and technology may be recalled due to unexpected customers' situations. Only a few global manufacturers that engage in the high application of scientific knowledge for industrial purposes and their organizational structures could survive in the ruthless race among competitors. To satisfy field requirements such as reliability and developmental period, the manufacturer needs to have an organized, reliable method closely bound to its developmental procedure. However, no structured reliability way—embedded in product development process, such as parametric ALT, can assess the robustness of design on the reliability. After an engineer finds the design defects by means of parametric ALT, manufacturers might properly incorporate new designs into systems.

This book provides basic concepts and tools, such as parametric ALT, as a well-ordered reliable way to evaluate the design of mechanical systems in a reliability-oriented development process. A mechanical product uses power to apply the mechanical advantages from mechanisms. Consequently, a mechanical system is exposed to repeated stress. Because design defects eventually bring about insufficient stiffness (or strength) in the structure, the product may be unsuccessful sometime during its anticipated life. A quantum-transported failure model and sample size will be used for this process. The organized steps

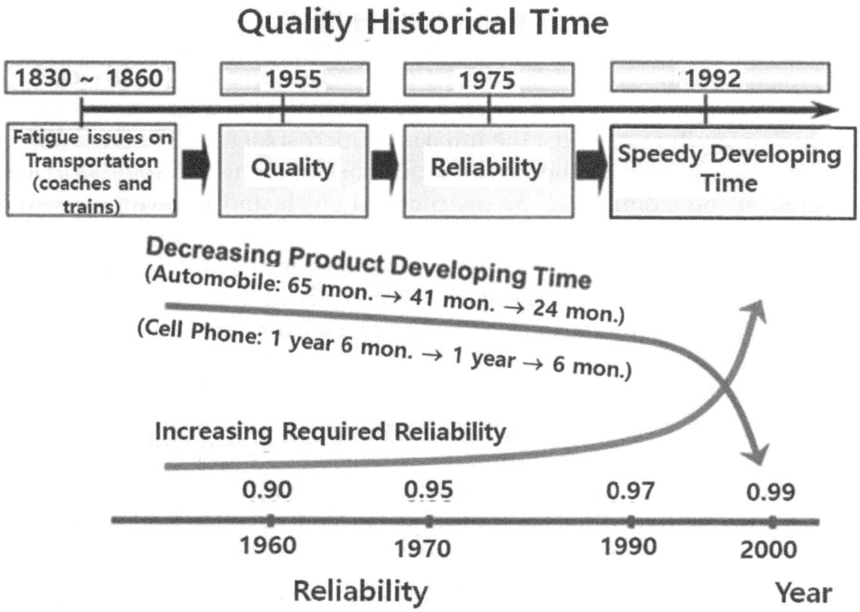

FIGURE 1.3
Contemporary trends for product quality.

for recognizing the sources of the failure mode and enhancing reliability are explained in order to inspect the proper design solutions and correct failure in the field: 1) an ALT plan created on a BX lifetime, 2) a load examination, 3) ALTs with the alterations, and 4) an evaluation of whether the design(s) fulfills the intended BX lifetime. To explain the design of the mechanical product, a broad understanding is required of probability/reliability ideas, many mechanisms, load examination, strength design, failures based on quantum mechanics, etc.

1.2 Design for Mechanical Product Such as a Refrigerator

Mechanical products such as a heat engine or refrigerator transfers (generated) power to carry out a task that needs forces and movements [2]. It ultimately causes mechanical advantages by properly adopting the mechanisms. The word "mechanical" comes from the Latin word *machina*, which may be described as being attached to machine. A mechanical system is formed from 1) a power source and some actuators that create forces and movements, 2) organized mechanisms that devise the actuator input to fulfill a specified execution of output forces and movements, and 3) a controller with sensors.

1.2.1 Design Process in Mechanical Engineering

The design procedure of a mechanical product is the act of devising a scheme to help a mechanical engineer in generating a new system. The fundamental ideas of mechanical engineering are intended to change resources in the most favorable way to satisfy an expressed customer requirement based on market investigation. The development of a mechanical product is a complete procedure: 1) identify the consumer requirements from the field, 2) construct the conceptual and physical design, 3) build a model (or prototype), 4) produce a demonstration or design product, and 5) assess the final product (Figure 1.4).

1.2.1.1 Feasibility Evaluation

The motive of a practicability evaluation is to decide whether the engineer's project may proceed to the design stage. The engineer questions whether this design project is constructed according to two criteria: 1) a successful engineer plan and 2) cost restrictions.

1.2.1.2 Setting Up the Design Requirements

The design requirements are an expression of an intended function that states how it might do or what it might carry out. There are the two kinds of needs: 1) practical needs—i.e., what the product might do?—and 2) nonfunctional needs—i.e., what limitations are there on the product and its evolution—data, user, usability, and environment?

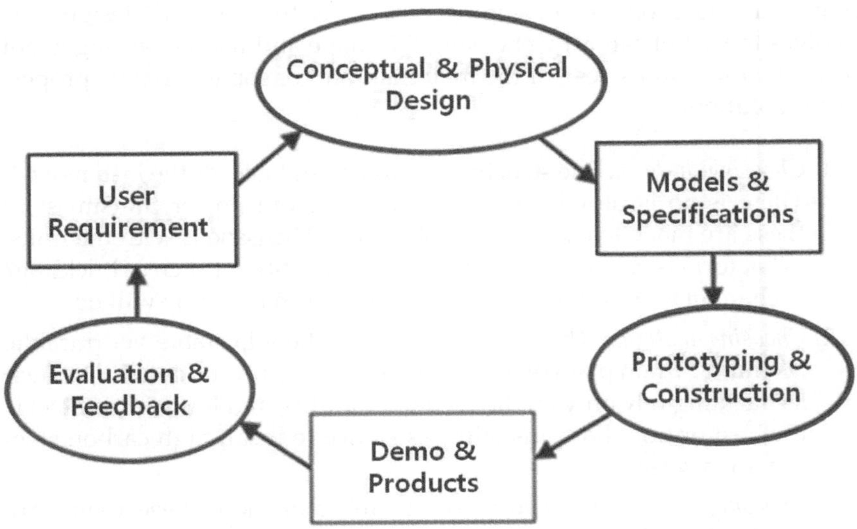

FIGURE 1.4
Iterative design summary of mechanical engineering.

1.2.1.3 Preliminary Design

In the introductory phase, from the specifications, engineers produce a design. They make an engineering prototype and its related design before generating engineering drawings and documents. Thus the preliminary design extends across the gap between the initial ideas and the detailed design. If the all-inclusive system arrangement in assigning a piece of work is stated, it may supply schematics, diagrams, and layouts as the early project arrangement. In the preliminary design, the engineer concentrates on creating the common substructure of the product.

1.2.1.4 Conceptual Design

The conceptual design is a stage of the project draft used to express how a new product may function and satisfy the demands in the specifications. It covers creating plans and considers the advantages and disadvantages of executing them. The engineer in this stage must recognize the troubles—the external load, manufacturing methods, limitations with regard to the shape and relative arrangement of the parts, etc. The project is intended to reduce the chance of mistake, to manage costs, to evaluate risks, and to assess the possible success of the planned project. As an engineering matter is stated, potential answers should be recognized. These answers may be discovered by means of an intellectual process—generating inventive ideas, recognizing and defining an issue, creating and assessing possible answers, and choosing the most promising solution.

1.2.1.5 Detailed Design

Central in the product developing process is the detailed design, which involves the act of carefully choosing the shape and relative arrangement of the parts (form and sizes), of the materials, and of some dynamic properties for the solution.

1. *Choosing the form*: The structure should be suitable for the load moved. There is no issue with the screw and pins in power transmission; these are made to conform to a standard. The issue is with the links. What form is effective for a compressed constituent such as buckling? A channel with roll-formed, double flat bar in the array will do.
2. *Choosing material*: The power screw may be adjustable yet durable. Medium carbon plain or alloy steel might be a fine option. Links such as buckling rely only on the section modulus, and low carbon steel is the best option. For pins with resistance to wear, high carbon steel is a good selection.
3. *Choosing dimensions*: The method differs on how these issues are resolved. If all the dimensions are given, the actual stresses in an element are controlled.

1.2.2 Mechanical Products

1.2.2.1 Refrigerator (or Air Conditioner)

To stock frozen (or fresh) food, cooled air in a refrigerator is provided from the evaporator to the refrigerator and the freezer sections. To satisfy these requirements, a domestic refrigerator employs vapor-compression refrigeration (VCR) cycle and can lower the temperature of a foodstuff to less than that of its environment. For transferring heat from the low to the high temperature, the VCR cycle applies some work to eliminate heat, expressed as the cooling load from a low-temperature environment in the refrigerator. The capacity can be expressed as the amount of refrigeration is similar to the heat necessitated for melting 1 ton of ice daily. To create a continual cooling effect, the VCR cycle in a refrigerator works at two pressures—a condensing (high) level and an evaporating (low) level. The main parts are the 1) compressor, 2) condenser, 3) capillary tube, and 4) evaporator (Figure 1.5).

There are two types for the VCR cycle: 1) vapor-compression refrigerating systems and 2) vapor-absorption refrigerating systems. The main implementations are 1) central air conditioning, 2) food storage in refrigerator, 3) ejection of heat in the zones of chemical processes, and 4) cryonics projects.

Another product— the air conditioner—functions by means of a VCR cycle, like a refrigerator. To increase the comfort of residents, air conditioning consists of a series of steps that remove the heat and moisture from the interior of an inhabited area. That is, the aim of a heat pump is to remove the heat to a high-temperature space, expressed as the heating load. Some of the applications for this utilization are 1) improving human comfort in hotels, offices,

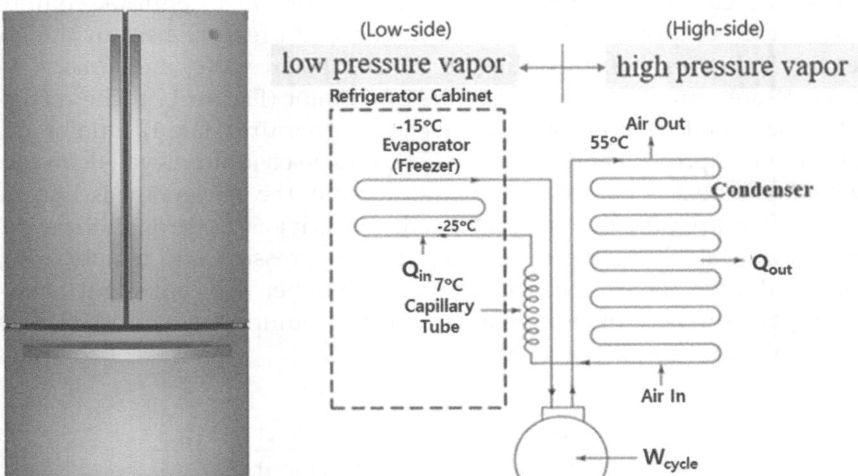

FIGURE 1.5
Refrigerator and its diagram.

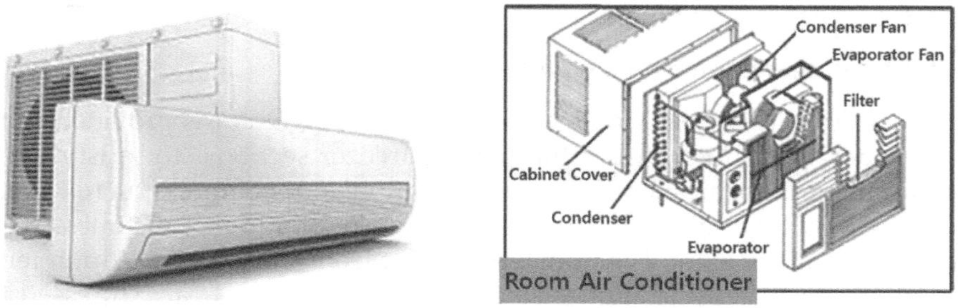

FIGURE 1.6
Typical air conditioner.

cars, buses, etc. and 2) food processing, pharmaceuticals, and machine tools in industries, chemicals, or printing.

For enhancing human comfort in spaces, presume that a room regularly remains at a temperature of 25°C. In the air conditioner, the room air is dragged by a fan and moved through a cooling coil, the surface of which is kept at a temperature of 10°C. By moving through the coil, the air is chilled (e.g., to 15°C) before being released into the room. As the room heat is absorbed, the air is moved through the cooling coil at 25°C (Figure 1.6).

1.2.2.2 Vapor-Compression Refrigeration (VCR) System

The VCR cycle consists of four processes: 1) *Process 1–2*: isentropic compression in the compressor, 2) *Process 2–3*: constant-pressure heat in the condenser, 3) *Process 3–4*: throttling in an expansion valve, and *Process 4–1*: constant-pressure heat is added to the evaporator (Figure 1.7). The refrigerant is sucked into the compressor inlet at low pressure (state 1). It flows away from the compressor and comes to the condenser at an elevated pressure (state 2). As heat is passed to the environment, the refrigerant is liquefied (state 3). The pressure of the liquid is reduced as it passes through the expansion valve (state 3). As the refrigerant at a low pressure and temperature is converted into vapor in the evaporator, the chamber will cool down (state 4).

The performances of refrigerators and heat pumps are expressed as the coefficient of performance (COP):

$$\mathrm{COP}_R = \frac{\text{Desired output}}{\text{Required input}} = \frac{\text{Cooling effect}}{\text{Worked input}} = \frac{\dot{Q}_\mathrm{L}}{W_{\mathrm{net,in}}} \tag{1.1a}$$

$$\mathrm{COP}_{HP} = \frac{\text{Desired output}}{\text{Required input}} = \frac{\text{Heating effect}}{\text{Worked input}} = \frac{\dot{Q}_\mathrm{H}}{W_{\mathrm{net,in}}} \tag{1.1b}$$

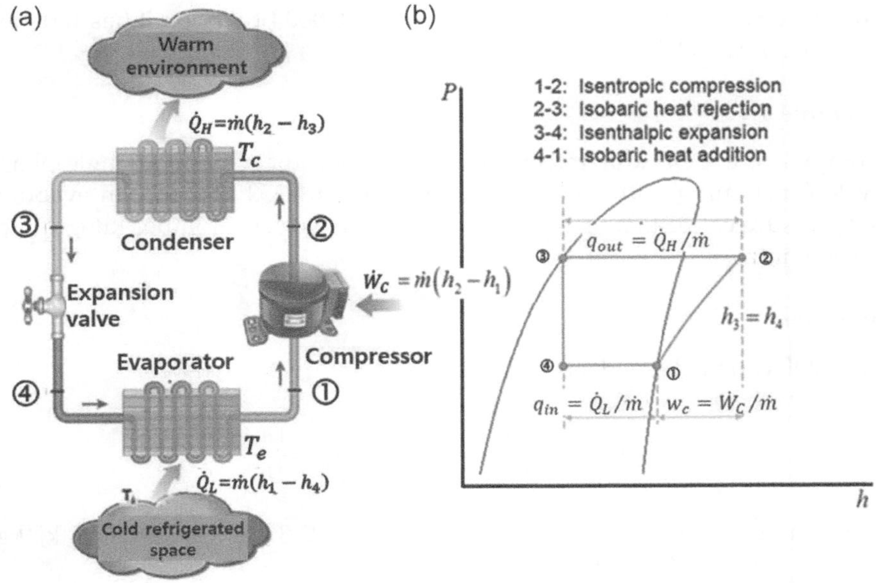

FIGURE 1.7
VCR cycle: (a) refrigeration cycle; (b) P-H diagram.

Both COP_R and COP_{HP} will be larger than 1. Under similar functioning circumstances, the COPs are defined as:

$$COP_{HP}=COP_R+1 \tag{1.2}$$

The cooling coefficient of performance (COP_C) is defined as

$$\eta_E=COP_C=\frac{\dot{Q}_{in}}{W_{cycle}}=\frac{\dot{m}(h_1-h_4)}{\dot{m}(h_2-h_1)}=\frac{(h_1-h_4)}{(h_2-h_1)} \tag{1.3}$$

where $\dot{m}$ = mass flow rate; compressor work: $W_{cycle}=\dot{m}(h_2-h_1)$; refrigeration effect: $\dot{q}_{in}=h_1-h_4$; and refrigeration capacity: $\dot{Q}_{in}=\dot{m}(h_1-h_4)$.

We notice that η_E will be >1. Thus a refrigeration system for cooling is called a refrigerator.

Cooling capacity is expressed as the quantity of heat lowered relative to the surrounding environment, which is defined in tons. This is identical to the heat necessitated for the daily melting of 1 ton of ice. One ton of refrigeration is the steady-state heat transfer rate necessitated to melt 1 ton of ice at 0°C (32°F) per day. Namely, it may be defined as

$$1\text{ ton}=144\text{ Btu/lb}\times 2{,}000=288{,}000\text{ Btu/day}=12{,}000\text{ Btu/hour}\,(=3.516\text{ kW})$$

For instance, if a system has a capacity of 1,200,000 Btu/hour, it has 100 tons (= 1,200,000/12,000).

Example 1.1

Compute the COP and refrigerating efficiency for a perfect single-phase cycle functioning with a condenser pressure of 1,253 kPa and an evaporator pressure of 201 kPa using R-134a. The cycle work is 31 kJ per kilogram of refrigerant.

Solution

The COP can be defined as

$$\eta_E = COP_C = \frac{\dot{Q}_{in}}{W_{cycle}} = \frac{\dot{m}(h_1 - h_4)}{\dot{m}(h_2 - h_1)} = \frac{(h_1 - h_4)}{(h_2 - h_1)}$$

From the thermodynamic table, because h_1 = 392.8 kJ/kg, h_4 = 268.5 kJ/kg, $W_{cycle} = 31$ kJ/kg, COP_C is

$$\eta_E = COP = \frac{\dot{Q}_{in}}{W_{cycle}} = \frac{(392.8\,kJ/kg - 268.5\,kJ/kg)}{31\,kJ/kg} = 4.0$$

A Carnot cycle working between the saturation temperatures in a manner conforming with the defined condenser and evaporator pressures has a COP as follows:

$T_e = -10 + 273 = 263$ K

$T_c = 48 + 273 = 321$ K

Thus we have

$$COP_C = \frac{263}{321 - 263} = 4.5$$

The refrigerating efficiency is

$$\eta_R = \frac{COP}{COP_C} = \frac{4.0}{4.5} = 0.89 = 89\%$$

Example 1.2

Refrigerant-134a is the working refrigerant in a conceptual VCR cycle. The fluid comes into the evaporator at –20°C and has a condenser pressure of 0.9 MPa. The mass flow rate is 3 kg/min. Find COP_R and $COP_{R,\,Carnot}$ for the same T_{max} and T_{min}, and the tons of refrigeration.

Solution

Using the refrigerant—R-134a Tables, we have

$$\begin{bmatrix} State\,1 \\ Compressor\ inlet \\ T_1 = -20^o\,C \\ X_1 = 1.0 \end{bmatrix} \begin{Bmatrix} h_1 = 238.41\dfrac{\text{kJ}}{\text{kg}} \\ s_1 = 0.9456\dfrac{\text{kJ}}{\text{kg}\cdot\text{K}} \end{Bmatrix} \begin{bmatrix} State\,2 \\ Compressor\,exit \\ p_{2s} = p_2 = 900kPa \\ s_{2s} = s_1 = 0.9456\dfrac{\text{kJ}}{\text{kg}\cdot\text{K}} \end{bmatrix} \begin{bmatrix} \begin{Bmatrix} h_{2s} = 278.23\dfrac{\text{kJ}}{\text{kg}} \\ T_{2s} = 43.79^o\,C \end{Bmatrix} \end{bmatrix}$$

$$\begin{bmatrix} State\,3 \\ Condenser\,exit \\ P_3 = 900\,kPa \\ X_3 = 0.0 \end{bmatrix} \begin{Bmatrix} h_3 = 101.61\dfrac{\text{kJ}}{\text{kg}} \\ s_3 = 0.3738\dfrac{\text{kJ}}{\text{kg}\cdot\text{K}} \end{Bmatrix} \begin{bmatrix} State\,4 \\ Throttle\,exit \\ T_4 = T_1 = -20^o\,C \\ h_4 = h_3 \end{bmatrix} \begin{bmatrix} \begin{Bmatrix} x_4 = 0.358 \\ s_4 = 0.4053\dfrac{\text{kJ}}{\text{kg}\cdot\text{K}} \end{Bmatrix} \end{bmatrix}$$

$$\begin{bmatrix} COP_R \dfrac{\dot{Q}_L}{\dot{W}_{net,in}} = \dfrac{\dot{m}(h_1 - h_4)}{\dot{m}(h_2 - h_1)} = \dfrac{h_1 - h_4}{h_2 - h_1} \\ = \dfrac{(238.41 - 101.61)\dfrac{\text{kJ}}{\text{kg}}}{(278.23 - 238.41)\dfrac{\text{kJ}}{\text{kg}}} \\ = 3.44 \end{bmatrix}$$

The tons of refrigeration, defined as the refrigeration effect, are

$$\begin{bmatrix} \dot{Q}_L = \dot{m}(h_1 - h_4) & COP_{R,Carnot} = \dfrac{T_L}{T_H - T_L} \\ = 3\dfrac{kg}{\text{min}}(238.41 - 101.61)\dfrac{kJ}{\text{kg}}\dfrac{1ton}{211\dfrac{kJ}{\text{min}}} & = \dfrac{(-20+273)K}{(43.79-(-20))K} \\ & = 3.97 \\ = 1.94\,ton & \end{bmatrix}$$

Another scale of the usefulness of the refrigeration cycle is how much input power to the compressor, in horsepower, is necessitated for each ton of cooling.

1.2.3 Short History of Refrigeration

Roughly 2,000 years ago, residents in Egypt produced ice by keeping water in porous pots outside the home at night. Roughly 1000 BCE, China used ice for chilling beverages. In the 4th century, East Indians produced ice by liquefying salt in water. In 1834, Jacob Perkins (American) developed a hand-powered refrigeration system employing ether as the refrigerant (Figure 1.8).

To cure people with high fevers by making ice, Dr. John Gorrie of Florida in 1851 attained the earliest American patent of a chilled air machine. David Boyle in 1872 initiated refrigeration that was operated by compressed ammonia (NH_3). Four years later, Carl Linde created the earliest efficient and good ammonia refrigerator. In 1904, a roughly 450-ton cooling machine was set for utilization on the New York Stock Exchange. In 1915, the earliest two-stage modern compressor was developed. In 1859, to make the first absorption refrigeration product, Ferdinand Carré, a French engineer, utilized ammonia's affinity for water (Figure 1.9).

Until the 1920s, the development of refrigeration products was restricted to developments in VCR systems and cold-air equipment. After the 1920s, there was great diversification the add-ons of refrigeration systems, managing to new developments such as thermo-electrics, vortex tubes, steam jets, pulse tubes, and centrifugal compression systems. In the 1930s, the most innovative thing was the refrigerant adaptation by Midgley (American), that is, chlorofluorocarbons (CFCs), or Freons.

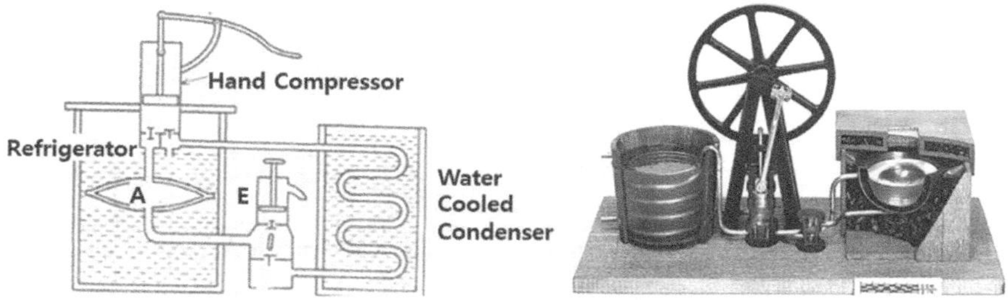

FIGURE 1.8
Hand-powered refrigeration machine (Jacob Perkins).

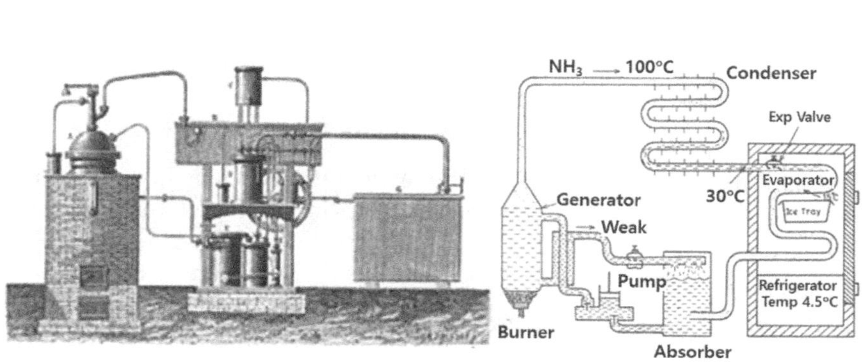

FIGURE 1.9
Vaporabsorption machine (Ferdinand Carré).

1.2.4 Refrigerant

A refrigerant in refrigeration is the working fluid. It functions in the evaporator (boiling) and the condenser (condensing) as the two phases, and it has the thermodynamic attributes of low-temperature-boiling fluids. Choosing refrigerant depends on the ability to remove heat, density, toxicity, cost, availability, etc. It is categorized as 1) main refrigerants and 2) secondary refrigerants.

Primary refrigerants are materials that function in compression, condensation, expansion, and evaporation in the VCR cyclic procedures. Ammonia, R12, R22, and carbon dioxide will cover in this category of refrigerants. The medium does not go through the cyclic procedure in a refrigeration system and is employed only as a means for heat transfer. This class consists of water, brine solutions of sodium chloride, and calcium chloride (Figure 1.10 (a)).

According to the American Society of Refrigerating Engineers (ASRE), definite conventions have been developed for employing dissimilar kinds of refrigerants. ASRE uses the three-digit nomenclature for this application. Thus the first digit represents the number of carbon atoms in the compound minus 1, the second digit stands for the number of hydrogen atoms plus 1, and the third digit stands for the number of fluorine atoms. The current atoms are still chlorine. For instance, there are refrigerants such as ammonia,

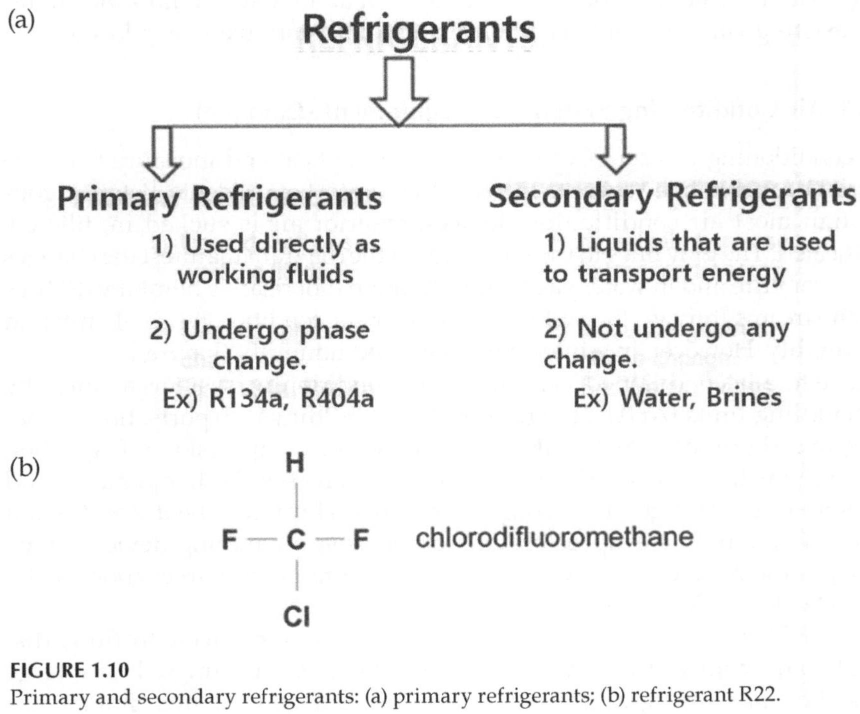

FIGURE 1.10
Primary and secondary refrigerants: (a) primary refrigerants; (b) refrigerant R22.

carbon dioxide, R12, R22, etc. Most refrigerants are halogenated hydrocarbons. The naming convention held by ASHRAE is defined:

$$\mathbf{R}(a{-}1)(b{+}1)d = \mathbf{C}_a\mathbf{H}_b\mathbf{Cl}_c\mathbf{F}_d \tag{1.4}$$

where $c = 2(a + 1) - b - d$.

Example 1.3

Compute the chemical formulation of refrigerant R22, which has three-digit nomenclature.

Solution

R22 (R022) is

Number of C atoms in R22: $a - 1 = 0 \rightarrow a = 1$,

Number of H atoms in R22: $b + 1 = 2 \rightarrow b = 1$,

Number of F atoms in R22: $d = 2$, $c = 2(a + 1) - b - d = 2\,(1 + 1) - 1 - 2 = 1$

Thus we may compute the chemical action of devising R22 (Figure 1.10 (b)).

Because there is one carbon atom, we know that the compound has started from the methane series (CH_4). From the action of the mathematical calculation, it can be seen that there is one hydrogen atom and two fluorine atoms. The existing valence bond of carbon shall still be stabilized by chlorine.

1.2.5 Air Conditioning Systems and Equipment (Example)

Air conditioning is a series of steps of removing heat and moisture from the interior of an inhabited space to increase the comfort of residents. Before it goes through most air conditioning devices, exterior air is sucked in, filtered, and heated. The gray lines in the lower part of the diagram manifest the changes in temperature and in water vapor concentration (not relative humidity (RH)) as the air streams through the system. In summer, air conditioning shall chill and dehumidify. However, in winter, it will heat and humidify (Figure 1.11).

To be totally conditioned, centralized air conditioning systems adapted by air handling units (AHUs) are used in large buildings, airports, hotels, theaters, and shopping malls. It also utilizes a vapor-compression refrigeration system, which is a form of compressor that increases the temperature and pressure of the refrigerant; a condenser, which eliminates heat wand is put in the system by the evaporator and compressor; a metering device, which manages the refrigerant movement to the evaporator; and an evaporator, by which heat is sucked up from the room.

An AHU is a piece of equipment that can be employed to flow, distribute, and control the conditioned air for HVAC buildings. It is a large metal box possessing a blower, heating or cooling elements, filter racks

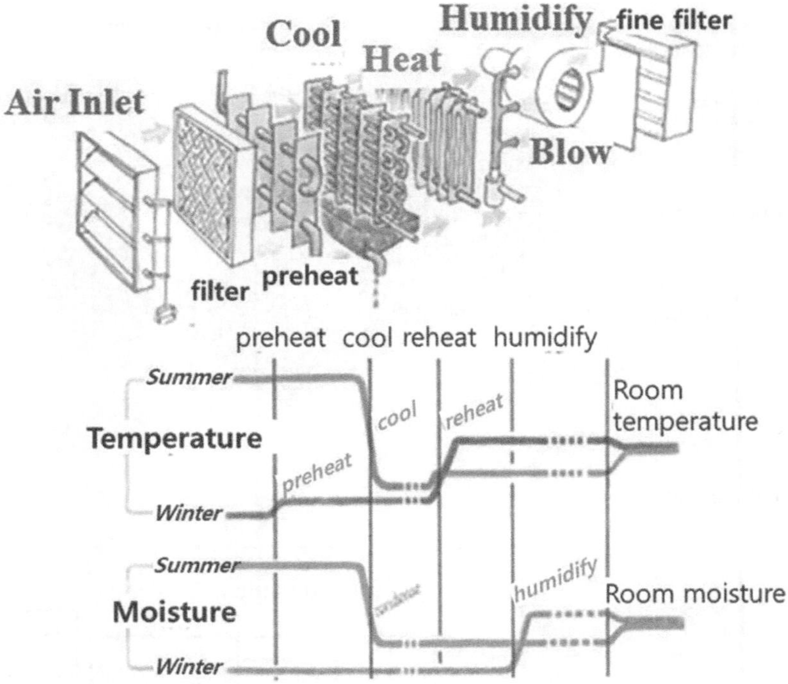

FIGURE 1.11
Total air conditioning.

or chambers, and humidity (45–55%) and temperature (19–23°C) control loops. The parts for an AHU are 1) a heating unit to warm the air to the managed temperature; 2) a cooling unit/dehumidifier to cool the air to the required temperature or to remove moisture from the air; 3) a humidifier to conduct the air to the managed humidity, if too low; 4) filters to eliminate particles of predetermined dimensions and/or microorganisms; 5) ducts to move the air; 6) a control damper to adjust the volume of air; 7) a silencer to lessen the noise created by air circulation, and 8) a weather louvre to stop insects, leaves, dirt, and rain from entering (Figure 1.12).

Centralized air conditioning systems can be all-air systems and air-and-water systems. An all-air system may be defined as a system providing the entire sensible/latent cooling capacity in the cooled air supplied by the system. It needs no additional cooling and falls into one of two categories: 1) A single duct system provides either cooling or heating using the same duct, but not both heating and cooling at the same time. Thus it has a constant volume that is operated by single-zone systems, a constant volume with multiple-zone systems, and variable volume systems. 2) Dual duct systems may provide both cooling and heating simultaneously. Thus they have a dual duct operated by constant volume systems and/or variable volume systems.

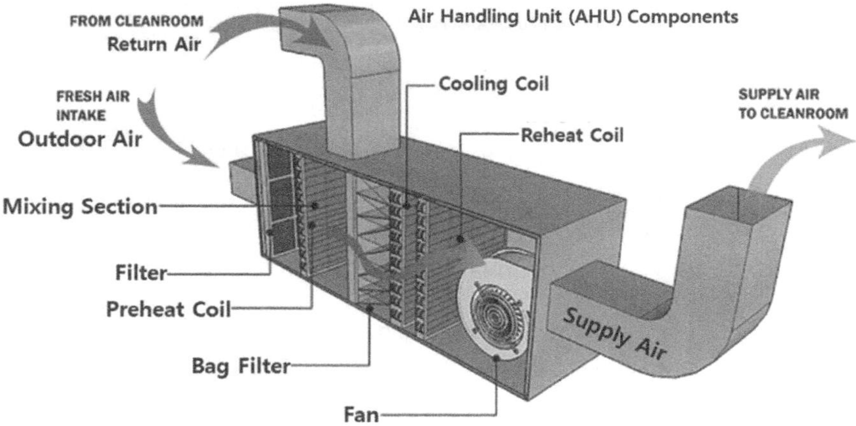

FIGURE 1.12
Air handling unit.

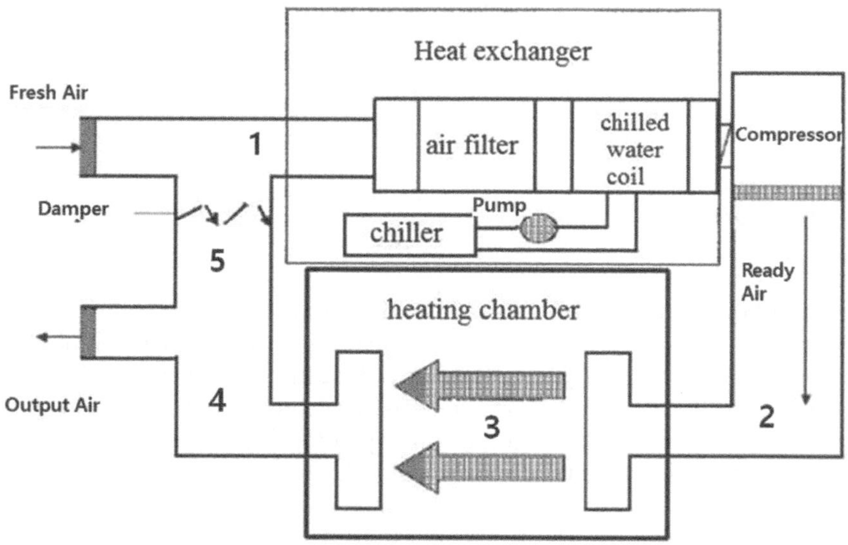

FIGURE 1.13
Constant volume, single zone system.

For example, a single-zone system is an uncomplicated shape of an all-air system, which has a single conditioner in the service of one temperature zone. It is relevant to small-scale department stores/single shops in shopping centers, single-class rooms, etc. (Figure 1.13).

As it is costly to devote a separate system for each zone in a big building, multiple-zone systems are utilized. For such instances, the basic control system idea is expanded to fulfill the cooling and heating requirements of the multi-zone. There are two kinds: 1) constant air volume systems with terminal reheat zones and 2) variable air volume systems with dual ducts (Figure 1.14).

As air is the medium that can be employed to stabilize the load, the engineer may choose to vary the size either by supplying a constant volume of air temperature or by adjusting the volume as the space load changes. Namely, a variable air volume (VAV) system may be relevant for inside or perimeter zones with common or separate fan systems, common or separate air temperature control, and with or without supplementary heating devices. The variable volume idea may be applicable to volume variation in the main system. Variations of flow under control of a space thermostat may be attained by placing an uncomplicated damper or volume adjusting device in a duct or a pressure reducing device at the terminal diffuser or grill.

To control the heating or cooling by the VAV system, a single stream of cool air services all the zones, and a thermostat in each zone modifies a damper to manage the flow rate of cooled air into the zone. The application is appropriate in an office building without heating loads and with only cooling loads (Figure 1.15).

The dual duct system controls all the air in the central equipment and passes it to the controlled rooms through two ducts. One duct passes cooled air, and the other passes warm air, thus providing air sources for both cooling and heating. In the controlled zone, in response to a room thermostat, warm/cold air is mixed in proportionate to attain the existing heat load of the space. The dual duct system responds rapidly to changes in zone load and may accommodate heating in some zones and cooling in others (Figure 1.16).

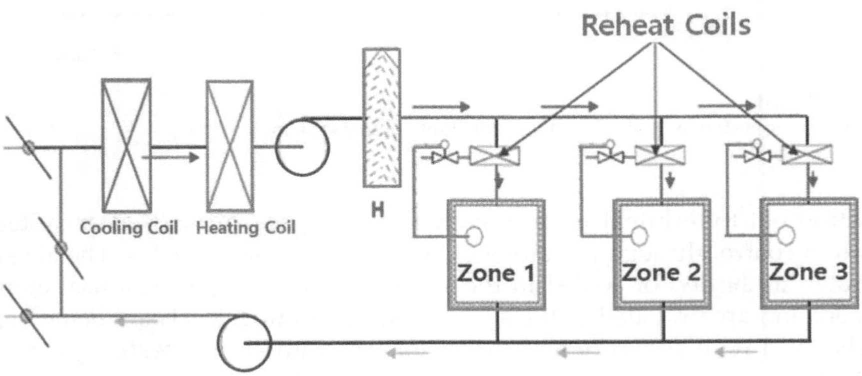

FIGURE 1.14
Multi-zone system with terminal reheat zone.

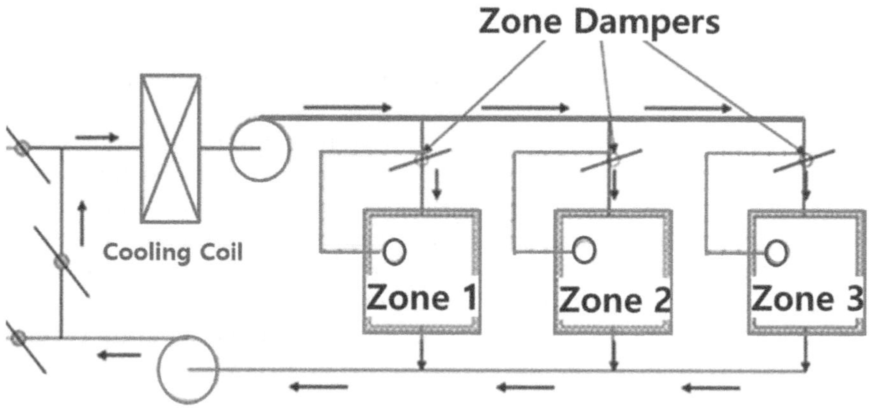

FIGURE 1.15
Multizone system with a single duct and a variable volume system.

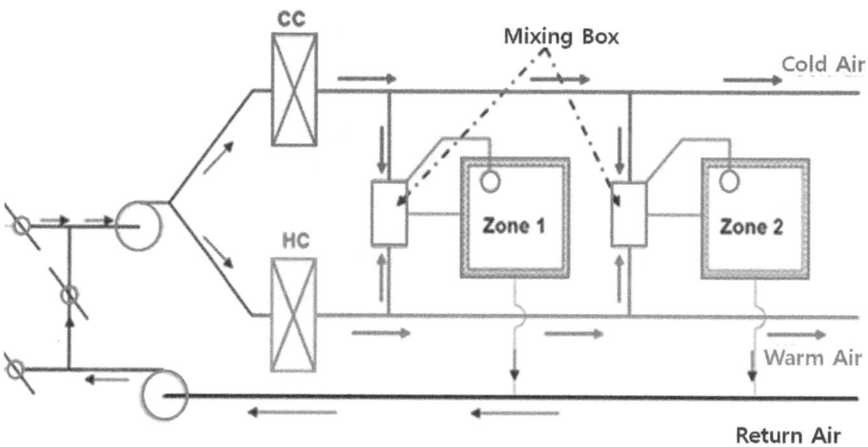

FIGURE 1.16
Multi-zone system with dual duct and constant volume system.

Run out to terminal units mounted in the space, air-and-water systems attain control through the usage of both air and water sources. The air and water are heated or cooled in the spaces of the central mechanical equipment and are provided to the space in which comfort levels are being kept. The most regular sort of terminal employed with air and water systems is the high-pressure induction unit. The air distributed to the induction unit is defined as primary air. It is supplied from the room of central mechanical equipment at high velocities between 10 and 24 m/sec. The water supplied to the induction unit is termed secondary water.

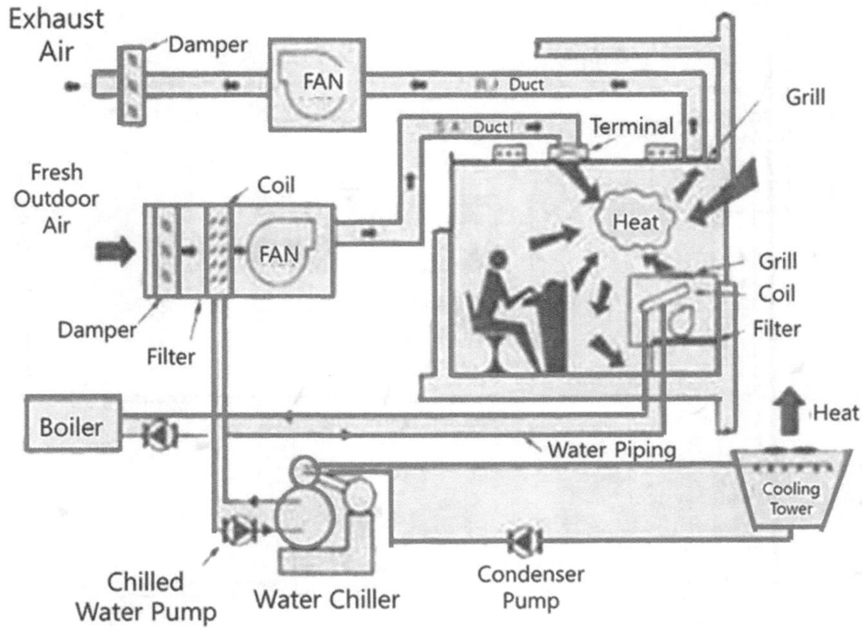

FIGURE 1.17
Basic air-water system.

Air-and-water systems are basically pertinent to a kind of multizone space in buildings where a broad extent of sensible loads occurs and where cautious control of humidity is not needed. Hospitals, apartment houses, schools, research laboratories, etc. are relevant as examples. The major subsystems are central air conditioning apparatus, duct distribution and water distribution systems, and a room terminal. The supplied air is normally a constant volume and provides outside air for ventilation (Figure 1.17).

Ultimately, from the AC central refrigeration system through cooling coils in terminal units distributed in building inhabited spaces, all-water systems attain both latent and sensible space cooling by running chilled water. The most normal terminals are fan coil units, unit ventilators, and valence units. The all-water system provides single room control, without cross contamination of recalculated air from one room to another (Figure 1.18).

1.2.6 Major Equipment of Vapor-Compression Refrigeration Systems

The major apparatus in the VCR system plays several roles: 1) The compressor increases the pressure and temperature of the refrigerant. Upon exit from the evaporator, it suctions low-pressure and -temperature refrigerant vapors and squeezes it. Through the discharge line, it releases the refrigerant vapors of high temperatures and pressure to the condenser. 2) The condenser removes

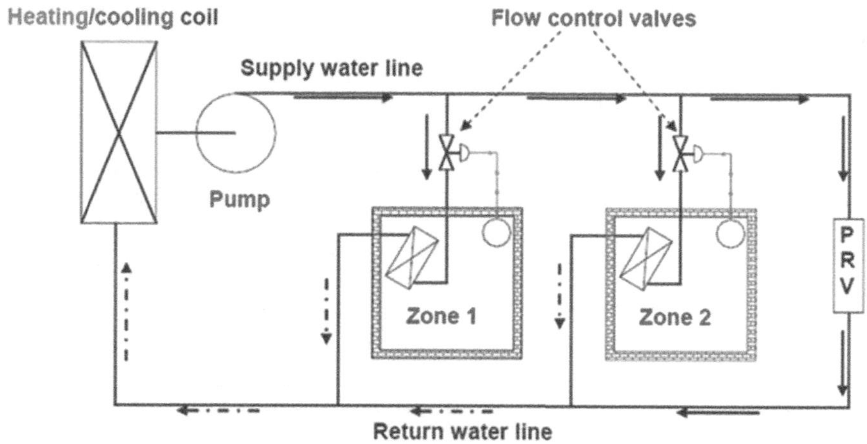

FIGURE 1.18
Two-pipe, all water system.

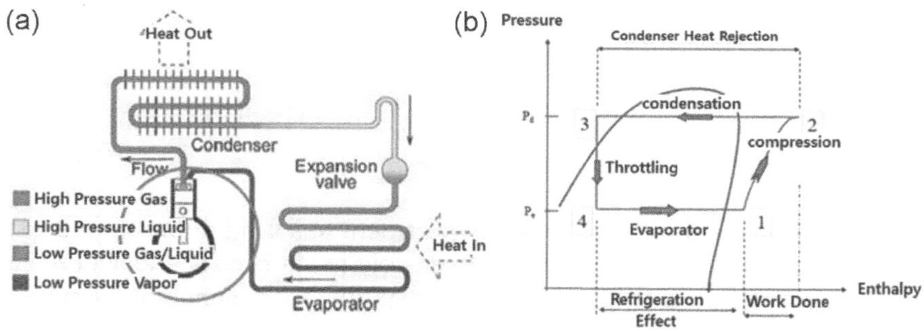

FIGURE 1.19
Key equipment of refrigeration: (a) major refrigeration equipment; (b) P-H diagram.

heat that was collected by the system by the evaporator and compressor. It is a heat exchanger that eliminates heat from the hot vapor refrigerant released from the compressor. 3) The metering device manages the refrigerant flow to the evaporator. It can be mainly categorized as either a capillary tube or a thermostatic expansion valve. 4) The evaporator, in condensing the refrigerant, sucks up heat from the room. Its role is to suck heat from the environment (Figure 1.19).

1.2.6.1 Compressors

Compressors can be categorized as positive displacement or dynamic. There are two kinds of positive displacement: reciprocating and rotary. For the dynamic kind, there are axial and centrifugal kinds (Figure 1.20).

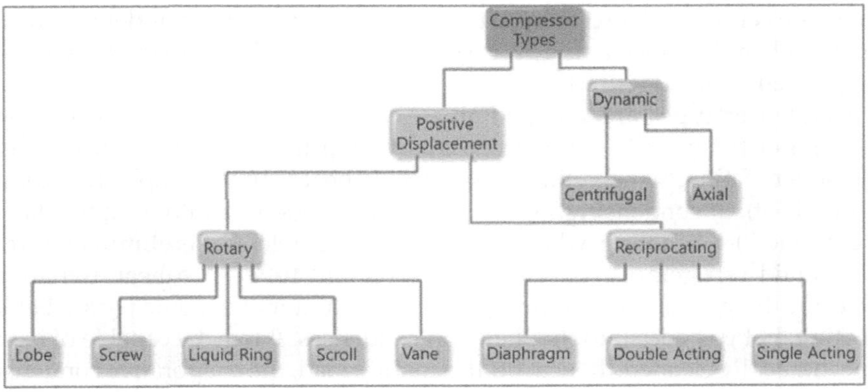

FIGURE 1.20
Compressor types.

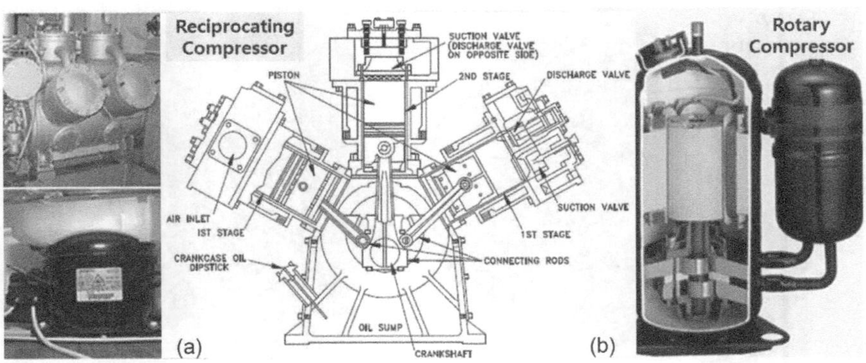

FIGURE 1.21
Compressor types—(a) reciprocating and (b) rotary.

Reciprocating compressors are furnished with cylinders and pistons and with suction and discharge valves. By the pressure difference across it, the valve position is adjusted. They may be categorized as open, semi-hermetic, or completely hermetic. The rotary compressor utilizes circular movement to attain compression. As refrigerant is enclosed by the revolving vanes, the refrigerant is squeezed as volume is reduced (Figure 1.21).

A helical rotary compressor uses a pair of rotors—a female and a male. The volume of the refrigerant lessens, and a continuous, flowing output is generated as it runs through the compressor. Compressor capacity is also adjusted by a slide valve. The scroll compressor utilizes the pair's machined scrolls in precisely the same way. Namely, one scroll does not move, and the other orbits. As the nesting of the scroll traps vapor, gas is set up from the outer edge, and refrigerant is let out from the center. A centrifugal compressor relies on centrifugal force. There are no pistons, valves, or cylinders. By using an impeller, the capacity is governed by inlet vanes. It is utilized for very big implementations (Figure 1.22).

Condenser types are grouped as air cooled (fin and tube), water cooled (shell and tube), and evaporative,. Air cooling (fin and tube) consists of 1) de-superheating, which eliminates sensible heat from the vapor refrigerant (superheat); 2) condensing, where the vapor condenses into a liquid (latent heat); and 3) subcooling, whereby additional sensible heat is eliminated from the liquid refrigerant. Water cooled—shell and tube—is a heat exchanger. Specifically, as water flows through the tubes and makes some passes before exiting, hot gas goes into the shell and condenses. It may be capable of being mechanically cleaned. To heat off the cooling load, the evaporative condenser uses air and water. Air circulates over the coil, and water may also flow over the coil if necessary. Transfers of latent and sensible heat occur as water is heated and evaporated (Figure 1.23).

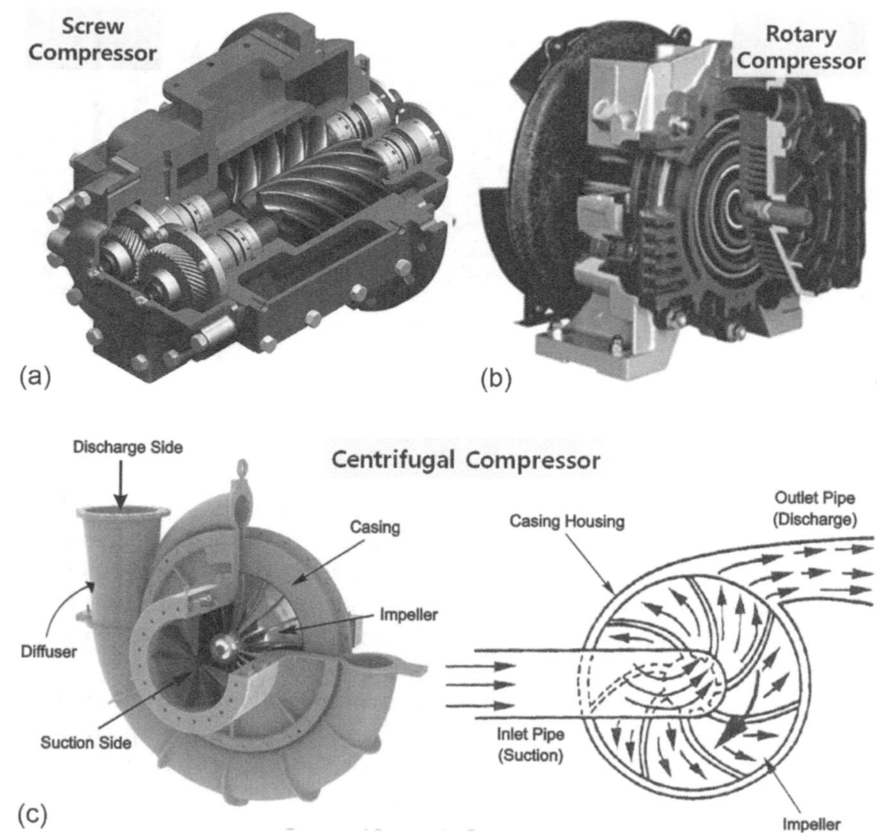

FIGURE 1.22
Compressor types: (a) helical rotary (screw type); (b) scroll type; (c) centrifugal type.

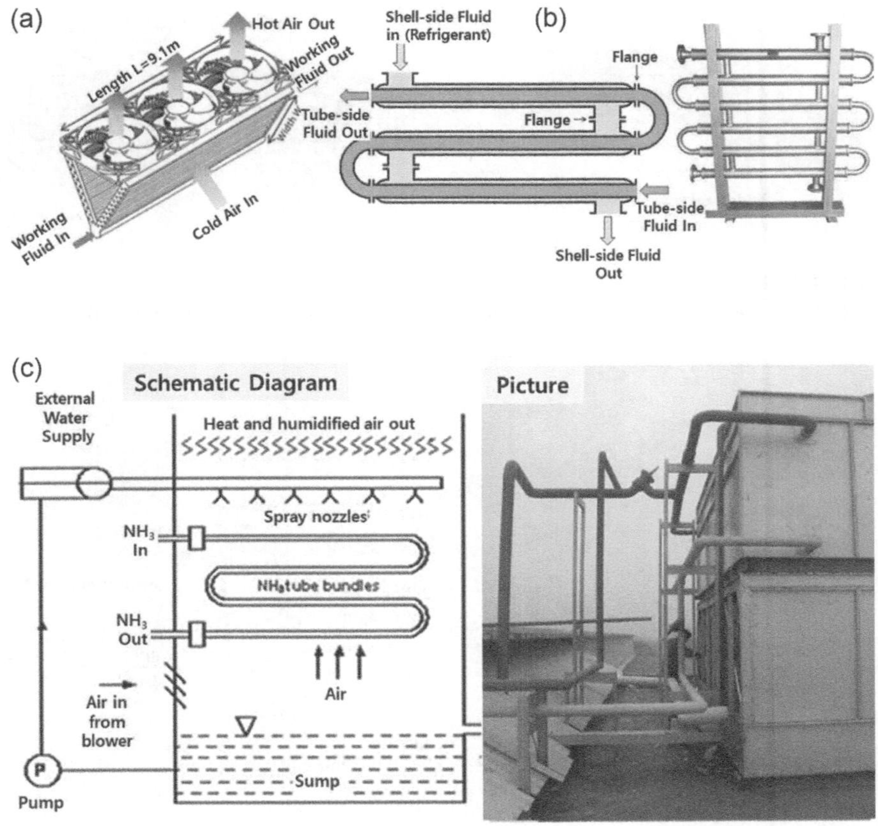

FIGURE 1.23
Condenser types: (a) air cooled (tube and fin); (b) water cooled (shell and tube); (c) evaporative condenser.

1.2.6.2 Evaporators

The evaporator is categorized as a direct expansion chilled water evaporator and flooded shell and tube. In a direct expansion chilled water evaporator, as refrigerant circulates in the tubes, water fills the shell. However, for the flooded shell and tube, as refrigerant enters the shell, water runs in the tubes (Figure 1.24).

1.2.6.3 Metering Devices

A metering device is mainly categorized as capillary tube and thermostatic. The thermostatic expansion valve continues superheating in the evaporator. Three pressures manage the valve position by bulb, spring, and

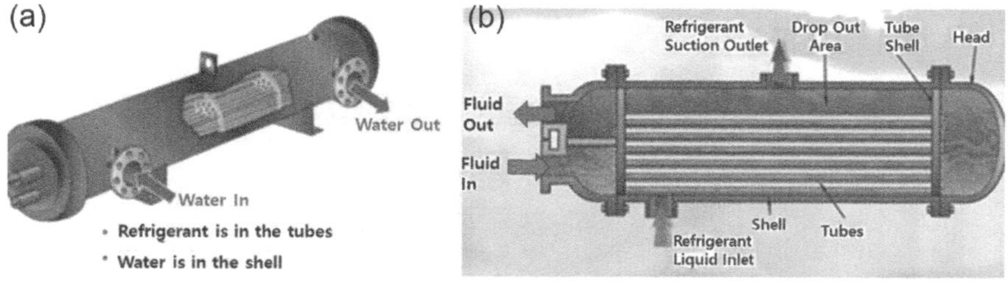

FIGURE 1.24
Evaporator types: (a) direct expansion chilled water evaporator; (b) flooded shell and tube.

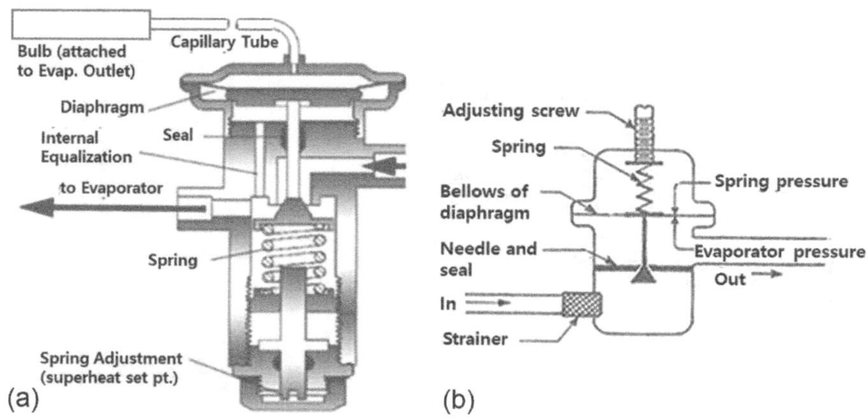

FIGURE 1.25
Evaporator types: (a) thermostatic expansion valve; (b) constant pressure expansion valve.

evaporator. As the load enlarges, the valve is open. As the load lessens, the valve is closed. The expansion valve keeps a constant evaporator pressure. A growth in evaporator pressure may cause the valve to close. A decrease in evaporator pressure will cause the valve to open. It is often utilized on systems with constant loads. Capillary tubes are seamless tubing with a small inside diameter and no moving components. A high side float valve guides the refrigerant to the evaporator at exactly the same speed as it departs.

A float on the high side keeps the flow right. Orifice plates are utilized fundamentally on centrifugal chillers. Two plates have holes in them. The pressure drops across the plate changes with the load. For a high load, liquid flows across the plates. For low loads, some liquid flashes into a vapor (Figure 1.25).

1.3 Reliability Disasters and Their Assessment Significance

1.3.1 Quality and Reliability

As mentioned, a customer chooses a product because of its greater performance, ease of use, lack of troubles (or reliability), and technological simplicity. Reliability is the product's potential to function under a group of conditions for a decided-upon period. From the viewpoint of a consumer, the reliability of the product is one of the sustainable qualities, irrespective of both the sort of product and the number of parts attached to the system. As a company makes a product with better quality—in terms of reliability and performance—the brand image may survive in the field.

The expression "quality of product" in everyday use is roughly intended to be an indication of the intrinsic level of the product's superiority. Namely, it may be defined as "conformance to specification at the starting of product usage." As quality defects appear in deficient products or components, they may be scrutinized and removed through the normal distribution by the standard specifications at product introduction. Recalls appear in the field when consumers begin to utilize a new product. From the examination of the basic causes of product recalls, we identify that they come from the design defects not considered in the product developing process. Reliability defects are evidence of unsuccessfulness in the future and are mainly analyzed through the exponential distribution. Thus the approach for reliability should be associated with the design field, not with manufacturing.

If the company doesn't assure the quality of product life for recently developed functions, it will be banished from the global market. We may infer the idea of reliability from the subsequent inquiry: How many of these samples meet their specifications at the close of the one-year assurance period? (See Table 1.1.)

TABLE 1.1

Quality Flaws and Failures

	Quality Defects	Failure
Concept	Out of specifications	Physical failure (e.g., fatigue)
Barometer	Defect rate	Lifetime, failure rate
Measure	Percent, ppm	Percent/year, year
Area	Manufacturing	Design
Distribution	Normal distribution	Exponential/Weibull
	$f(x)=\frac{1}{\sigma\sqrt{2\pi}}e^{-\frac{(x-\mu)^2}{2\sigma^2}}$	$F(t)=1-R(t)=1-e^{-\lambda t}$

1.3.2 Reliability Disasters

Due to design defects, a disaster such as an oil spill, a car recall, or a nuclear plant problem is a failure of a system's intended function, which can result in fatalities and have environmental impacts that go beyond the forecasting capability of the community or society to manage its own resources. That is, such disasters also have no forecasting ability of the society or community to manage its own resources because people may identify its consequence after revelation. Therefore, people acquire a knowledge of lessons from their past errors so that similar disasters may be anticipated by identifying their basic causes beforehand.

For instance, the RMS *Titanic* in 1912 sank in the Atlantic, and more than 2,200 passengers were injured or lost. That is, as the crew discerned the floating icebergs, they couldn't rotate the ship swiftly enough. This situation induced the ship to strike an approaching iceberg on its right-hand side. It only took about three hours to become submerged, which cost the deaths of 1,500 people. The reason behind this occurrence is that the metal iron in the ship, in the freezing seawater of the Atlantic Ocean, was too breakable to support the iceberg impact. There were no other ideas for saving the ship, which sank quickly in a few hours (Figure 1.26).

From the viewpoint of material, the *Titanic* sinking was essentially brought on by the expanded brittleness of the steel employed to construct the ship hull due to immersion in the freezing seawater of the North Atlantic. Though the collision of a large floating mass of ice detached from a glacier had a comparatively a small impact on the vessel, it suffered a huge quantity of damage in the structure. Consequently, the bolts clasping on the steel plates fractured, which contributed to the malfunction of the vessel hull. Instantly after the sinking of RMS *Titanic*, every ship was mandated to have a correct emergency evacuation scheme.

FIGURE 1.26
RMS *Titanic* of the White Star Line sinking (Wikipedia).

Thus we list a few common instances of disasters.

- The Versailles rail accident happened in 1842 on the railroad between Versailles and Paris. Following King Louis Philippe I's celebrations at the Versailles Palace, a train returning to Paris departed from its tracks possibly at Meudon. As the major car cracked an axle, the carriages behind piled into it and caught fire. With about 200 deaths, including that of the explorer Jules Dumont d'Urville, it was noted as one of the first of fatal railway mishaps in the world. Because most of the dead passengers were wearing their seat belts, the disaster led to the abandonment of the practice of locking passengers in their carriages, and it started the study of metal fatigue exposed to repetitive loads such as S–N curves (Figure 1.27).
- As the earliest commercial jet engine in airplane, the de Havilland DH 106 Comet was redesigned for people to cross the Atlantic. The first Comet plane had an aerodynamic design with four turbojet engines in pair of wings, an aerodynamic fuselage, and large square-kind windows. It first flew on July 27, 1949, and, as of 1952, it provided a silent and pleasant passenger cabin. In 1953, the Comets began to undergo design issues, and three airplanes broke up in flight. Caused by metal fatigue in its airframe, a design defect in the corners of the square windows, exposed to repetitive stresses, was found in the Comet. As an outcome, it was redesigned with oval windows, constructional reinforcement, and other changes (Figure 1.28).
- Situated in the West Virginia/Ohio frontier, the Point Pleasant Bridge collapsed and plummeted in 1967. This catastrophe caused 46 deaths, and the root cause was the metal fatigue of a critical eyebar in the suspension bridge (Figure 1.29).

FIGURE 1.27
Versailles rail disaster of 1842 (Wikipedia).

FIGURE 1.28
De Havilland DH 106 Comet (1954) (Wikipedia).

FIGURE 1.29
Point Pleasant Bridge disaster (Wikipedia).

- Because welded designs are speedier and cheaper than riveted designs, the initial Liberty ships were created by all-welded hulls. As computed by the Charpy impact tests, the steel utilized to construct the Liberty ships also had low toughness. The Liberty ships sprung hull and deck cracks. During WWII, there were almost 1,500 instances of significant brittle fractures; of the 2,710 Liberties, 12 ships instantly shattered in half. The SS *John P. Gaines*, for example, broke in two and sank on November 24, 1943, because the low-grade metal in the hull became brittle in the frigid northern waters, costing the death of ten people (Figure 1.30).
- The Space Shuttle Challenger catastrophe occurred in 1986, killing all of the crew. The principal cause of this calamity was a series of design defects—rubber O-rings under a cooled winter environment in southern Texas. Simultaneously, engineers from the

FIGURE 1.30
Early Liberty ship disaster (Wikipedia).

FIGURE 1.31
Space Shuttle Challenger disaster: (a) phenomenon; (b) structure (Wikipedia).

booster manufacturer noticed a feasible issue with the O-ring seals and suggested that the launch period might be delayed. However, they had no proof data to persuade the National Aeronautics and Space Administration (NASA) officials. Numerous Americans were crushed in pride because of the Space Shuttle Challenger's spectacular failure (Figure 1.31).

- In 1986, the nuclear reactor detonation occurred in Chernobyl, in the then Soviet Union. Some 31 people passed away, and this region is still contaminated due to the existence of radioactive materials. The disaster was the result of design defects, such as a troublesome switch in the design of the reactor (Figure 1.32).
- In 2007, due to a leak at the McKee Refinery in 2007, propane gas got away from the McKee Refinery's Propane Deasphalting Unit in Sunray, Texas. Three workers were injured with critical burns, and the refinery was closed for two months. Gas prices increased

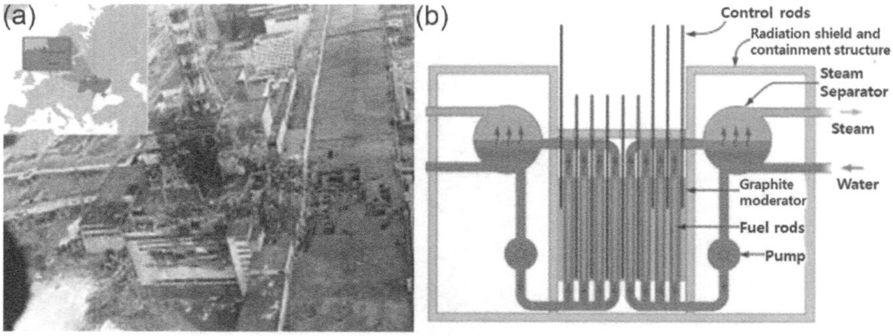

FIGURE 1.32
Chernobyl disaster: (a) phenomenon; (b) structure (Wikipedia).

FIGURE 1.33
Leak at McKee Refinery in 2007 (Wikipedia).

rapidly by nine cents per gallon in the West. The leak was created by a cracked pipe likely due to water freezing inside it, which then quickly released a large flammable vapor cloud when the temperature rose and the ice melted (Figure 1.33).

References

1. Garvin, D.A. Competing on the Eight Dimensions of Quality, *Harvard Business Review.* **1987**, *65*(6), 101–109.
2. Halliday, D.; Resnick, R. *Fundamentals of Physics*, 10th ed. John Wiley & Sons: Hoboken, NJ, 2014.

2

Probability, Statistical Distribution, and Reliability Definitions

2.1 Fundamentals of Probability

An event can be expressed as the probability of how likely something is going to happen. That probability will be located somewhere between impossible (zero) and certain (one) [1]. Namely, because the probability of an event will not be less than zero, zero probability is impossible. On the other hand, because the probability of an event cannot be more than one, one is certain—known for sure. The probability is familiar that 1) the sun will arise again, 2) I will not have to commit mathematics to memory at school, 3) when a coin is thrown, it will be heads-down or heads-up, and 4) we can select a red ball from a bag with three green balls and one red ball (Figure 2.1).

Choosing a specified subset $A_i \in S$ makes an occurrence. Certain set theory has particular terms and definitions when used in the probability conditions.

1. A subset A or B of S is defined as an occurrence.
2. For any occurrence A, $B \in S$:
   - The complement occurrence A^c is the occurrence that is not in A or not in A'.
   - Union $A \cup B$: a result if it is either in A or B.
   - Intersection $A \cap B$: a result if it is in both A and B.
   - $A \subseteq B$: The occurrence of A signifies the occurrence of B.
   - $A \cap B = \varphi$: A and B are disjoint occurrences with mutual action.

In probability, if the sample space comprises a limited numeric of probable results, $S = \{A_1, A_2, A_3, \ldots\}$, the probability can traditionally be expressed:

$$P(A) = \frac{\text{number of outcomes in } A}{\text{number of outcomes in } S} = \frac{|A|}{N} \tag{2.1}$$

DOI: 10.1201/9781003451471-2

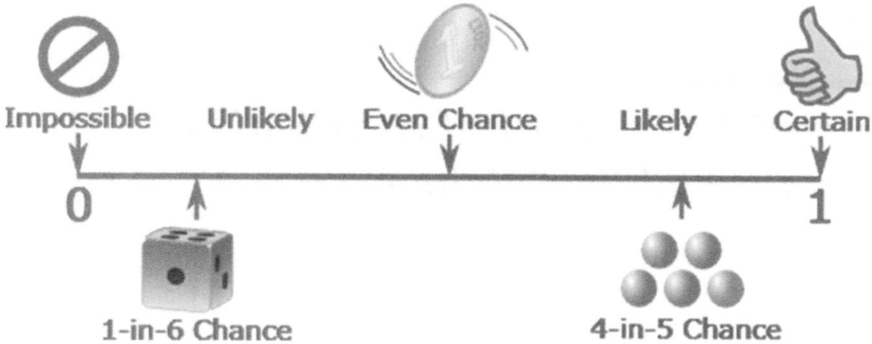

FIGURE 2.1
Probability line.

This traditional expression for probability in Equation (2.1) enables us to think, understand, and form judgments by a process of logic without conscious thought. However, it requires some empirical assumptions on the existence of the maximum at the period under consideration and the fact that the ratio value reaches to only one value for all feasible orders of trial results. Or, if the experimental number n grows with no limit, the relative frequency $\frac{r}{n}$, where r is the frequency for occurrence A, may advance toward $P(A)$. As $n \to \infty$, the possibility of a tail in the single flip of a two coin is $P(A) = lim \frac{r}{n} = \frac{1}{2}$.

Instead of making these assumptions, the favored current hypothesis of probability describes a more basic set of axioms about the essential part of the probability measure P. Utilizing the following axioms, we may attain more complex results, involving the instinctive definition of probability as a fraction of occurrence. The possibility of an occurrence A is a number $P(A)$, which is allocated to A, which fulfills the axiom conditions:

1. $0 \le P(A) \le 1$ for sample space S and occurrence E
2. $P(S) = 1, P(\emptyset) = 0$ (2.2)
3. For any order of occurrences $A_1, A_2, \ldots, A_n$, which are mutually complete, fulfill

$$P(A_1 \cup A_2 \cup \cdots \cup A_n) = P(A_1) + P(A_2) + \cdots + P(A_n) \tag{2.3}$$

The first axiom states that the probability of an occurrence is a numeric between 0 and 1. The second axiom prescribes that the occurrence indicated by the entire sample space has a possibility of 1. The third axiom describes how to combine the possibilities of complete occurrences in a mutual relationship.

- *Permutation*: From a group of n components, the numeric of manners to attain an ordered subset of k elements is expressed:

$$_nP_k = n(n-1)(n-2)\cdots(n-r+1) = \frac{n!}{(n-r)!} \tag{2.4}$$

- The number of permutations of n individual things is n factorial, stated as $n!$, which signifies that the multiplication of all positive integers is less than or equal to n. Namely,

$$_nP_n = n(n-1)(n-2)\cdots 1 = n! \tag{2.5}$$

- *Combination*: In permutation, the sequence is crucial. On the other hand, combination is an option of entries from a group, so that the order of an act of selecting does not have any significance. For example, allowed three fruits—an apple, an orange, and a pear—three combinations of two could be removed from a group of things: an apple and a pear, an apple and an orange, or a pear and an orange. Namely, it is the numeric of ways for choosing k unordered results from n possibilities. It is also close to the choice number or binomial coefficient, which may read "n choose k."

$$_nC_k = \binom{n}{k} = \frac{_nP_r}{k!} = \frac{n!}{(n-k)!} \tag{2.6}$$

- *Addition theorem of probability*: The probability of occurrence A or B is achieved by adding the probabilities of both occurrences, A and B, and deducting any intersection of the two occurrences:

$$P(A \cup B) = P(A) + P(B) - P(A \cap B) \tag{2.7}$$

- If $A \cap B = \emptyset$, $P(A \cup B) = P(A) + P(B)$; that is, occurrences A and B are exclusive occurrences with mutual action.
- *Conditional probability*: The conditional possibility of an occurrence B is the possibility that the occurrence will occur, given the understanding that an occurrence A has already happened. This probability is defined as $P(B|A)$, notation for the possibility of B in case of given A. In the process of dealing with a particular situation where occurrences A and B are independent (where occurrence A has no effect on the probability of occurrence B), the conditional probability of occurrence B, if given occurrence A, is uncomplicatedly the probability of occurrence B, that is, $P(B)$. If occurrences A and B are

relying on each other, the probability of the intersection of A and B (the possibility that both occurrences occur) is defined:

$$P(A \cap B) = P(A)P(B \mid A) \tag{2.8}$$

From Equation (2.8), the conditional probability $P(B|A)$ is straightforwardly attained by dividing by $P(A)$:

$$P(B \mid A) = \frac{P(A \cap B)}{P(A)} \tag{2.9}$$

- *Multiplication theorem of chance*: As $P(A) > 0$ and $P(B) > 0$,

$$P(A \text{ and } B) = P(A \cap B) = P(A)P(B \mid A) = P(B)P(A \mid B) \tag{2.10}$$

- *Complement probability*: As A is an occurrence and A^c is its complement, the probability of A is identical to one minus the possibility of A^c:

$$P(A^c) = 1 - P(A) \tag{2.11}$$

It may be relevant to all occurrences and their complements.

2.2 Brief History of Probability and Statistics

The mathematical theory of probability emerged from analyzing games of chance, with key contributions from Gerolamo Cardano in the 16th century and Pierre de Fermat and Blaise Pascal in the 16th century. On the other hand, statistics is the discipline that concerns the action of collecting data, planning, analysis, elucidation, and presentation of data. The reason of statistics is to develop an organized way for attaining empirical comprehension or specific information from both data and testing. Its fundamental part in data inspection and statistical reasoning is also very feasible in data collection (design of experiments and studies) and in leading correct scientific inference. Statistics is neither accurately a science nor an area of mathematics.

As the science of statecraft extending over the collection and utilization of data by the state, a German professor Gottfried Achenwall early utilized the word "statistics" in the mid-18th century. The term "statistics"—from the Latin *status*, Italian *statista*, German *statistik*, or French *statistique*—indicating a political state, originally meant information functional to the state, such as information on the magnitudes of the population (animal, human, products, etc.) and armed forces.

In a manner conforming with explorer/statistician Yule, the term "statistics" occurred first in the book *The Element of Universal Erudition* by Baron (1770). In 1787, a broader meaning was used by E.A.W. Zimmermann in "A Political Survey of the Present State of Europe." It appeared in the encyclopedia of Britannica in 1797 and was employed by Sir John Sinclair in Britain in volumes successively published between 1791 and 1799, providing a statistical description of Scotland. In the 19th century, the term "statistics" attained a broader explanation, extending over numerical data of almost any topic and the interpretation of data through acceptable examination. In brief, the history of probability and statistics is as follows:

1. 17th century—probability hypothesis
2. 18th century—inferential statistics
3. 19th century—statistics applied to education and sociology
4. 20th century—regression and correlation; computer analyses

As the first probability and statistics hypotheses evolved in the 19th century, statistical reasoning and probability models by social scientists also began with the recent sciences of exploratory psychology and sociology. It was applicable to thermodynamics and statistical mechanics by physical scientists in the last century. At an initial stage, Pierre-Simon Laplace proposed the central limit theorem. He also studied Gauss's achievement and built the thoroughgoing link between the central limit theorem and least squares estimation. The Gauss–Laplace coalescence is regarded as one of the pivotal milestones in the history of science. Karl Pearson (1857–1936) created the *P* value, Pearson's correlation coefficient, Pearson's chi-square test, and main component examination, among many other things.

At the start of the 20th century, Ronald Fisher (1890–1962) executed statistical steps in the design of scientific tests such as ANOVA. Fisher also outlined his statistical attempt in *Statistical Methods and Scientific Inference* (1956) [2]. Established on a compact collection of data, William Sealy Gosset set up procedures for making decisions. In the manufacture of mechanical products, there is a chance of recalls from the field. Based on this previous achievement on the probability and its statistics, a new structured reliability method could be evolved in the future (Figure 2.2).

2.3 Statistics Terminologies

Statisticians gather data by implementing particular experimental designs and survey samples. Typical sampling is convincing that inferences and conclusions can stretch out from the sample to the population. Before getting

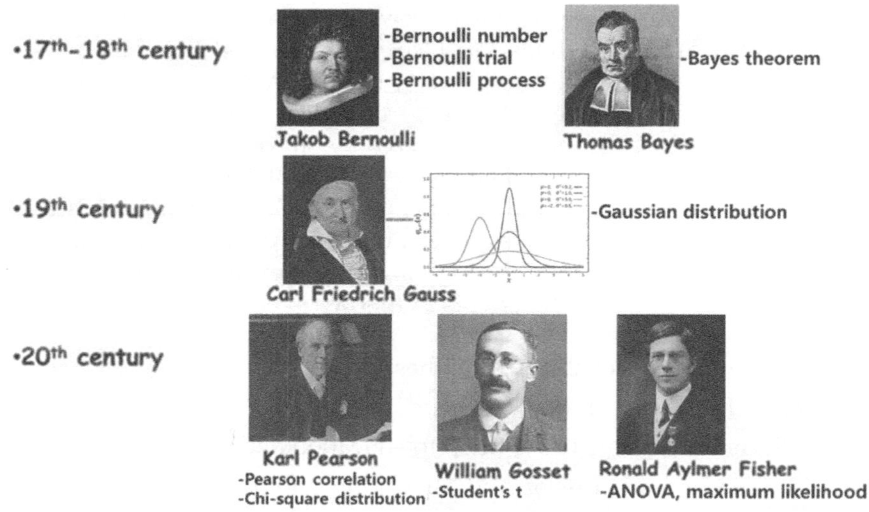

FIGURE 2.2
Short history of statistics.

further knowledge or information, engineers should be familiar with statistical terms such as sample, population, parameter, and statistics (Figure 2.3).

- A *population* is a collection of all items or things under examination in a statistical investigation. Population is attributed with the set of things or persons in which an engineer is interested and from which the engineer desires to reach deductions. For instance, if an engineer is about to design a car engine, the entire population may be connected to the engine that will be or was produced in the lot of the same year.
- A *sample* is the subset of the population that is openly being observed or learned. The grounds to select a sample from the population are that 1) population examination is unfeasible because of the period and economics, 2) investigation outcomes might be supplied in period, and 3) there is the chance of incorrectness in the whole sample due to examination mistake.
- *Random sampling* is a way to choose a sufficient representation sample from a population. For a sample to be exact, it could be random with an alike possibility of each representative of the population possessing a probability of being chosen. This may decrease bias.
- A *parameter* is a quantitative feature of the population in estimating or testing, such as a population mean or proportion. For example, if you desire to decide on the classes of freshmen selected at a college,

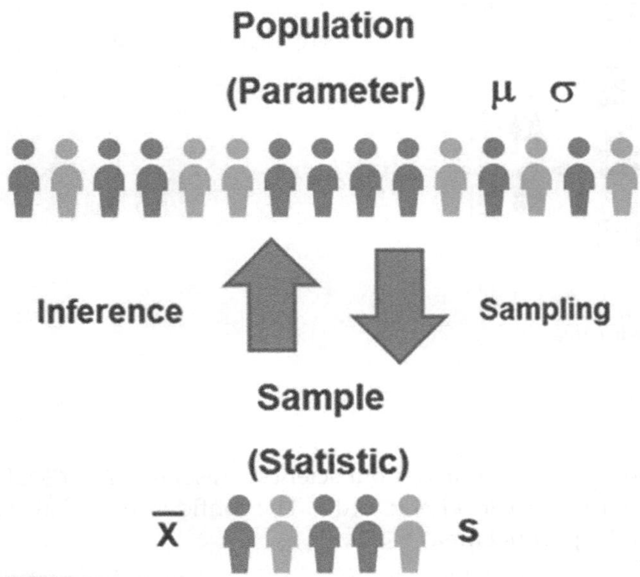

FIGURE 2.3
Population and sampling.

you may question everyone (via email) and attain a parameter such as their mean height (or weight).

- A *statistic* is a quantitative feature of a sample that can estimate or test the population parameter, such as a sample mean or proportion.
- A *confidence interval* (*CI*) describes the issue that have a good chance of happening to detain the population parameter of interest. Specifically, it symbolizes the frequency of practical CIs that detain the accurate merit of the population parameter not described in detail (Figure 2.4).
  - A CI provides an extent of values for unspecified parameters of the population by calculating a statistical sample. It is manifested as the level of confidence of the parameter in the interval. Hypothesis testing is where scientists declare the population by inspecting a statistical sample. By design, there is the unspecified quality of being unpredictable in this process. This is expressed as a level of significance.
  - A CI is discerned as the possibility that a random value lies in a definite span. CI is defined as a percentage. For example, because a 90% CI indicates that, in 90 out of 100 occurrences, the noticed value falls in this certain interval. As an unquestionable sample is selected, the population parameter is either or not in the interval attained. The compulsory level of confidence is decided by

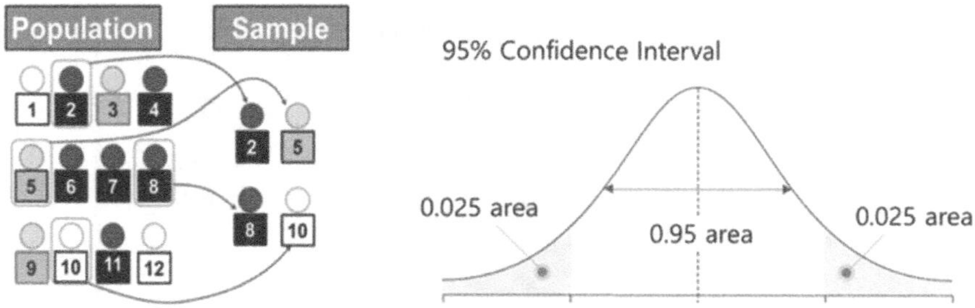

FIGURE 2.4
Idea of confidence interval.

the person who carries out scientific research. A 90% CI signifies a significance level of $\alpha = 10\%$. The confidence level also depends on the product area.

- Some 60% CI is extensively utilized as an international standard organization (ISO).
- For the International Electrotechnical Commission (IEC) and General Motors (GM), 50% CI is utilized.
- The MIL standard adjusts 60–90% CI.

- A random variable *X* is one whose practical values are the numerical results of a random event from a statistical test. A random variable is a function from a sample space *S* into the real numbers (Figure 2.5).
    - There are the two sorts of random variables: discrete and continuous. A discrete random variable is one that puts on only a countable number of certain values such as 0, 1, 2, 3, 4. . . . Instances of discrete random variables are as follows: 1) the numeral of mistakes in written writings on a page; 2) the numeral of children in a family; 3) the numeral of problematic light bulbs in a box, and 4) the numeral of problematic products per lot in manufacturing. A continuous random variable is one that sets on an infinite number of possible values in a real interval. Examples are 1) the leaving period of airplanes, 2) the period, height, and weight necessitated to go a mile.
    - A random variable X is continuous if there is a function $f(x)$, such that for any $a \le b$, we have

$$P(a \le X \le b) = \int_a^b f(x)dx \tag{2.12}$$

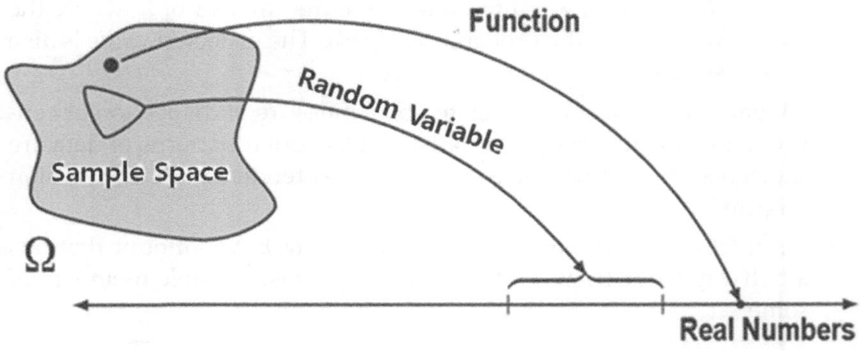

FIGURE 2.5
Random variable.

- The function *f*(*x*) is stated as the probability density function (*pdf*).
- The *pdf* always accomplishes the properties as follows:

$$1.\quad f(x) \geq 0\,(f \text{ is not negative}) \tag{2.13}$$

$$2.\quad \int_{-\infty}^{\infty} f(x)dx = 1(\text{This is identical to } P(-\infty < X < \infty) = 1) \tag{2.14}$$

- *f* (*x*) is also stated as the probability distribution function. Presume X is a continuous random variable with range [*a*, *b*] and probability density function *f*(*x*). The expected value of X is defined as

$$E(X) = \int_{a}^{b} xf(x)dx \tag{2.15}$$

- However, the expected value of X for a discrete random variable is

$$E(X) = \sum_{i=1}^{n} x_i f(x_i) \tag{2.16}$$

- The discrete formula puts a weighted process of adding the values, x_i, of X, where the weights are the probabilities $f(x_i)$. $f(x)dx$ indicates the probability that X is in an extremely small range of width dx around x. Thus we can describe the formula for $E(X)$ as a weighted integral of the values x of X, where the weights are the probabilities $f(x)dx$. The expected value is also stated as the mean or average.

- All gathered data could be stated by a measure of dispersion or central tendency in descriptive statistics. Dissimilar patterns of data are shortened by distinct measures of central tendency and dispersion (Figure 2.6).
- *Mean (also called the arithmetic mean or average)*: An amount that has a halfway value between those of the greatest possible members of some set.

$$\overline{x} = \frac{1}{n}(x_1 + x_2 + \cdots + x_n) = \frac{\sum_{i=1}^{n} x_i}{n} \tag{2.17}$$

- *Median (the central of a data set)*: The median, in statistics, is the middle point value of the specified list of data, when arranged in an order.
- *Mode*: The number that emerges most regularly in a group of numbers.
- *Range*: The span of the variable is the dissimilarity between the lowest and highest values in a data set.
- *Interquartile range (IQR)*: This is an index of statistical dispersion, which is the scattering of the data. The most frequently used percentiles are quartiles. The quartiles of the variable break the noticed

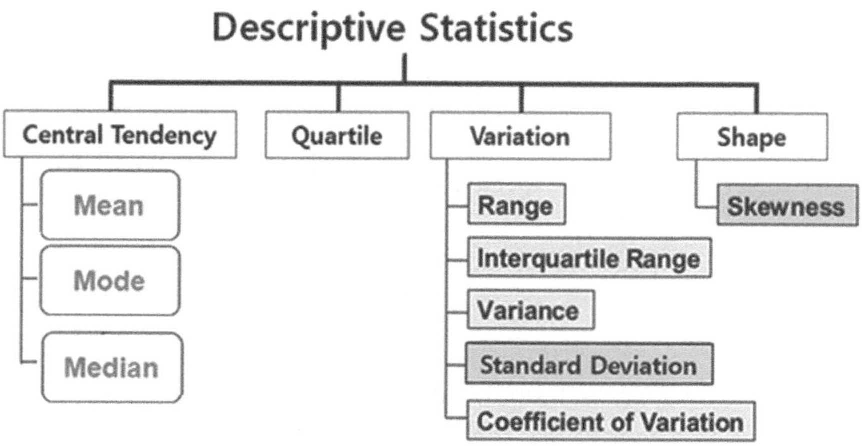

FIGURE 2.6
Measures summary of central tendency and variation.

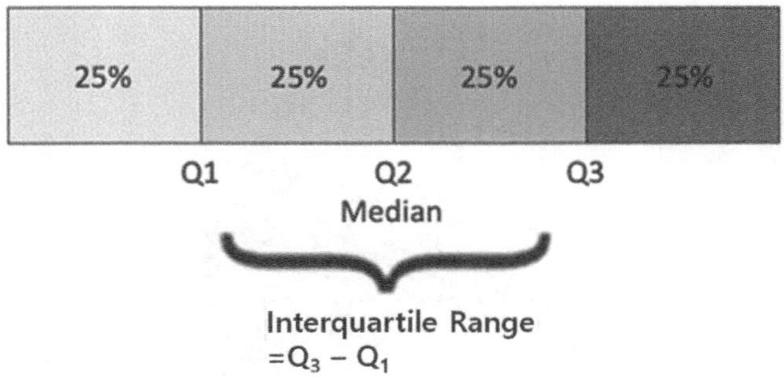

FIGURE 2.7
Interquartile range (IQR).

values into quarters. The three quartiles, Q_1, Q_2, and Q_3, may be expressed: The first quartile Q_1 is at a particular place $(n + 1)/4$, the second quartile Q_2 (the median) is at a particular place $(n + 1)/2$, and the third quartile Q_3 is at a particular place $3(n + 1)/4$. Thus the interquartile range, signified as IQR, is the dissimilarity between the first and third quartiles of the variable, namely, $IQR = Q_3 - Q_1$. The IQR gives the extent of the central 50% of the noticed values (Figure 2.7).

- *Standard deviation (SD)*: Standard deviation is one of the basic methods of statistical inspection, which is the positive square root of the variance. For a variable x, the sample standard deviation, signified by s_x, is

$$s_x = \sqrt{\frac{\sum_{i=1}^{n}(x_i - \overline{x})^2}{n-1}} = \sqrt{\frac{\sum_{i=1}^{n} x_i^2 - n\overline{x}^2}{n-1}}. \quad (2.18)$$

  - 68% of all data lies in 1 SD of a mean, and 95% of all data lies in 2 SDs of a mean (Figure 2.8).

2.4 Statistical Distributions

2.4.1 Binomial Distribution

In modern life, the binomial distribution can be easily found at a few common instances: 1) vote computation for the two unlike candidates in a vote, 2) the numeric of heads/tails following coin flips in a specific order, 3) the

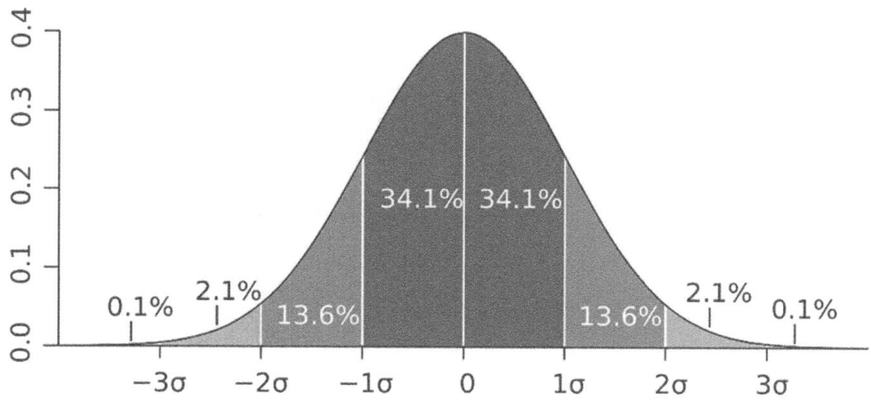

FIGURE 2.8
Standard deviation in normal distribution.

numeric of lucky sales calls, 4) the numeric of male/female workers in an office, and 5) the numeric of troublesome products in a production line [3].

Some basic presumptions describe a binomial distribution: 1) n prearranged statistical examinations are carried out; 2) each experiment has one of the two results—a success or a failure (Bernoulli tryout); 3) the probability of "success" p is the same for each result; 4) the results of single tests are independent; and 5) we are interested in the whole numeric of successes in these n tests.

Under these presumptions, if the random variable X is the entire numeric of successes, the probability distribution of X is expressed as the binomial distribution:

$$P(X=x)=\binom{n}{x}p^x(1-p)^{n-x}=\frac{n!}{x!(n-x)!}p^x(1-p)^{n-x} \quad \text{for } x=0,1,2,\ldots,n \quad (2.19)$$

The binomial average and variance are defined:

$$\mu=E(X)=np, \mathrm{Var}(X)=np(1-p) \quad (2.20)$$

where the values of n and p are the parameters of the binomial distribution.

As $p = 0.5$, the binomial distribution is consistent—the mean and median are similar. Even when $p < 0.5$ (or $p > 0.5$), the bigger the value of N is, the more consistent the shape of the distribution is. As the binomial distribution is complicated, there are rough calculations of the binomial distribution, which is not difficult to use when N is big (Figure 2.9).

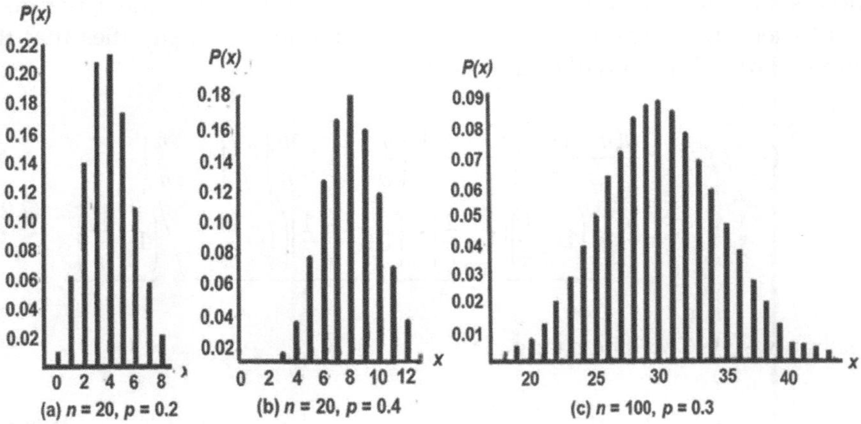

FIGURE 2.9
Form of the binomial distribution according to n and p: (a) $n = 20, p = 0.2$; (b) $n = 20, p = 0.4$; (c) $n - 100, p = 0.3$.

2.4.2 Poisson Distribution

Poisson distribution appears how frequently one might anticipate an occurrence to happen in an allotted period. The Poisson distribution was coined after Simeon Poisson (1781–1840), a French mathematician, and employed in situations where big reductions occur in a time with a specific mean rate, irrespective of the elapsed period. More exactly, this distribution is employed when the numeric of feasible occurrences is big, but the occurrence possibility over a specific time is tiny. The probability distribution originates from a Poisson trial: 1) each test outcome will be grouped as a success or failure; 2) the average numeric of successes (μ) that occur in a specific area is identified; 3) the probability that a success will occur is proportional to the size of the area; and 4) the probability that a success will occur in a minimal size district is very close to zero [4].

The Poisson distribution happens as follows: 1) the numeric of hurricanes striking Hawaii every year, 2) the every hour numeric of customers visiting at a bank, 3) the numeric of battery failures and renewal, 4) the numeric of hummingbirds perceived while attending school every morning, 5) the everyday numeric of mishaps on a particular area of highway, 6) the daily numeric of urgent calls in a town, 7) the every month numeric of employees who had a failure to attend in a large office, 8) the numeric of a typographical error in a book, and 9) the hourly numeric of entrances to a particular web server.

The Poisson distribution also has some premises: 1) independence: occurrences should be independent; 2) homogeneity: the mean numeric of targets achieved is assumed to be the same for all groups; and 3) time (or space) should be set. The Poisson distribution is a kind of setting as a limit case of the binomial distribution when the numeric of tests n is big and the possibility of success p is small. That is, if the probability p is minute and the experiment is sufficient, the Poisson probability from Equation (2.19) signifies that the binomial distribution will be assessed.

$$
\begin{aligned}
{}_nC_x p^x (1-p)^{n-x} &= \frac{n(n-1)\cdots(n-x+1)}{x!}\left(\frac{m}{n}\right)^x \left(1-\frac{m}{n}\right)^n \Big/ \left(1-\frac{m}{n}\right)^x \\
&= \frac{m^x}{x!}\underbrace{1\left(1-\frac{1}{n}\right)\left(1-\frac{2}{n}\right)\cdots\left(1-\frac{x-1}{n}\right)}_{A}\underbrace{\left(1-\frac{m}{n}\right)^n}_{B} \Big/ \underbrace{\left(1-\frac{m}{n}\right)^x}_{C}
\end{aligned} \tag{2.21}
$$

As n grows, A~C will be changed.

$$A = 1\left(1-\frac{1}{n}\right)\left(1-\frac{2}{n}\right)\cdots\left(1-\frac{x-1}{n}\right) \underset{n\to\infty}{\to} 1 \tag{2.22(a)}$$

$$B = \left(1-\frac{m}{n}\right)^n \underset{n\to\infty}{\to} e^{-m} \tag{2.22(b)}$$

$$C = \left(1-\frac{m}{n}\right)^x \underset{n\to\infty}{\to} 1 \tag{2.22(c)}$$

Thus, the outcomes can be abridged as follows (Figure 2.10):

$$P(X = x) = \frac{(m)^x e^{-m}}{x!} \tag{2.23}$$

2.4.3 Poisson Process

In probability hypothesis, the Poisson process is one of the major random procedures. It is widely used to model random "points" in period and space. Some evaluative probability distributions usually come from the Poisson process. Namely, the exponential distribution with a steady failure rate in the bathtub curve could be defined by a Poisson process (Figure 2.11).

- Period Uniformity: $P(k,\tau) =$ Probability of k arrivals in interval of duration τ
- The numeric of the action of arriving in disjoint period intervals are independent
- Small interval probabilities: $P(k,\tau) = \begin{cases} 1-\lambda\delta \; if \; k=0 \\ \lambda\delta \; if \; k=1 \\ \lambda\delta \; if \; k>1 \end{cases}$, where λ is the arrival rate

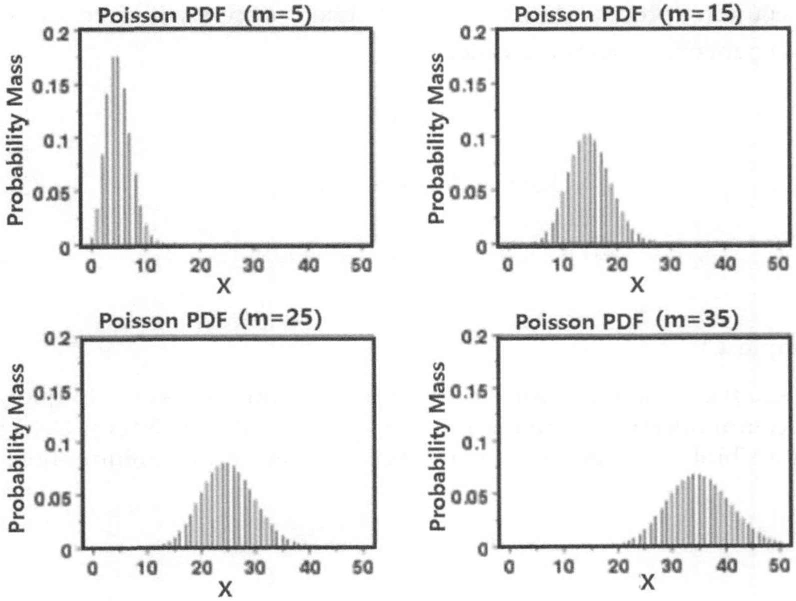

FIGURE 2.10
Form of the Poisson distribution as stated by *m*.

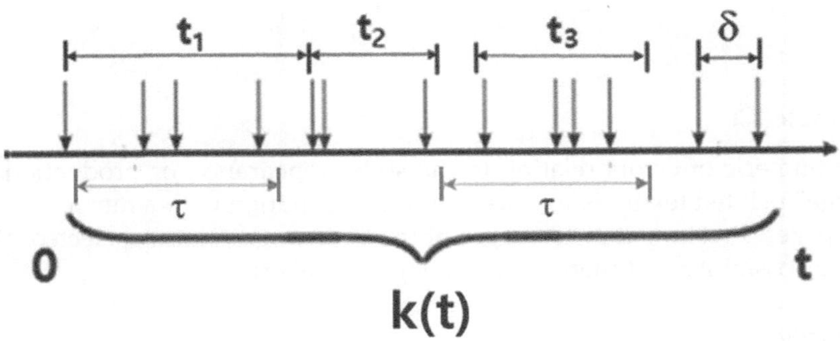

FIGURE 2.11
Definition of the Poisson process.

Namely, the process is defined by the presumed counter process $k(t)$. The counter informs the numeric of failures which have happened in the intervening time $(0, t)$ or, more usually, in the intervening time (t_1, t_2).

$N(t)$ = numeric of failures in the intervening time $(0, t)$ (the stochastic process)

$N(t_1, t_2)$ = numeric of arrivals in the intervening time (t_1, t_2) (the increment process $N(t_2) - N(t_1)$).

The counting process $\{N(t),\ t \geq 0\}$ is a Poisson process with rate λ if all the subsequent expressions are valid:

$$N(0) = 0, \tag{2.24(a)}$$

$$N(t) \text{ has independent increments,} \tag{2.24(b)}$$

$$N(t) - N(s) \sim \text{Poisson}\big(\lambda(t-s)\big) \text{for } s < t. \tag{2.24(c)}$$

Example 2.1

You email in a manner conforming with a Poisson process at a frequency of $\lambda = 5$ communications per hour. You check your email every 30 min. Determine the possibility of no new communication and one new communication.

$$m = \lambda t = 5 \cdot \frac{1}{2} = 2.5$$

$$P(0, 1/2) = e^{-2.5} \frac{(2.5)^0}{0!} = 0.082$$

$$P(1, 1/2) = e^{-2.5} \frac{(2.5)^1}{1!} = 0.205$$

Example 2.2

The numeric of errors relating to the style, appearance, or production of a "large" printed textbook follows Poisson distribution with a mean of 1.5 per 100 pages. Presume that 100 pages of the book are randomly selected. What is the possibility that there are no typographical errors?

Solution

$$P(X = 0) = e^{-m} \frac{m^x}{x!} = e^{-1.5} \frac{1.5^0}{0!} = 0.2231$$

Example 2.3

In a determined region, the mean numeric of traffic mishaps happens in one per two days. Determine the possibility of $x = 0, 1, 2$ accidents happening in a stated day.

Solution

One per two days signifies the average of traffic mishaps, $m = \lambda t = 0.5$,

$$X = 0,\ f(0) = \frac{(0.5)^0 e^{-0.5}}{0!} = 0.606$$

Accident days = 365 days × 0.606 = 221 days

$$X = 1,\ f(1) = \frac{(0.5)^1 e^{-0.5}}{1!} = 0.303$$

Accident days = 365 days × 0.303 = 110 days

$$X = 2,\ f(2) = \frac{(0.5)^2 e^{-0.5}}{2!} = 0.076$$

Accident days = 365 days × 0.076 = 27 days

2.4.4 Exponential Distribution

In the probability hypothesis and statistics, the exponential distribution is the statistical distribution of the interval between occurrences in a Poisson point process, i.e., a process when occurrences happen independently and continuously at a continual mean rate [5]. In reliability engineering, the exponential distribution with only one parameter not stated explicitly is the simplest of all life distribution prototypes. Numerous engineering modules manifest a steady failure rate in the product lifetime if they fit the exponential distribution. Moreover, it is relatively simple to carry out reliability examination. The analytic equations for the exponential distribution are defined:

$X(t)$ = the period it takes for one added arrival, presuming someone comes at period t

By definition, the subsequent circumstances are similar:

$$(X(t) > x) \equiv \left(N(t) - N(t+x)\right) \tag{2.25}$$

The event on the left-hand side of Equation (2.25) takes the possibility that no one has come in the period interval $[t, t + x]$, which specifies that the numeric of arrivals at period $t + x$ is identical to the numeric of arrivals at period t, which is the likelihood on the right side.

From the complement rule, we also have

$$P(X(t) \le x) = 1 - P(X(t) > x) \tag{2.26}$$

Employing the equality of the two occurrences that we just indicated, we will rewrite Equation (2.26):

$$P(X(t) \le x) = 1 - P(N(t+x) - N(t) = 0) \tag{2.27}$$

where $P(N(t + x) - N(t) = 0) = P(N(x) = 0)$.

Employing the Poisson distribution Equation (2.23), where λ is the average numeric of arrivals per period unit and x is the quantity of period units, the right side in Equation (2.27) is defined:

$$P(N(t+x) - N(t) = 0) = \frac{(\lambda x)^0}{0!} e^{-\lambda x} \tag{2.28}$$

Interchanging Equation (2.28) into Equation (2.26), we have

$$P(X(t) \le x) = 1 - e^{-\lambda x} \tag{2.29}$$

Namely, let X_1 be the period of the first failure. reliability function, $R(t)$:

$$R(t) = P(X_1 > t) = P(\text{no failure} \quad \text{in } (0,t]) = \frac{(m)^0 e^{-m}}{0!} = e^{-m} = e^{-\lambda t} \tag{2.30}$$

Thus the accumulative distribution function as a complement $F(t)$ is also expressed:

$$F(t) = 1 - e^{-\lambda t} \tag{2.31}$$

As the cumulative distribution function is observed, the density function in probability is attained:

$$f(t) = \lambda e^{-\lambda t} \quad t \ge 0, \lambda > 0 \tag{2.32}$$

Failure rate $\lambda(t)$ is defined:

$$\lambda(t) = \frac{f(t)}{R(t)} = \frac{\lambda e^{-\lambda t}}{e^{-\lambda t}} \tag{2.33}$$

Think that the failure rate decreases the constant λ for any time. The exponential distribution has a steady failure rate. Further, the other name for

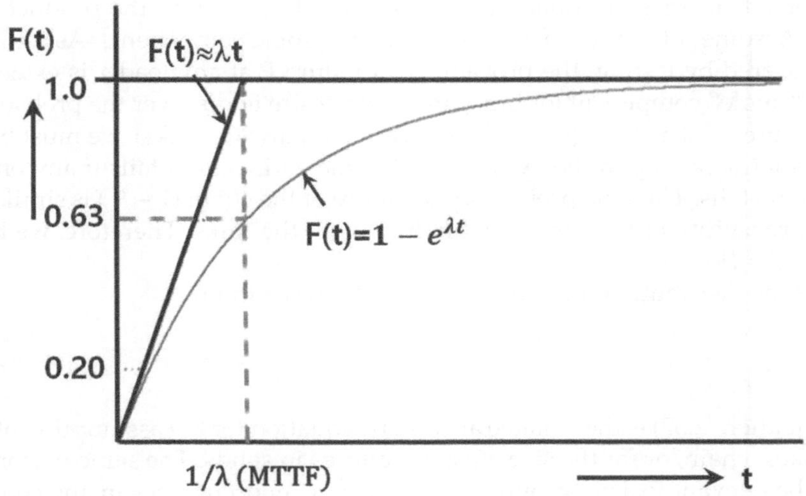

FIGURE 2.12
Cumulative distribution function *F*(*t*) of the exponential distribution.

the exponential mean is the average period fail, or MTTF, and we have MTTF = $1/\lambda$. Normally, if a product follows the exponential distribution, its MTTF is 0.63 at $1/\lambda$ (Figure 2.12).

2.5 Weibull Distributions in Reliability Testing

As a random variable *X* is distinguished as the singles of a population, the distribution function of *X*, nominated *F*(*x*), may be defined as the numeric of all singles having an $X \le x$, split by the whole numeric of singles. This function also takes the probability *P* of choosing at random a single possessing a value of *X* is similar to or less than *x*, and thus we have

$$P(X \le x) = F(x) \tag{2.34}$$

Any distribution function shall be defined:

$$F(x) = 1 - e^{-\phi(x)} \tag{2.35}$$

Though this seems to be a difficulty, this formal change depends on the relationship

$$(1-P)^n = e^{-n\phi(x)} \tag{2.36}$$

The advantages of Equation (2.36) may be described by the product failure. Assume that we have a product creating some components. As we have discovered by testing, the probability of failure P at any load x is extended to a "single" component (or link), and if we desire to discover the probability of failure P_n of a chain (or product) creating n parts (or links), we must build our subtraction upon the hypothesis that the entire chain fails if any one of its parts fails. Thus the probability of failure of the chain $(1 - P_n)$ is similar to the probability of the concurrent failure of all the links. Therefore, we have $(1 - P_n) = (1 - P)^n$.

As the distribution function of a single link, we attain

$$P_n = 1 - e^{-n\phi(x)} \tag{2.37}$$

Equation (2.37) is the suitable arithmetic equation for the assumption of the weakest chain, or for the size effect of failures in solids. The same reasoning can be relevant to a big group of issues, where an occurrence in any component of an object is said to have occurred in the object as a single unit—e.g., electrical insulation breakdowns, static or dynamic strengths, the phenomena of yield limits, the life of electric bulbs, or even the death of a man—as the possibility of not possessing failed from numerous dissimilar sources [6].

In this case, we must express the function $\varphi(x)$. The only necessary circumstances this function usually has to satisfy is to be a positive, increasing, or remaining function. The straightforward function achieving this state is

$$F(x)\left(=1 - R(t)\right) = \int_0^{\infty} f(x)\,dx = 1 - e^{-\left(\frac{x}{\eta}\right)^{\beta}} \tag{2.38}$$

where η is characteristic life, and β is shape parameters.

This is mainly due to the product's weakest chain propositions. However, the other causes for its increasing failure rate are, for instance, the age of its components and diversity of distribution forms. The increasing failure rate describes to some extent the fatigue failures. The density function relies on the shape parameter, β. For low β values ($\beta < 1$), the failure characteristics will be similar to the exponential distribution. For $\beta > 1$, the density function begins at $f(t) = 0$ and grows to a maximum as the life goes, and then it lessens steadily again. A two-parameter fit is more usual in reliability testing and more well-structured with the same sample size. As the shape parameter β is assumed to be familiar, it will decrease the fit to one parameter (Figure 2.13).

By using the maximum likelihood, we estimate the characteristic life:

$$\hat{\eta} = \left[\sum_{i=1}^{N}\left(\frac{t_i^{\beta}}{r}\right)^{\frac{1}{\beta}}\right] \tag{2.39}$$

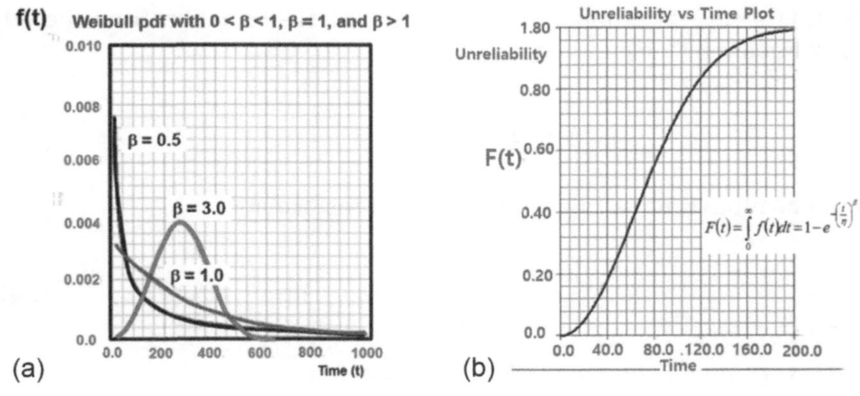

FIGURE 2.13
(a) Probability density and (b) cumulative distribution function on the Weibull distributions.

where t is the testing period, r is the numeric of unsuccessful samples, and N is the entire numeric of failures.

Two methods are commonly used, which could roughly calculate the life distribution parameters from a specific data set: 1) roughly graphical calculation in the Weibull plot and 2) maximum likelihood estimation (MLE) and median rank regression (MRR) [7]. Weibull plotting is a graphical way for unofficially inspecting the presumption of the Weibull distribution prototype and for assessing the two Weibull parameters—characteristic life η and shape parameter β.

For a Weibull probability chart, draw an aligned line from the y-axis to the fitting line at the 62.3 percentile point. The roughly calculated line cuts in half the line running through the points at a period that is close to the characteristic life η. To roughly calculate the incline of the fitting line (or the shape parameter β), select any two points on the fitting line and divide the change in the y variable by the change in the x variable.

The way of median ranks is used to roughly calculate the unreliability for each failure.

First, we examine the fitting of the two-parameter Weibull prototype employing the MRR way. MRR determines the best-fit straight line by least squares regression curve fitting. This approach starts as follows:

Categorize the times-to-failure in a manner conforming with the growing order $t_1 < t_2 \ldots < t_n$

i	1	2	3	$r-1$	r
t_i	t_1	t_2	t_3	t_{r-1}	t_r

The accumulative distribution function $F(t)$ has an S-like curve (Figure 2.13b). As the function $F(t)$ is marked in a Weibull probability paper, it is feasible to evaluate the life of a mechanical system in reliability testing.

Setting a reverse number and logarithmic transformation from the reliability Equation (2.38):

$$\ln(1-F(t))^{-1}=\left(\frac{t}{\eta}\right)^{\beta} \tag{2.40}$$

After one more time setting a logarithmic transformation:

$$\ln\left(\ln\frac{1}{1-F(t)}\right)=\beta\cdot\ln t-\beta\ln\eta \tag{2.41}$$

As F is very small, Equation (2.41) can be redefined:

$$\ln\left(\ln\frac{1}{1-F(t)}\right)\cong\ln F(t)=\beta\cdot\ln t-\beta\ln\eta \tag{2.42}$$

Equation (2.42) is identical to a linear equation:

$$y=ax+b \tag{2.43}$$

with the variables

$$a=\beta \quad \text{(slope)} \tag{2.44(a)}$$

$$b=-\beta\ln\eta \quad (\text{axis intersection}) \tag{2.44(b)}$$

$$x=\ln t \quad (\text{abscissa scaling}) \tag{2.44(c)}$$

$$y=\ln(-\ln(1-F(t))) \quad (\text{ordinate scaling}) \tag{2.44(d)}$$

The two-parametric Weibull distribution could be expressed as a straight line on the Weibull probability paper. The slope of its straight line is the shape parameter β (Figure 2.14).

Calculate the median ranks: Categorize the failure periods in a growing order. Mean ranks are less precise for the slanted Weibull distribution; thus, median ranks are better. Median ranks can be computed:

$$\sum_{k=i}^{N}\binom{N}{k}(MR)^{k}(1-MR)^{N-k}=0.5=50\% \tag{2.45}$$

Bernard employed an estimation of it:

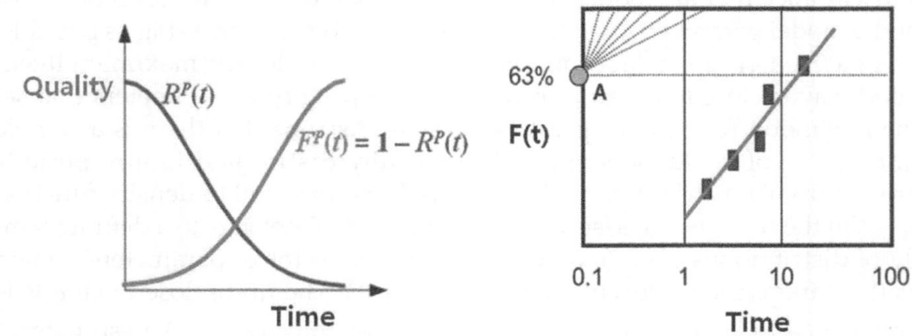

FIGURE 2.14
Chart of Weibull probability paper.

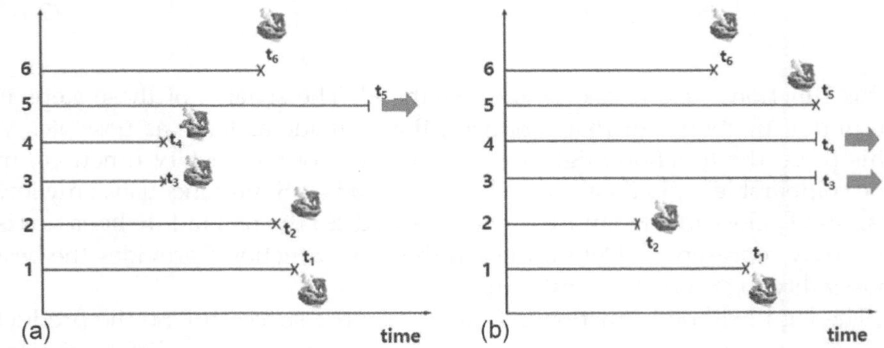

FIGURE 2.15
Clarified demonstration of Type I (a) and Type II censoring (b).

$$F(t_i) \approx \frac{i - 0.3}{n + 0.4} \times 100 \tag{2.46}$$

where i is the failure-order number, and N is the whole sample size.

As seen in Figure 2.15, engineers find out that the test criteria of reliability can be expressed in units of period (Type I censoring) or of failure number (Type II censoring). In Type I censoring, the items can be put to the test for a definite quantity of period, whereas in Type II censoring, the items can be put to the test until a definite numeric of failures is reached. As numerous surviving components are ordinary in period-terminated tests, engineers tend to use the failure number (Type II censoring).

To predict the system lifetime, MLE is an advocated manner that can be found on the noted data of reliability testing. MLE is a statistical manner for estimating the parameters of a statistical model—some undescribed mean and variance that are associated with a data set. It selects a group of values of the model parameters that make the likelihood function as big as possible.

For a fixed group of data and a basic statistical model, the maximum likelihood way is close to a group of values of the prototype parameters that set the likelihood function as great as possible. Assume that there is a sample $t_1, t_2, \ldots, t_n$ of n independent and identically distributed failure periods arising from a distribution with an undisclosed probability density function $f_0(\cdot)$. On the other hand, assume that the function f_0 belongs to a definite family of distributions $\{f(\cdot|\ \theta), \theta \in \Theta\}$ (where θ is a vector of parameters), stated as the parametric prototype, so that $f_0 = f(\cdot|\ \theta_0)$. The undisclosed value θ_0 is anticipated to be an accurate value of the parameter vector. An estimator $\hat{\theta}$ could be very near to the accurate value θ_0. The noticed variables t_i and the parameter θ are vector parts:

$$L(\Theta;t_1,t_2,\ldots,t_n) = f(t_1,t_2,\ldots,t_n\,|\,\Theta) = \prod_{i=1}^{n} f(t_i\,|\,\Theta) \tag{2.47}$$

This function is expressed as the likelihood. The scheme of these steps is to find a function f for that product L that is made as large as possible. At this place, the function might have high values of the density function f in the comparable region with some failure periods t_i. Simultaneously, only low values of f close to few failures are discovered. So the real failure behavior is precisely represented. Determined in this way, function f provides the best possibility to describe the test samples.

The log-likelihood function is frequently more successful. So the product equation becomes an addition equation, which elucidates the differentiation. Because the natural log is a monotonic function, this step is reasoning by means of mathematics:

$$ln\,L(\Theta;t_1,t_2,\ldots,t_n) = \sum_{i=1}^{n} ln\, f(t_i\,|\,\Theta) \tag{2.48}$$

Differentiating Equation (2.48), the maximum of the log-likelihood function and thus the statistically optimal parameters Θ_l can be discovered:

$$\frac{\partial\, ln(L)}{\partial \theta_l} = \sum_{i=1}^{n} \frac{1}{f(t_i;\Theta)} \cdot \frac{\partial f(t_i;\Theta)}{\partial \theta_l} = 0 \tag{2.49}$$

Because these mathematical problems may be nonlinear, it is commonly practical to seek the parameter estimates in the Weibull distribution by a numerical series of steps such as the Newton–Raphson method. By the likelihood function value, the probability is specified to estimate the quality of the alternation of a distribution to the failure data. The bigger the value in the likelihood function is, the better will be the incontrovertible distribution function, which is regarded as the real failure behavior. However, established on MLE, the characteristic life η_{MLE} from the reliability testing can be estimated in the Weibull chart.

2.6 Reliability and Bathtub Curve

2.6.1 Hazard Rate Function

If the survival period T for individuals in a population have a density function $f(t)$, the identical distribution function is the fraction of the population that fail by period t:

$$F(t) = \int_{-\infty}^{t} f(s)ds \tag{2.50}$$

The complementary function, $R(t) = 1 - F(t)$, expressed as the reliability function, is the fraction remaining alive at period t. The failure rate function $\lambda(t)$ computes the immediate risk, in which $\lambda(t)\delta t$ is the probability of being unsuccessful in the subsequent small-scale interval δt stated survival to period t. From the deduction,

$$\text{pr}(\text{survival to } t + dt) = \text{pr}(\text{survival to } t)\ \text{pr}(\text{survival to } \delta t \mid \text{survival to } t) \tag{2.51}$$

where *pr* is the probability.

We have the equations:

$$1 - F(t + \delta t) = \{1 - F(t)\}\{1 - \lambda(t)\delta(t)\} \tag{2.52}$$

$$\delta t F'(t) = \{1 - F(t)\}\lambda(t)\delta(t) \tag{2.53}$$

so that the failure rate function is expressed as:

$$\lambda(t) = \frac{f(t)}{\{1 - F(t)\}} \tag{2.54}$$

A distribution for survival times might have a failure rate function with real properties. Thus, for a big period t, the failure rate function may increase because, at a certain point in the period, the probability of breakdown usually increases. For a small-scale t, numerous shapes may prove to be reasonable, involving the one that decreases with t at first, for such a distribution could indicate the behavior of a mechanical component with a settling-in time, where reliability increases once the starting period is over.

The simple failure rate function requires an exponential distribution of survival periods with a Poisson process. For if T has the density

$$f(t) = \lambda e^{-\lambda t} \quad t \geq 0, \tag{2.55}$$

then

$$F(t) = 1 - e^{-\lambda t}, \tag{2.56}$$

From Equations (2.55) and (2.56), we may discover the hazard rate function:

$$\lambda(t) = h(t) = \frac{f(t)}{\{1 - F(t)\}} = \frac{\lambda e^{-\lambda t}}{e^{-\lambda t}} = \lambda \tag{2.57}$$

2.6.2 Bathtub Curve

To stretch out justifiable conclusions from a test's results, mechanical engineers could examine data after a new product is put to the test under consumer utilization (or environmental circumstances) for its lifespan. To design a product, an engineer should decide on the meaning and explanation of numerical and diagrammatic expressions needing data. At a period, we depend on the basic tabular and graphic skills for exhibiting the middle tendencies and span of data. We may discover another tendency, not a representative histogram, that is defined as data and its number. Numerous product failures happen at first. As the period goes on, the numeric of failures may lessen (Table 2.1). We realize that the curve has a right-skewed

TABLE 2.1

Quality Defects and Failures

Time	Number of Failures	$\lambda(t)$
1	5	5/100
2	10	10/95
3	35	35/85
4	30	30/50
5	15	15/20
6	2	2/5
7	2	2/3
8	1	1/1
Total	100	

shape. The fundamental question is how to successfully give an account in words on its tendencies. The last response is the bathtub curve. Namely, if the failure rate appears in the period, we may discover another tendency of the product lifetime—the bathtub curve [8].

On the bathtub curve, we may constitute the product lifetime with clear and vividly explicit details. Namely, it consists of three divisions: (1) an untimely mortality period with a lessening failure rate, (2) followed by a typical life period (also called useful life) with a small and steady failure rate, and (3) coming to the conclusion with a wear-out time that indicates an increasing failure rate. The bathtub curve thus gives a profile of how untimely mortality, usual life failures, and wear-out failures are integrated to build the widespread product failure distributions (Figure 2.16).

As a product is designed and first manufactured, the life trends of the troublesome product, such as a vacuum tube, may be explained as the bathtub curve because a vacuum tube has a high failure rate and short life. It is manufactured by working the initial failure rate as "embryonic mortality" failures ($\beta < 1$), the random failure rate with a steady failure rate for its "useful life" ($\beta = 1$), and in the end the rate of "wear-out" failures ($\beta > 1$), which follow the Weibull distribution.

Premature possible sources of inceptive failure, such as storage, handling, and installation error, are controlling in bathtub curves. The aging test can instantly remove them. In the middle of the curve, the failure rate of a product is steady because it pursues an exponential distribution. In this time, the product may undergo random failures due to, for instance, excessive use by consumers and exposure to overstress. The product may undergo wear failure due to extended utilization in the end region of the bathtub curve. But if any design issues are in the product, its failure rate could grow immediately and greatly during its useful life. Thus this region may be examined by the Weibull distribution.

As a product follows the bathtub curve, it will have difficulties succeeding in the market. Because of the quality issues—high failure rates and

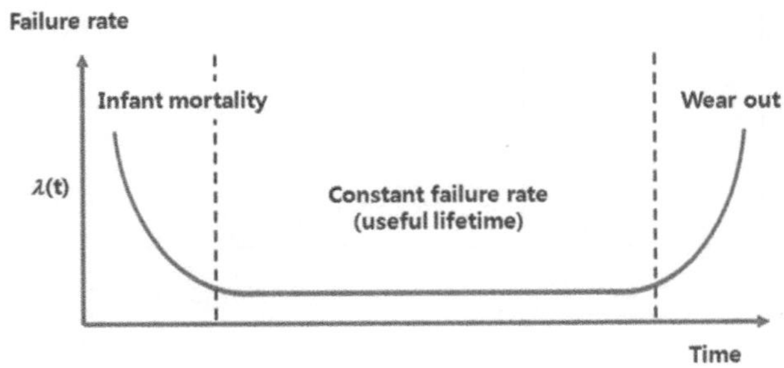

FIGURE 2.16
Bathtub curve.

short lives—due to design flaws, manufacturers will undergo economic losses throughout the entire product life. Thus they should clearly improve the design of a product by planning reliability aims for new products to 1) decrease untimely failures, 2) lessen random failures for the product functioning period, and 3) enhance product life.

2.6.3 Cumulative Distribution Function $F(t)$

As seen in Figure 2.17, reliability is the probability that a mechanical product may acceptably function for a design life under the operating/environmental circumstances. As T is the random variable designating the period to failure, the reliability at period t could be defined as

$$R(t) = P(T > t) \tag{2.58}$$

The failure rate in a period interim $[t_1, t_2]$ could be stated as the probability that a failure rate unit period occurs in the interval stated and that no failure has occurred prematurely to t_1, the beginning of the interval.

So the failure rate is expressed as

$$\frac{R(t_1) - R(t_2)}{(t_2 - t_1)R(t_1)} \tag{2.59}$$

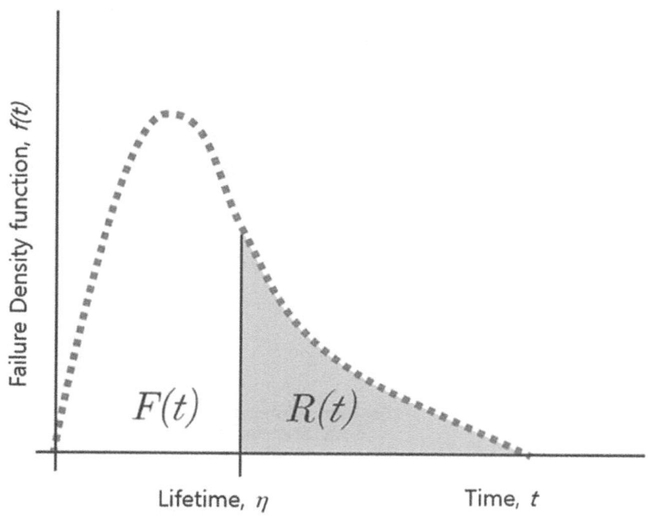

FIGURE 2.17
Reliability function $R(t)$ and accumulative distribution function $F(t)$.

If we replace t_1 by t and t_2 by $t + \Delta t$, we can define Equation (2.59) again:

$$\frac{R(t)-R(t+\Delta t)}{\Delta t R(t)} \tag{2.60}$$

As Δt proceeds toward zero, the instantaneous failure rate can be expressed:

$$\lambda(t)=\lim_{\Delta t\to 0}\frac{R(t)-R(t+\Delta t)}{\Delta t R(t)}=\frac{1}{R(t)}\left[-\frac{d}{dt}R(t)\right]=\frac{f(t)}{R(t)} \tag{2.61}$$

A hazard function is utilized to inspect the expected full length of period until one or more occurrences happen, such as a failure in mechanical products. The cumulative hazard rate function $\Lambda(t)$ is defined:

$$\Lambda(t)=\int_0^t \lambda(x)dx \tag{2.62}$$

Presume that the failure rate $\lambda(t)$ is identified. So $f(t)$, $F(t)$, and $R(t)$ may be attained:

$$f(t)=\frac{dF(t)}{dt}=-\frac{dR(t)}{dt}\Rightarrow \lambda(t)=-\frac{\frac{dR}{dt}}{R} \tag{2.63}$$

If Equation (2.63) is integrated, the reliability function may be attained:

$$R(t)=\exp\left[-\int_0^t \lambda(\tau)d\tau\right] \tag{2.64}$$

Thus the accumulative distribution function and density function are defined again:

$$f(t)=\lambda(t)\exp\left[-\int_0^t \lambda(\tau)d\tau\right] \tag{2.65}$$

$$F(t)=1-\exp\left[-\int_0^t \lambda(\tau)d\tau\right] \tag{2.66}$$

The relationship between the cumulative distribution function $F(t)$ and reliability function $R(t)$ in reliability engineering may be summarized, as shown in Figure 2.18.

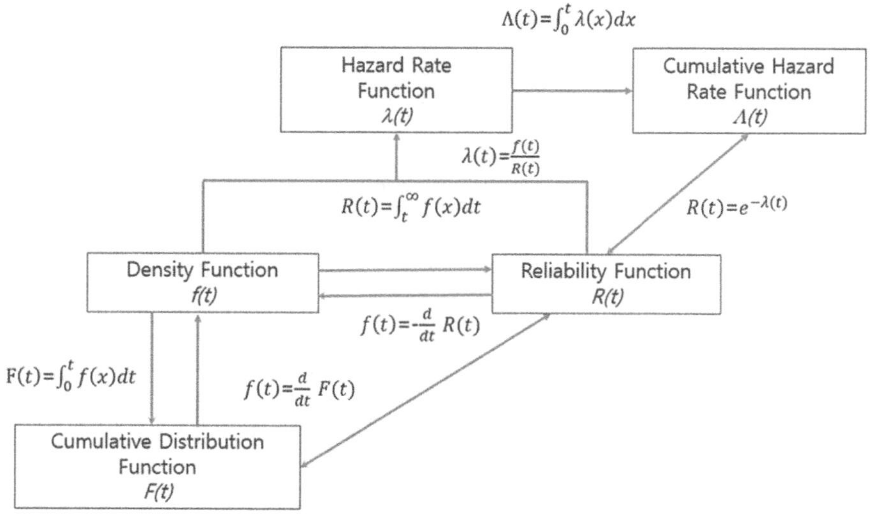

FIGURE 2.18
Relationship between the reliability function *R*(*t*) and cumulative distribution function *F*(*t*).

2.7 Lifetime Metrics for Design

As a consumer uses a product, failure is revealed in the fragile component with a design defect. As the period goes on, the tendency of product reliability will be expressed as the failure rate in the bathtub curve. For the useful life, the product follows an exponential distribution with a continual failure rate. Engineers normally describe a design's life to mean an engineer's specifications of product use under which reliability will be demonstrated. Normally, the design's life (or useful life) is 2–3 times longer than the BX lifetime (qualification life) (Figure 2.19).

2.7.1 Mean Time to Failure (MTTF)

MTTF is the length of period that a product is expected to stop working. MTTF is a basic life measure of reliability to express the lifetime of products broken beyond repair such as "one-shot" tools in light bulbs. A nonrepairable system is one for which failed items are removed permanently from the population. It is the mean period until a part of a larger piece of equipment fails. MTTF is the mean over a lengthy period of time with a large unit (Figure 2.20):

$$\mathrm{MTTF} = \frac{t_1 + t_2 + \cdots + t_n}{n} \tag{2.67}$$

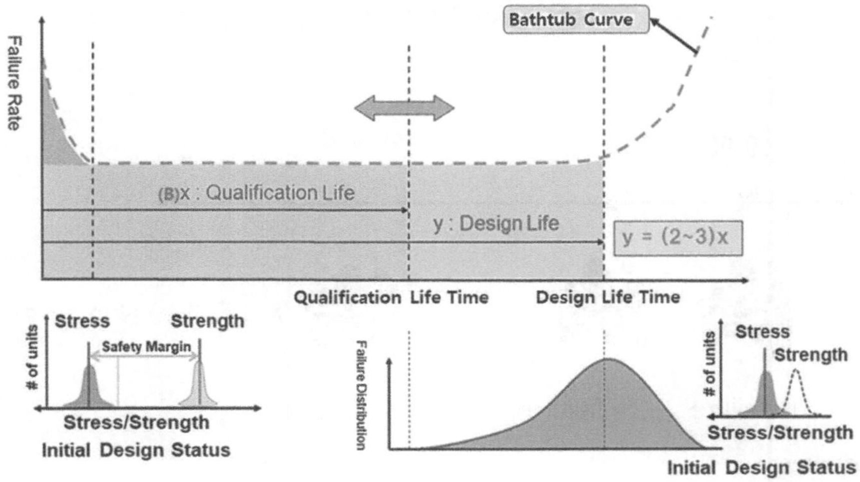

FIGURE 2.19
BX life (qualification life) and design life.

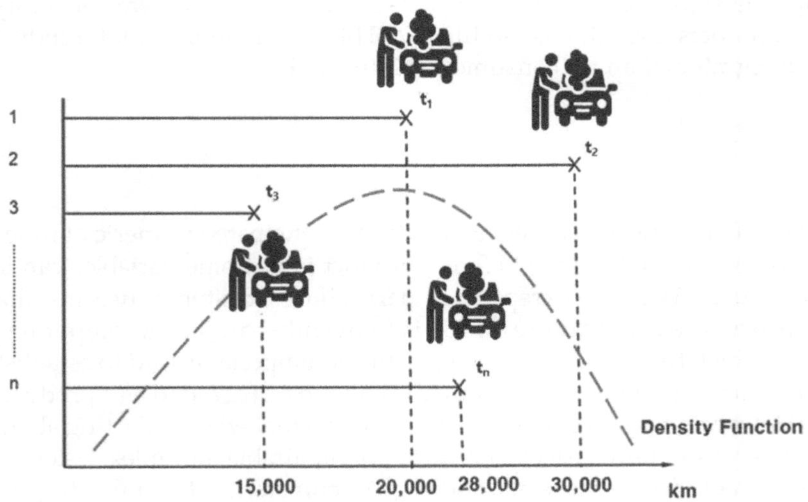

FIGURE 2.20
Plan of mean time to failure.

2.7.2 Mean Time Between Failures (MTBF)

MTBF is the lapsed time between the failures of a product to stop functioning. That is, MTBF may be computed as the mean period between the failures of a product. MTBF is a reliability measure employed to describe the mean life of repairable parts such as refrigerators, automobiles, appliances, construction machines, and airplanes. A repairable system is one that can

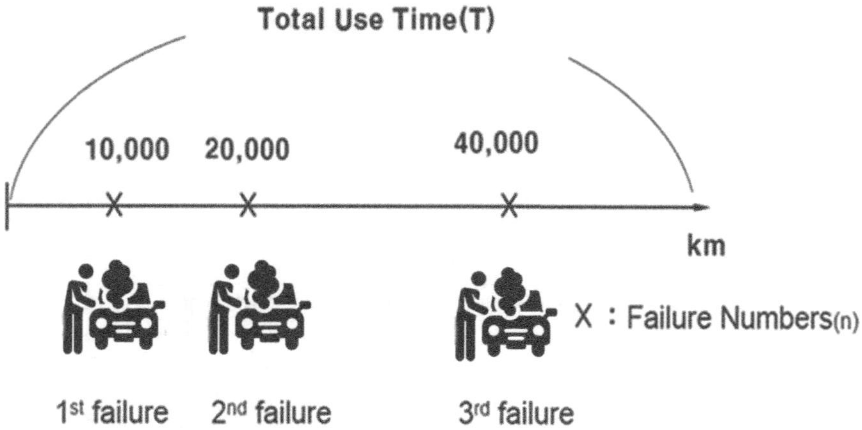

FIGURE 2.21
Notion of MTBF.

be restored to adequate operation by component replacements or changes to adaptable settings. Thus MTBF is a basic standard of system reliability for most products, though it is modified. MTBF is more important for industries and integrators than for consumers (Figure 2.21).

$$\text{MTBF} = \frac{T}{n} \tag{2.68}$$

The MTBF value is in agreement with the anticipated numeric of functioning hours (service lifetime) before a product fails. Some variables can influence failures. With the exception of part failures, customer use/installation can also evolve into a failure. MTBF is frequently based on a computing process in which the factors in all of a product's components lead to establishing the product's lifetime cycle in hours. MTBF can be regarded as a product failure. However, it is still examined as a practical tool when reflecting about the installation and purchase of a product. For repairable multiplex systems, failures are regarded as those out-of-design circumstances that take the product out of service and put in for repair. Strictly speaking, MTBF is employed only in relation to a repairable component, while MTTF is used for nonrepairable components such as electric parts.

2.7.3 BX Life

A real index of product lifetime, such as BX lifetime, should be chosen for carrying out parametric ALT. BX lifetime is expressed as the period at which $X\%$ of the samples in a population may fail. For instance, as the life of a product has a B20 lifetime of ten years, 20% of the population may fail after a ten-year functioning time. "BX life of *y* years" may acceptably describe the cumulative failure rate of a product and its life in accordance with field needs. The MTTF, as the reverse of the failure rate, may not be utilized for the lifetime because it corresponds to a B60 lifetime that is too large. BX life shows a more suitable approximation of life compared to MTTF.

The design lifetime of products, BX, will fail to agree with the life measure—MTTF or MTBF. The BX lifetime metric was originated by a ball and roller bearing company. However, it has become today a product life standard employed across the spectrum of manufacturing. It is mainly functional in expressing the warranty period for a system. The BX % life or "bearing life" is the life measure that is unsuccessful in achieving *X*% of the samples in a population. For example, as a sample has a B10 lifetime of 1,000 km, 10% of the population will be unsuccessful by 1,000 km of operation.

On the other hand, the B10% life indicates 90% reliability of a population at a specific termination in the product lifetime. At first, the B10 lifetime measure was used to express approval by industries when manufacturers products met requirements. At the moment, the B1, B10, and B50 life values work as measurements of product reliability. For example, if the MTTF of a product is 100,000 km, its B10 lifetime may be 10,000 km (Figure 2.22).

The specific design life gives a use or period setting for reliability tests, such as parametric ALT. Some companies could simply opt to design a product to be consistently good in quality or performance over an expressed warranty (qualification) time. As BX is identical to how long the product is anticipated to be utilized in the market, an advanced company could select its own design life. Relying on the designer's viewpoint, the life definition could be built on any of the design life calculations.

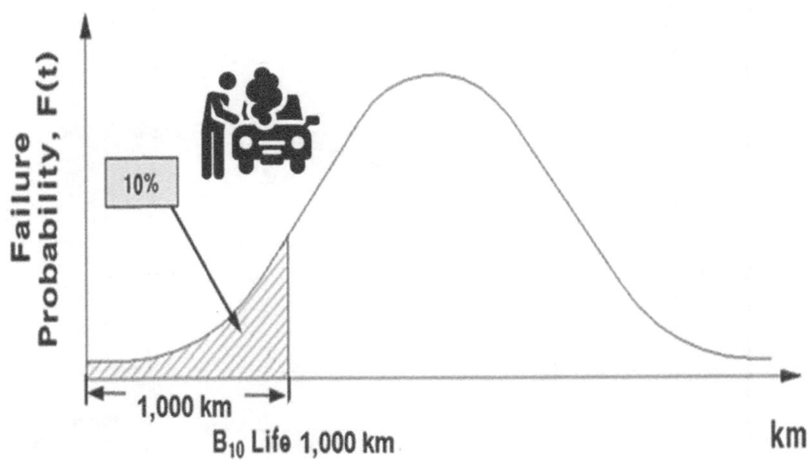

FIGURE 2.22
Idea of BX% life.

2.8 Robust Design by Taguchi Methods

Taguchi developed an early and influential robust design approach for enhancing reliability at low cost in a short-term period [9]. Robust design is a structured statistical engineering method for making the most effective use of any circumstances so that the product can quickly respond to numerous noise source variations (see Figure 2.23). Since the 1980s, the method has been exerted in a way that affects a large area to enhance the quality of innumerable products and processes.

Robustness is expressed as the capacity of a product to carry out its intended function invariably despite the existence of noise factors. At this time, noise factors are the variables that have unfavorable results in the intended function and that are unsuitable to control. Environmental stresses are the classic noise factors. This explanation is generally relevant in the area of quality engineering to describe inceptive robustness when the product service period is zero. As consumer satisfaction over a period is examined, the effect of the period might be considered along with other factors before reaching a decision. The reliability of a mechanical product may be recognized as robustness over a period or under environmental circumstances. A reliable product has high robustness under dissimilar usage circumstances. To fulfill high robustness, the Taguchi way suggests that the optimal designs are not sensitive to noise parameters.

Taguchi ways initially enhance product quality and lately have become applicable to ensure robust design in engineering. Statisticians have embraced

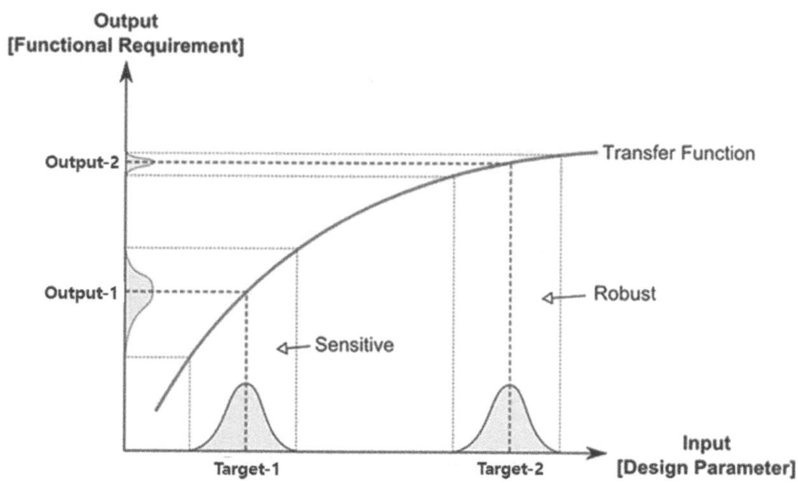

FIGURE 2.23
Robust design—inputs that make the outputs less sensitive.

the aims and advances brought about especially by Taguchi's evolution of designs for investigating variation but have found fault with the inefficiency of some of Taguchi's schemes. As a different way, parametric ALT, will be investigated in Chapter 7.

2.8.1 Specific Loss Function

To roughly calculate invisible quality costs, a Taguchi quality loss function (QLF) method has been put forward. Taguchi's proposal is different from the conventional proposal of quality costs. In the conventional outlook, if you have two products, one of which is within the particular limits and the other of which is exactly outside the particular limits, the difference is small. Though the difference is slight, the product within the limits is regarded as an acceptable product. On the other hand, the outside one is regarded as an unacceptable product. Taguchi disagrees with this point of view. According to Taguchi, when a product moves from its aim value, that motion produces a loss. It doesn't matter if the move drops inside or outside the stated limits. For this cause, Taguchi evolved the QLF to compute the loss that is linked with invisible quality costs. This loss occurs when a variation brings about the product to depart from its aim value.

QLF is a U-formed parabola. The horizontal axis is tangent with the parabola at the aim value. This is a quadratic loss function because it supposes that when a product is at its aim value (T), the loss is zero (Figure 2.24).

Quality characteristics may be grouped into three circumstances: (1) on-target (completely accurate), minimum-variation; (2) the smaller the better; and (3) the larger the better.

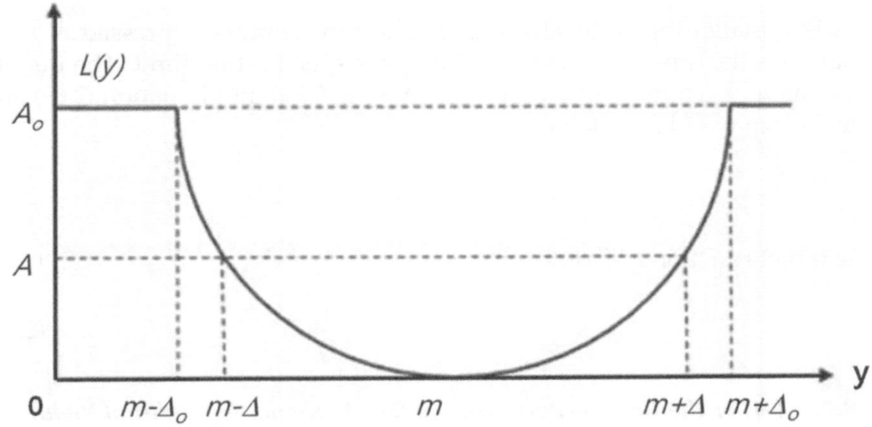

FIGURE 2.24
Taguchi quality loss function for a nominal-the-best attribute.

2.8.1.1 On-Target, Minimum-Variation (For Instance, a Mating Component in an Assembly)

In engineering design, we often come across on-target attributes. Due to production process variation, the characteristics are permitted to form within a range, suppose ±Δ, where *d* is entitled the tolerance. Equation (2.68) expresses the quality loss of this kind of attributes. The unit loss is decided by the mathematical relationship:

$$L(t) = k(y - T)^2 \tag{2.69}$$

where k = a proportionality constant dependent on the organization's failure cost structure, y = true value of quality characteristic, and T = target value of quality characteristic.

The value of k should be decided on early, before the loss may be roughly calculated. To decide the value of k:

$$k = c/\Delta^2 \tag{2.70}$$

where c = loss connected with the specification limit, and Δ = deviation of the specification from the aim value.

The target function is expressed as

$$f(y) = 1/(y - t)^2 \tag{2.71}$$

2.8.1.2 Smaller the Better—Variance (For Instance, Carbon Dioxide Emissions)

As y is a smaller-the-better characteristic, its range may be expressed as $[0,d]$, where 0 is the aim value and d is the upper specification limit. The quality loss function is attained by replacement with $T = 0$ into Equation (2.69) and may be expressed as follows:

$$L(t) = ky^2 \tag{2.72}$$

The target function is expressed:

$$f(y) = 1/y^2 \tag{2.73}$$

2.8.1.3 Larger the Better—Performance (For Example, Agricultural Yield)

As y is a larger-the-better characteristic, its range is $[d,\infty]$, where d is the lower limit. Because the reciprocal of a larger-the-better characteristic has the same quantitative way of behaving as a smaller-the-better one, the quality loss function may be attained by replacing with 1/y for y in Equation (2.74). Thus we have

$$L(t) = k\left(\frac{1}{y}\right)^2 \tag{2.74}$$

The target function is expressed as

$$f(y) = y^2 \tag{2.75}$$

2.8.2 Robust Design Process

Robust design is a structured statistical engineering method for reducing the performance variation of a product by selecting the optimal design circumstances to make the product's performance nonsensitive to noise factors. Taguchi discerned that the leading chance to remove variation is during the design of a product and its manufacturing process. Thus he evolved a quality engineering procedure that is made up of three phases—system design, parameter design, and tolerance design (Figure 2.25).

2.8.2.1 System Design

System design requires the best suitable action of carefully choosing technology and components for usage, the design of system architecture, evolution of a model that satisfies the consumer need, and the determination of the manufacturing process. System design has notable influences on cost, yield, reliability, maintainability, and numerous other forms of product performance. It also plays an important role in lessening product sensitivity to noise factors. If a system design is faulty, the ensuing parameter design and tolerance design targeted at robustness development will be unsuccessful. In recent years, some structured methods of system design have appeared, such as Theory of Inventive Problem Solving (TRIZ).

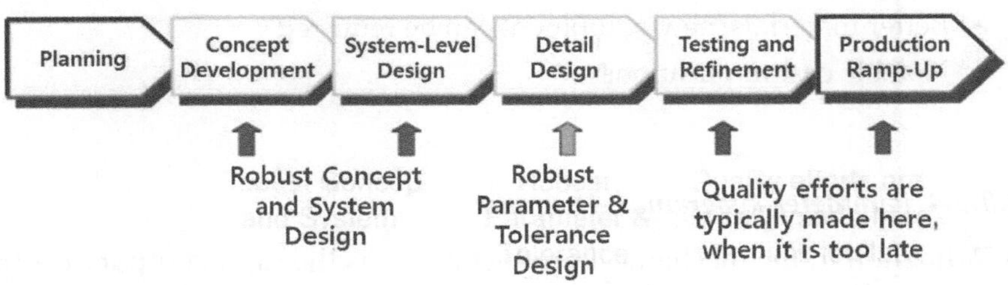

FIGURE 2.25
Steps of robust design.

This step is the conceptual design level, requiring creativity and innovation.

- Becoming involved in the "design space"
- Creating a feasible design
- Involves innovation

2.8.2.2 Parameter (Measure) Design

Parameter design aims to reduce the sensitivity of the product performance to noise factors by putting its design parameters at the optimal levels. In this step, designed tests are generally carried out to explore the connections between the design parameters and performance characteristics of the product. Utilizing such connections, one may decide the optimal positions of the design parameters.

If the idea is initiated, the nominal values of the several sizes and design parameters in the product should be set, as in the detail design stage of traditional engineering. In numerous situations, this permits the parameters to be selected so as to reduce the effects on performance from variation in environmental noise loads. As precisely as possible, parameter design should mean robust design.

- Optimizing in the "design space" (not altering anything basically)
- Settings for the factors recognized in product design

2.8.2.3 Tolerance Design

With a successfully completed parameter design, tolerance design is intended to select the tolerance of critical components to lessen the performance sensitivity to noise factors within cost constraints. Tolerance design can be carried out after the parameter design is finalized. If the parameter design cannot fulfill adequate robustness, tolerance design is finalized. In this step, the main components whose variability has the greatest results on product sensitivity are identified through experimentation. Thus the tolerance of these components is secured by utilizing higher-grade components established by the trade-off between the expanded cost and the reduction in performance variability.

- Securing tolerances on main factors
- Better materials/new equipment can be required
- Notable cost implications

2.8.2.4 Parameter Diagram (P-Diagram)

A P-diagram in the mechanical product explains the input, outputs, control factors, and noise factors. Figure 2.26 presents a general P-diagram. Signals

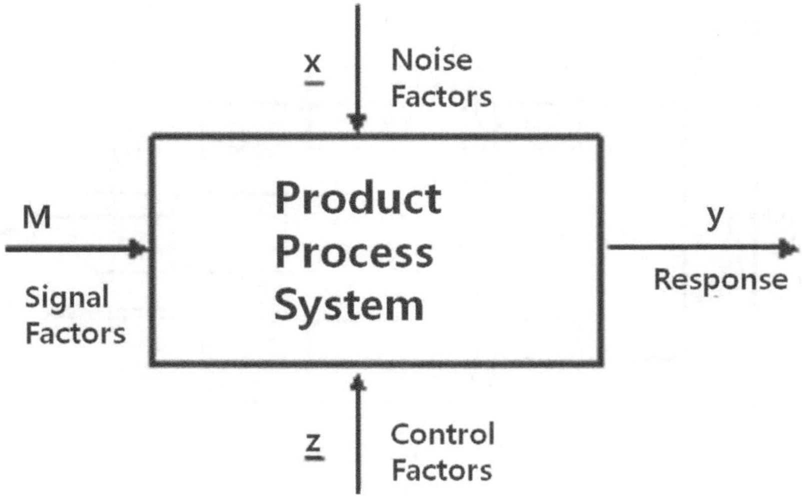

FIGURE 2.26
Steps of robust design.

are inputs from modules or subsystems to the system. The system converts the signals into practical responses. Signals are necessary to achieving the system function. Noise factors are variables that have unfavorable results on robustness. Normal instances in the mechanical system are a number of random loads.

Control factors are the design parameters whose degrees are stated by engineers. The aim of a robust design is to select the optimal levels of the parameters. In reality, design parameters play an important role in mechanical products even though they are not the main factors of robustness. So the key design parameters are involved in a robust design. These key design parameters are recognized by utilizing engineering judgment, analytical investigation, a preliminary test, or market data (Figure 2.27).

2.8.2.5 Inefficiencies of Taguchi's Designs

Robust design in a Taguchi approach utilizes parameter design to position the design where random "noise" does not bring about failure and to decide the real design parameters and their levels. However, for a straightforward mechanical construction, a large number of design parameters might be regarded as the Taguchi's robust design procedure. Those mechanical products with design defects can develop into recalls and loss of brand name.

Interactions are also part of the actual world. Interactions in Taguchi's arrays are confounded and hard to settle. Response surface methodology (RSM) is a "follow-up design" that solves only the confounded interactions. RSM design can be utilized to inspect possible high-order univariate effects

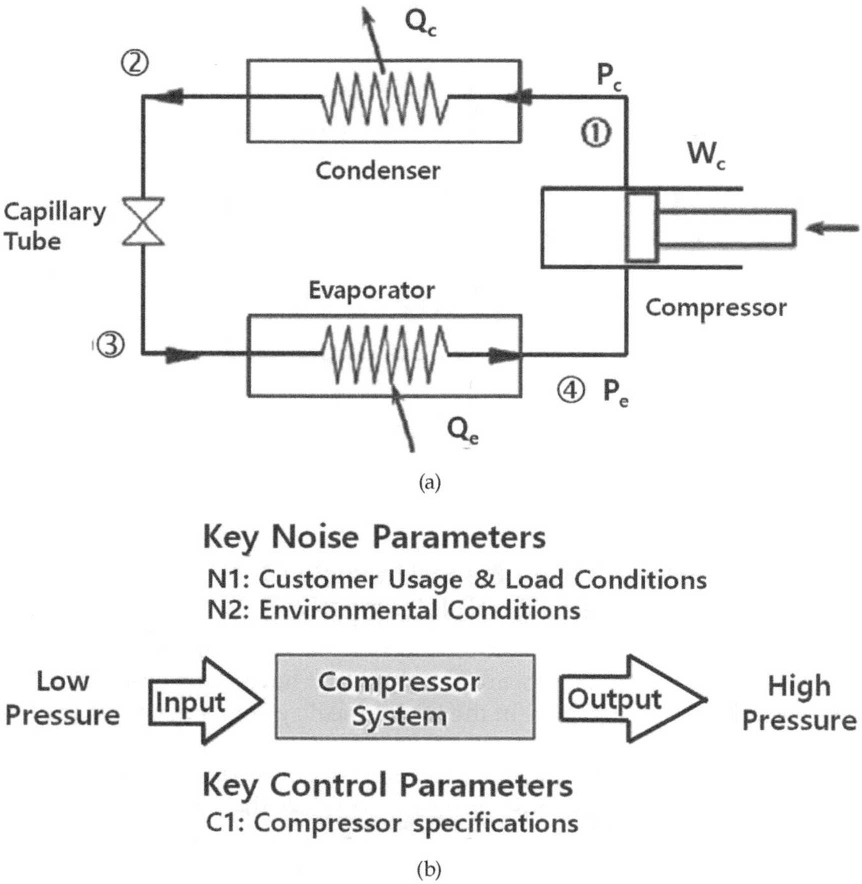

FIGURE 2.27
Schematic diagram for mechanical system (example: compressor system): (a) compressor in a vapor-compression refrigeration cycle; (b) parameter diagram of compressor system.

of the remaining variables and can lead to considerable statistical efficiency. The sequential designs of RSM decrease the number of experimental runs that Taguchi's designs would require. However, Taguchi's designs still demand a large number of experimental runs.

References

1. Feller, W. *An Introduction to Probability Theory and Its Applications*, Vol. 1, 3rd ed. John Wiley & Sons: New York, 1968.

2. Fisher, R.A.; Owen, A.R.G. An Appreciation for the Life and Work of Sir Ronald Aylmer Fisher. *Journal of the Royal Statistical Society. Series D (The Statistician).* **1962**, *12*(4), 313–319.
3. Jowett, G.H. The Relationship Between the Binomial and F Distributions. *Journal of the Royal Statistical Society, Series D (The Statistician).* **1963**, *13*(1), 55–57.
4. Haight, F.A. *Handbook of the Poisson Distribution*. John Wiley & Sons: New York, 1967.
5. Chiu, S.N.; Stoyan, D.; Kendall, W.S.; Mecke, J. *Stochastic Geometry and Its Applications*. John Wiley & Sons: New York, 2013.
6. Weibull, W. *The Statistical Theory of the Strength of Materials*. Ph.D. Thesis, Royal Technical University, Stockholm, 1939.
7. Rossi, R.J. *Mathematical Statistics: An Introduction to Likelihood Based Inference.* John Wiley & Sons: New York, 2018.
8. Klutke, G.; Kiessler, P.C.; Wortman, M.A. A Critical Look at the Bathtub Curve. *IEEE Transactions on Reliability.* **2003**, *52*(1), 125–129.
9. Taguchi, G. *Taguchi on Robust Technology Development*. ASME Press: New York, 1992, pp. 16–62.

3

Design of Mechanical System Including Mechanisms

3.1 Introduction

A mechanical system in power transmission converts (created) power to attain its purpose, which needs force (effort) and velocity (flow) and which supplies mechanical advantages through properly embracing the mechanisms. A mechanism in the product is a sort of kinematic chain that generates movements such as linear movement, oscillation, and rotary movement. And there are some procedures of power transmission in the mechanical system: 1) gear trains, 2) chain drive, 3) rope drive, and 4) belt drive. For example, an automobile utilized for transit is a four-wheeled vehicle that consists of engine, transmission, drive, and body structures [1]. Under the principle of Carnot cycle, an engine experiences four phases: 1) induction, 2) compression, 3) ignition, and 4) emission. That is, the automobile is designed to change fuel into mechanical energy through the engine, the slider crank mechanism, transmission containing gear system, rear drive shaft, rear differential, and vehicle axle. Finally, the car attains enough torque to go ahead (Figure 3.1).

In power transmission, a product which includes some modules may be exposed to various stresses due to repetitive loads. If there is a design flaw in the construction, it may cause structural failure, such as the (low/high) cycle fatigue. To design the mechanical system, an engineer might understand the product mechanism and the characteristics of systemic loading.

3.2 Mechanical Structure Including Mechanisms

3.2.1 Mechanical Mechanisms

Constructions such as frame, bridge, and building are an assembly of rigid bodies attached by joints without mobility and requiring no mechanism. They thus do not change the accessible energy into work and bear loads only.

DOI: 10.1201/9781003451471-3

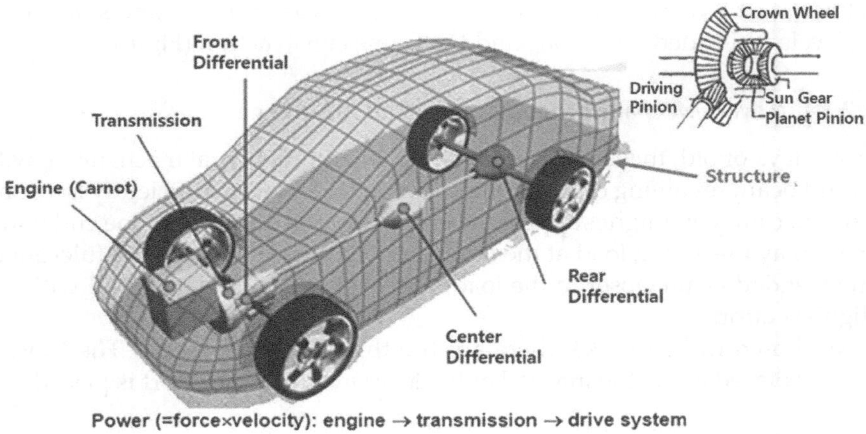

FIGURE 3.1
Automobile: cycle, structure, and mechanisms (example).

FIGURE 3.2
Golden Gate Bridge (Wikipedia).

The structure is designed to withstand loads for its own sake as a facility with some volume. For instance, numerous ancient structures, presided over by the Pharaohs of ancient Egypt and the Caesars of Rome, are still situated on ancient sites. In Europe, numerous bridges and buildings established during the Renaissance Period are still working. These constructions are constructed in thousands of forms and dissimilar dimensions in nature, which supplies them with built-in robustness (Figure 3.2).

On the other hand, a mechanism is a gathering of conveying components that carry out some tasks, producing movement by utilizing a slider crank. The mechanism and construction in a product are connected together as a positioning of components and supply sufficient strength and stiffness to support loads. Most mechanisms in the mechanical system are designed to supply a mechanical advantage. For example, a bicycle in transport is worked by a sizable driver gear wheel with pedals. When the back wheel rotates, the bicycle advances. A mechanism such as the chain and sprocket in a bicycle is decidedly built around the frame construction (Figure 3.3).

3.2.1.1 (Lever) Mechanisms and Mechanical Advantage

From days of old, the lever has been the most fundamental mechanism, with a rigid beam revolving on a fulcrum: "If you had a lever sufficiently long and a fit fulcrum, you might support the earth." A little effort from one end of the beam may raise a big load at the other end. That is, by moving the fulcrum of an extended beam closer to the load, the lever may lift a large load with the slightest effort.

As shown in Figure 3.4, there are the three kinds of levers. The Class 1 lever is placed along the interval of the lever in which the effort is placed. For

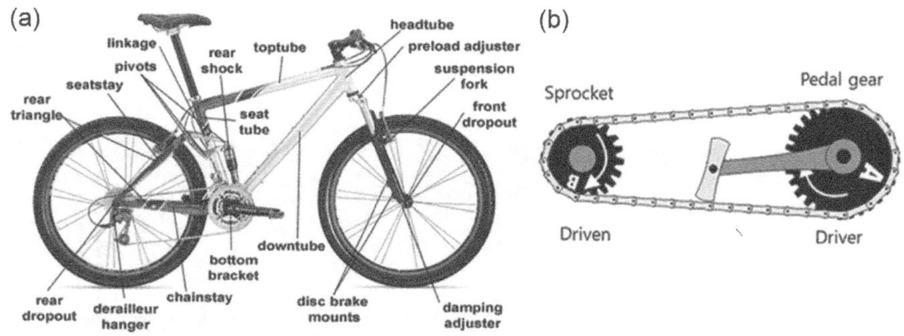

FIGURE 3.3
Bicycle and its mechanism: (a) bicycle (structure + mechanism); (b) chain and sprocket mechanism.

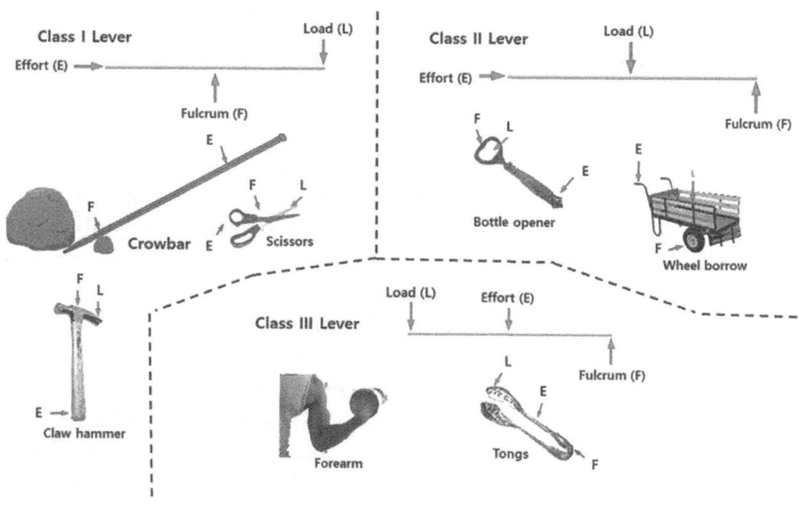

FIGURE 3.4
Three classes of levers.

instance, the crowbar is a kind of lever, and the pliers/scissors use two levers attached at the fulcrum. A bottle opener and wheelbarrow are instances of Class 2. The human forearm and tongs are instances of Class 3 levers.

Leonard Euler (1707–1783) was one of the first specialists to explore the mathematics of linkage design (synthesis). Most linkages are in the form of a plane, and their motion is confined to a flat surface. The common investigation of linkage movements, such as planar and spatial, is expressed as screw theory. Sir Robert Stawell Ball (1840–1913) is regarded as the pioneer of screw theory. During the Industrial Revolution, many of the weaving machines proposed complex apparatus to change a waterwheel's rotary movement into complicated movements. The design of the steam engine created an enormous need for new mechanisms and apparatus. Long linear motion travel was necessitated to robustly utilize steam power. In particular, although he did not comprehend it, James Watt (1736–1819) applied thermodynamics with rotary joints and long links to produce structured straight-line movement (Figure 3.5).

While utilizing a stick to lift a rock with a lever mechanism, the worker exerts a small effort at the end of the lever to change the rock (Class 1). Levers are used to alter the positioning, distance, velocity of movement in order to decrease the effort required to lift a load. This reducing effort is familiar as mechanical advantage. With its fulcrum at X, the lever is a bar of length AB, breaking the length of the bar into components L_1 and L_2. To lift a load W through a height of h, a force F might be utilized downward through a distance s. The triangles AXC and BXD are similar and proportional. In this case, we can ignore friction.

In the example, as the effort of 10 kg_f is required to lift a load of 50 kg_f, we may compute the mechanical advantage. Namely, the mechanical advantage

FIGURE 3.5
James Watt's industrial steam engine.

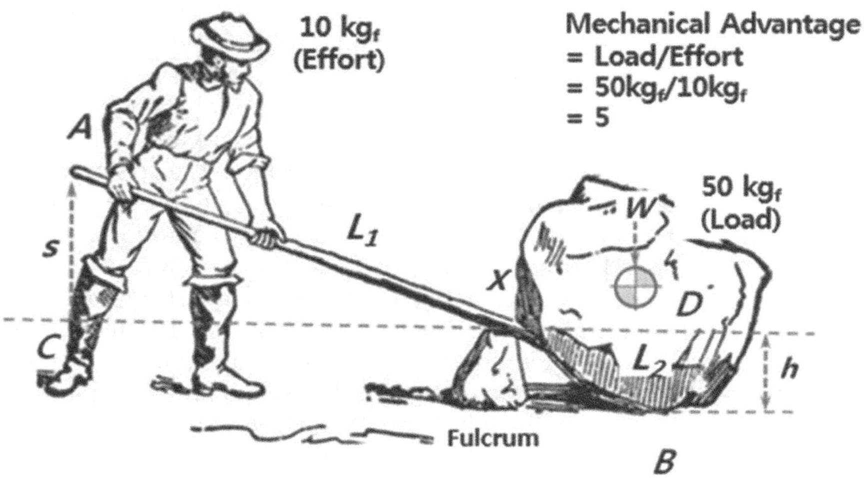

FIGURE 3.6
Mechanical advantage.

(MA) = load/effort = $s/h = L_1/L_2$ = 50 kg$_f$/10 kg$_f$. So this system has an MA of 5. There are no units for mechanical advantage because it is a proportion of the same units (Figure 3.6).

3.3 Design of Mechanisms

To achieve a group of performance requirements for the machinery, a mechanism design is necessitated to specify the dimensions, shapes, material layouts, and positioning of components by a kinematic analysis and synthesis. Mechanisms create a kinematic chain where no less than one link is fastened to a frame of reference (ground) and constructs four dissimilar movements, such as reciprocating motion, linear motion, rotary motion, and oscillation. Linear motion is a one-dimensional movement along a direct line. Reciprocating movement is again a back-and-forth movement. Rotary movement is the beginning point for numerous mechanisms, generally supplied by an engine or dc motors. Oscillation is a back-and-forth movement about a pivot location [2,3].

3.3.1 Classification of Mechanisms

There are three implementations of mechanisms: 1) path generation mechanisms: the path of a tracer point; 2) function generation mechanisms: the

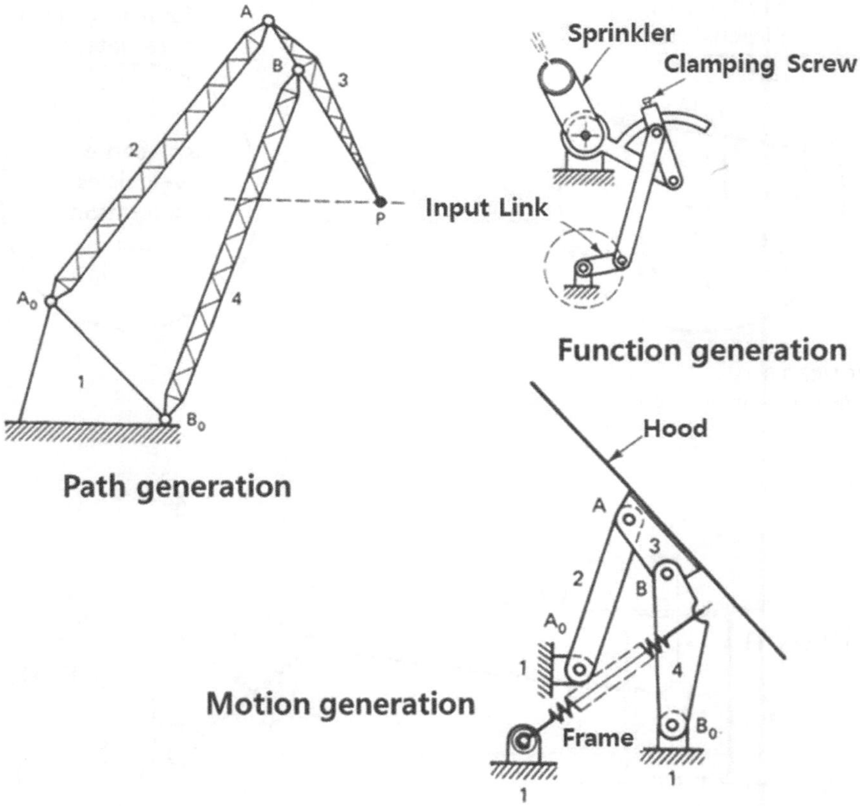

FIGURE 3.7
Classifications of mechanisms in application.

comparative movement between the links attached to the frame; and 3) motion generation mechanisms: the movement of the coupler link (Figure 3.7).

Analysis is a way of carrying out a particular task that allows an engineer to examine a formerly existing design to resolve its properness for the task. The inspection of mechanisms and machineries thus aims to perceive the relationship between the motions of the machinery parts (kinematics) and the forces that construct the motions (kinetics). In a mechanism design, kinematic analysis is required to decide the displacement, rotation, velocity, and acceleration of the mechanism. On the other hand, synthesis is a series of steps of running a scheme of carrying out an assigned aim. That is, it is a procedure of developing a mechanism to achieve a group of performance requirements for the machinery. Mainly in mechanisms, synthesis is the design of a linkage to produce a required output movement for a described input movement (Figure 3.8).

A kinematic linkage is a gathering of links and joints that provides an intended output movement as a reaction to a specified input movement. As

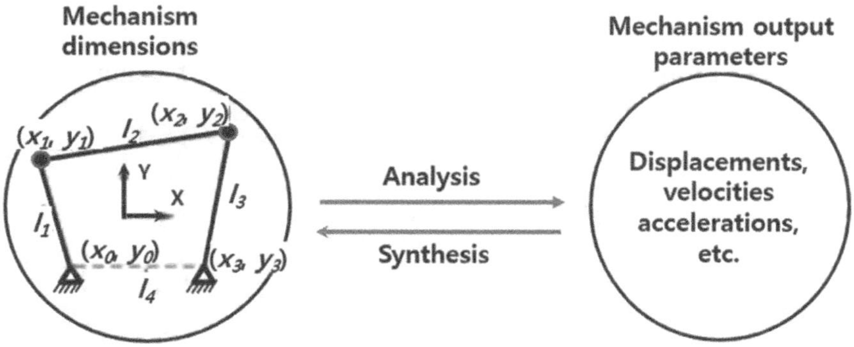

FIGURE 3.8
Kinematic synthesis versus analysis.

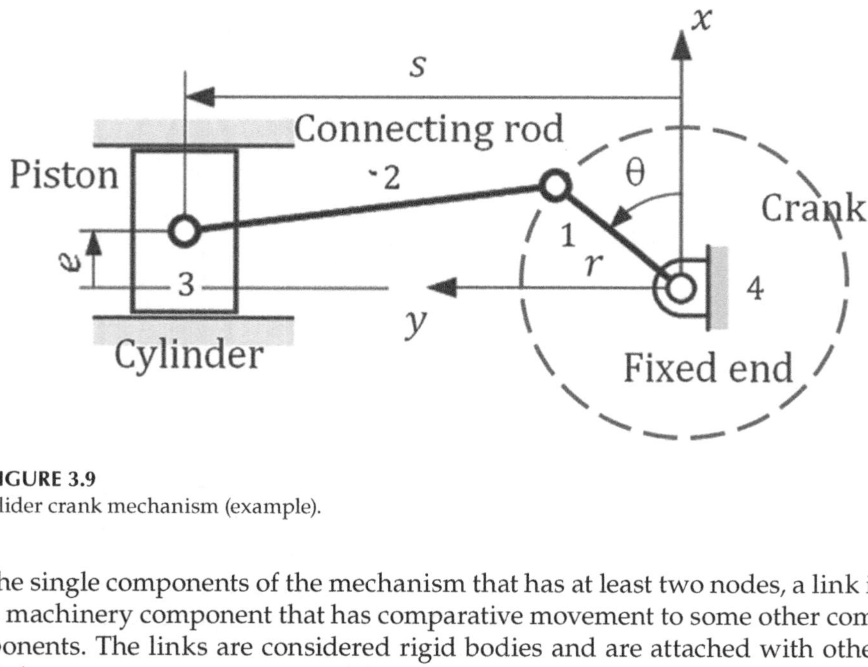

FIGURE 3.9
Slider crank mechanism (example).

the single components of the mechanism that has at least two nodes, a link is a machinery component that has comparative movement to some other components. The links are considered rigid bodies and are attached with other links to convey movement and force. It is a gathering of resistant structures fastened in such manner that they make a single component of it and have no comparative movement between them. Mainly, in a slider crank mechanism, the crank (Link 1) has movement relative to the attaching rod and piston (Links 2 and 3) and the frame (Link 4) (Figure 3.9).

Kinematic links are categorized as 1) lower joint and higher joint, 2) closed joint and unclosed joint, and 3) sliding and turning joint. A joint (or pair) is a portable attachment between links and permits relative movements between the links. As a joint in movement makes surface contact between its elements,

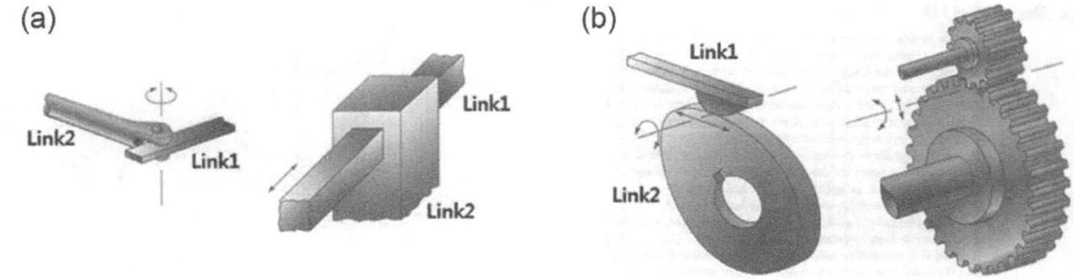

FIGURE 3.10
Kinematic links and joints: (a) primary joints: pin and sliding; (b) higher-order joints: cam joint and gear joint.

it is named a lower joint. For instance, a shaft rotates in a bearing. As a joint in movement has a line or point contact between its elements, it is called a higher joint; for instance, a ball and shaft bearing. Revolute and sliding joints are primary joints. If the elements of the joint in a mechanical system are clasped jointly, they make a closed joint; for instance, all lower joints. If the two elements of the joint in mechanical system are not clasped jointly, they make an unclosed joint; for instance, a cam and follower. Two elements in a joint have sliding motion relative to each other; for instance, a piston and cylinder joint. If two components are attached such that one element rotates around the other, it amounts to a turning point. It is also familiar as a revolute pair; an instance is a slider crank mechanism (Figure 3.10).

3.3.2 Kinematic Chain and Mobility

A kinematic chain is a merging of kinematic joints connected in such a manner that the relative movement between them is entirely restricted. It is categorized: 1) four bar chain, 2) slider crank chain, 3) double slider crank chain, etc. The degree of freedom (DOF) is the numeric of self-determining inputs required to precisely position all links of a mechanism with regard to the ground or the numeric of actuators needed for the mechanism to function. That is, it is provided by mobility M:

$$M = \text{degree of freedom} = 3(n-1) - 2jp - jh \tag{3.1}$$

where n is the whole numeric of links, jp is the numeric of primary joints, and jh is the numeric of higher-order joints.

3.3.3 Kinematic Model/Diagram

A kinematic model is a description of a mechanism that only shows the dimensions that influence the motion. To produce the kinematic prototype from a real mechanism, detach any redundant details and make the drawing

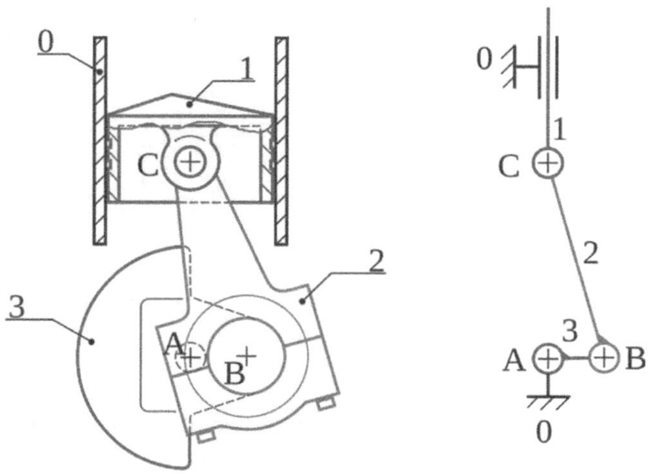

FIGURE 3.11
Engine and its kinematic mechanism.

of the mechanism simpler for additional inspection. Finally, perform the displacement, velocity, and acceleration analysis of a mechanism in machinery (Figure 3.11).

Example 3.1

As seen in Figure 3.12, a shear press is employed to cut the electronic circuit board laminates. Find a kinematic diagram.

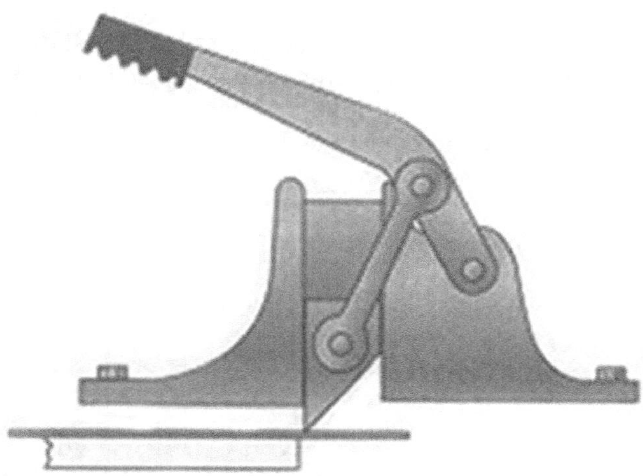

FIGURE 3.12
Shear press (example).

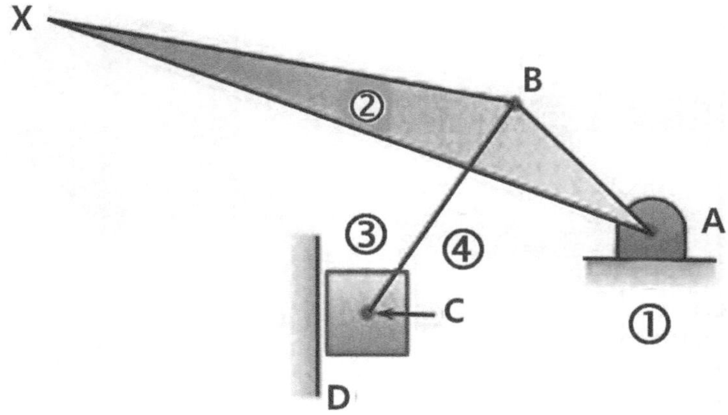

FIGURE 3.13
Kinematic diagram of the shear press.

Solution

The first step in constructing a kinematic diagram is to discover the link component that may be designated as the frame. The kinematic diagram is identified in Figure 3.13. The motion of all other links may be resolved relative to the frame. In some examples, its choice is clear as the frame is affixed to the ground. In this problem, the large base bolted to the table is nominated as the frame. The movement of all other links is determined relative to the base (frame).

The base is assigned as link 1. Careful watching manifests three other moving components: ② handle, ③ cutting blade, and ④ bar, which fastens the cutter with the handle. Pin joints are employed to fasten these three dissimilar components. Pin joints are tagged as *A–C*. The cutter also slides up and down along the base. This sliding joint is tagged as *D*. Ultimately, the movement of the end of the handle is expected. This is nominated as the point of interest *X*.

Example 3.2

Figure 3.14 shows the vice grips that are similar and are intended to be used together. Sketch a kinematic diagram.

Solution

The first step is to find out the component, which shall be nominated the frame. In this problem, no components are added to the ground. Thus the option of the frame is rather random. The top handle is nominated as the frame. The movement of all other links is resolved relative to the top handle.

The upper handle is assigned as link 1. Careful watching manifests three other moving components: ② the bottom handle, ③ the bottom jaw, and ④ the bar that fastens the upper and lower handles. Four-pin joints are utilized to fasten these dissimilar components. These joints are labeled A–D.

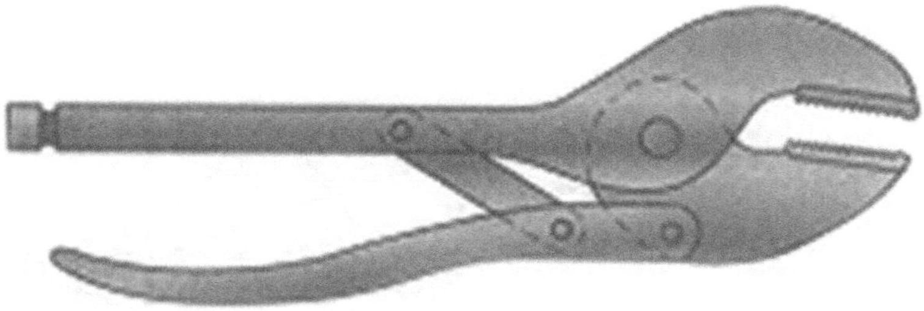

FIGURE 3.14
Pair of vice grips (example).

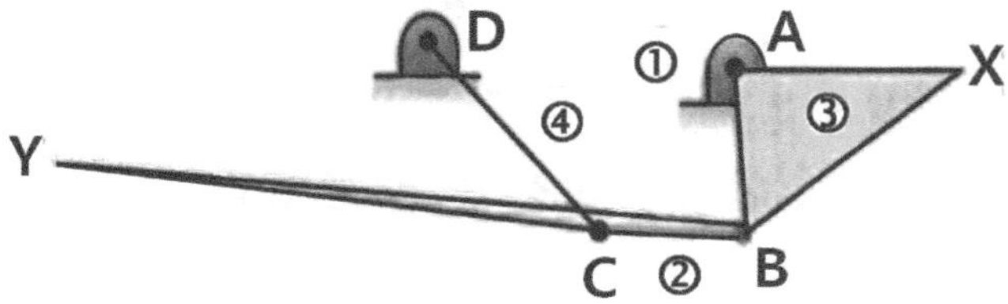

FIGURE 3.15
Kinematic diagram of vice grips.

Furthermore, the movement of the end of the bottom jaw is required. This is specified as the point of interest X. Eventually, the movement of the end of the lower handle is also required. This is appointed as the point of interest Y. The kinematic diagram is specified in Figure 3.15.

3.3.4 Grashof's Law and Some Inversion Mechanisms

Grashof's law expresses that, for a planar four-bar linkage system, the whole amount of the shortest and longest link lengths may not be greater than the whole amount of the existing two link lengths if there is to be a continual relative rotation between the two members. Algebraically, assume s = length of the shortest link, l = length of the longest link, p and q = the lengths of other links (Figure 3.16),

$$s + l \leq p + q \tag{3.2}$$

3.3.4.1 Inversions of Four Bar Mechanisms

Some main inversions of four bar mechanisms are the 1) coupled wheel locomotive, 2) beam engine, 3) pantograph, and 4) Watt mechanism. The coupled wheel locomotive consists of four links. A coupling rod or side rod attaches the driving wheels of a locomotive. Steam locomotives generally have them. The opposite links are identical in their lengths. Because Links 1 and 3 are cranks, the mechanism is also familiar as double crank mechanism (Figure 3.17).

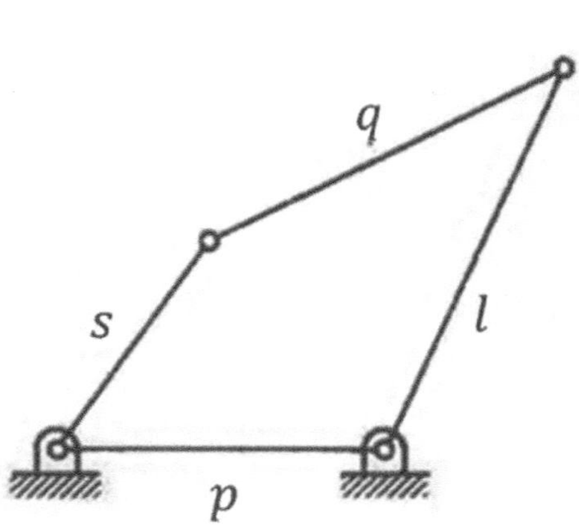

FIGURE 3.16
Planar four-bar linkage.

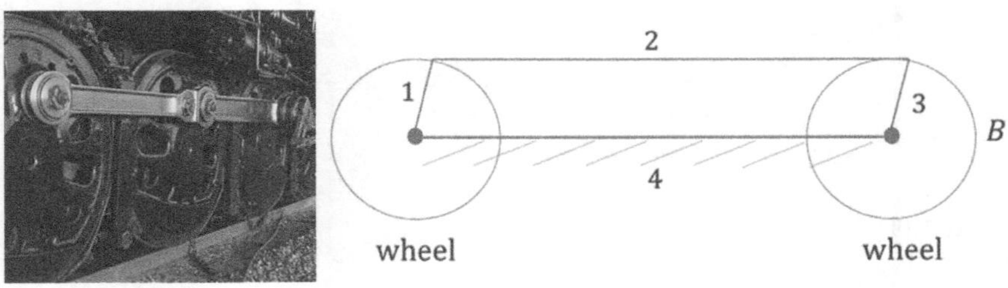

FIGURE 3.17
Coupled wheel locomotive.

A beam engine, familiar as a crank and lever mechanism, a sort of steam engine where a pivoted overhead beam is utilized to exert the force from an upright piston to an upright connecting rod. The mechanism is utilized to change rotary movement into reciprocating movement. Link 3 swings about point C. Link 4 is fastened (Figure 3.18).

As shown in Figure 3.19, pantograph is a mechanical linkage attached in a way established on parallelogram so that the motion of one pen, in tracking down an image, creates similar motions in a second pen. If a line drawing is tracked down by the first point, a similar, enlarged, or miniaturized copy may be drawn by a pen fastened to the other. That is, it may replicate a displacement in a lessened or enlarged scale and be utilized for copying the drawings, maps, plans, etc. This may be discovered in add-on arms for barrier-mounted mirrors, momentary fences, pantographic knives, scissor lifts, and other scissor mechanisms such as the pantograph utilized on electric locomotives and trams. All the four links are turning in nature, and the

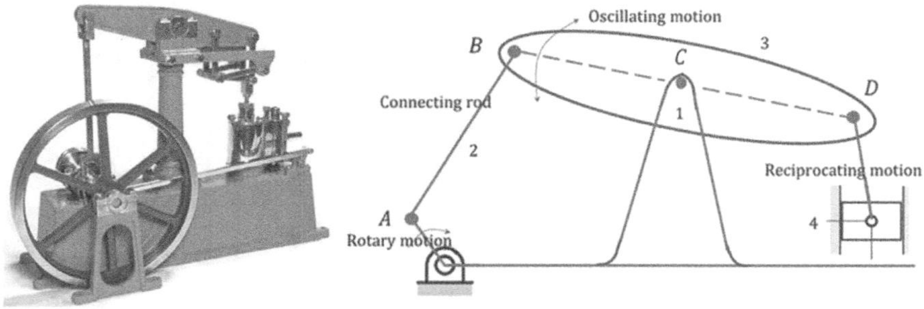

FIGURE 3.18
Beam engine.

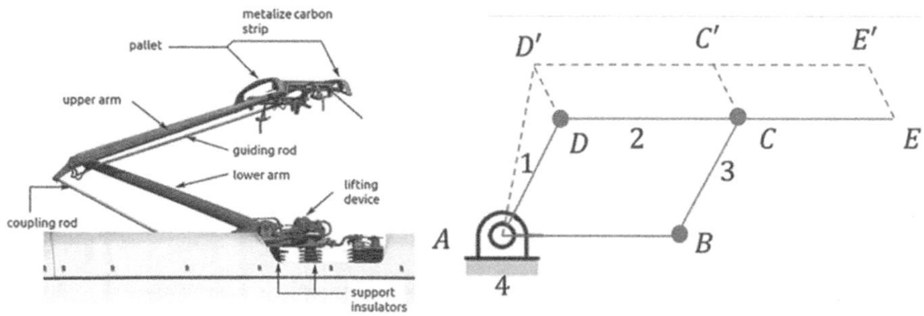

FIGURE 3.19
Pantograph.

mechanism is fundamentally a quadratic cycle in the shape of a parallelogram as stated by the geometry ($AC / AC' = AE / AE'$).

The Watt mechanism (the parallel linkage) is a sort of mechanical linkage originated by James Watt in which his steam engine may lead the piston rod, and the central moving point of the linkage is compelled from traveling on an estimation to a linear line. It is utilized in automobile suspensions, permitting the shaft of a vehicle to move vertically while stopping sideways movement. Lines OA and BC are parallel in the average location of the mechanism. They are the levers of the mechanism and attached with a Link AB. For small displacement of levers OA and BC, D may track down a roughly straight line on AB, such that ($AD / DB' = OA / BC'$) (Figure 3.20).

3.3.4.2 Inversions of Slider Crank Chain

Some main inversions of a slider crank chain are as follows: 1) slider crank mechanism, 2) hand pump mechanism, 3) oscillating cylinder engine,

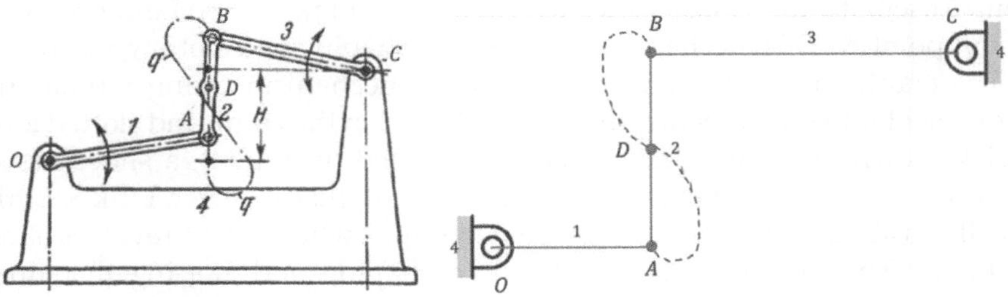

FIGURE 3.20
Watt mechanism.

4) crank and slotted lever quick return motion mechanism, 5) Whitworth quick return motion, and 6) rotary engine.

Inversion of slider crank mechanism is a positioning of mechanical components, which is designed to change rotary movement to straight-line movement. Link 2 in the slider crank mechanism is the crank. Link 3 is the connecting rod between Links 2 and 4, which is the slider. Link 1 is the frame fastened. Fastening Link 1 and revolving Link 2 will make Link 4 slide in Link 1, transferring rotary movement into oscillatory movement (Figure 3.21).

Hand pumps are a sort of manually worked positive displacement pump. They have been utilized for centuries, but this straightforward technology remains the central component of rustic water stocks in many countries. Hand pumps have a straightforward and robust design, need little maintenance, and are uncomplicated to work; this has led to their common implementation all over the world (Figure 3.22).

An oscillating cylinder engine is a straightforward steam engine design that needs no valve gear (Figure 3.23). On other hand, the crank and slotted lever quick return mechanism changes the rotary movement into reciprocating movement. Link 2 in the crank and slotted lever quick return motion mechanism is fixed. Link 4, a slider reciprocates in an oscillating lever, Link 3.

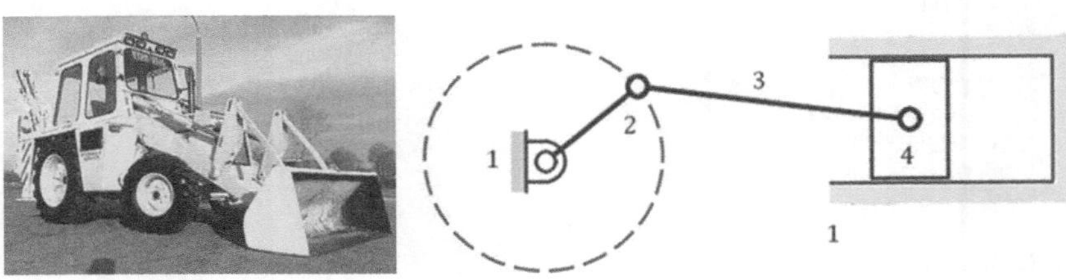

FIGURE 3.21
Inversion of slider crank mechanism.

Link 3 oscillates about O, pivoted to the fastened Link 2 when Link 1 revolves about point A. This mechanism is utilized in shaping and slotting machines.

The crank and slotted lever quick return mechanism changes rotational movement into reciprocating movement. Link 2 in the crank and slotted lever quick return movement mechanism is fastened. In Link 4, a slider moves backward and forward in a straight line in an oscillating lever, Link 3. Link3 oscillates about O, pivoted to the fastened Link 2 when Link 1 revolves about point A. This mechanism is utilized in shaping and slotting machines. The Whitworth quick return mechanism may change circular motion into

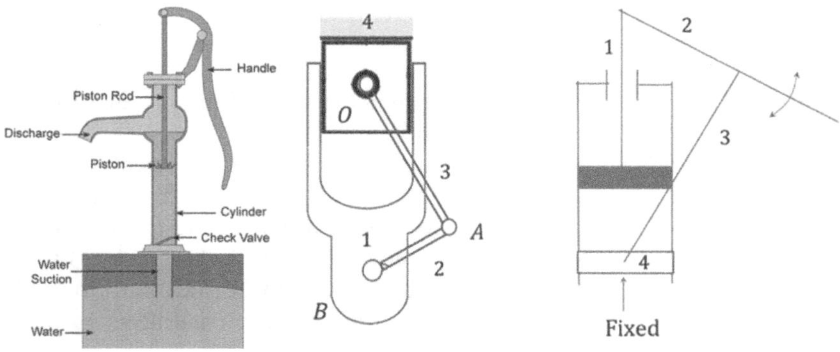

FIGURE 3.22
Hand pumps.

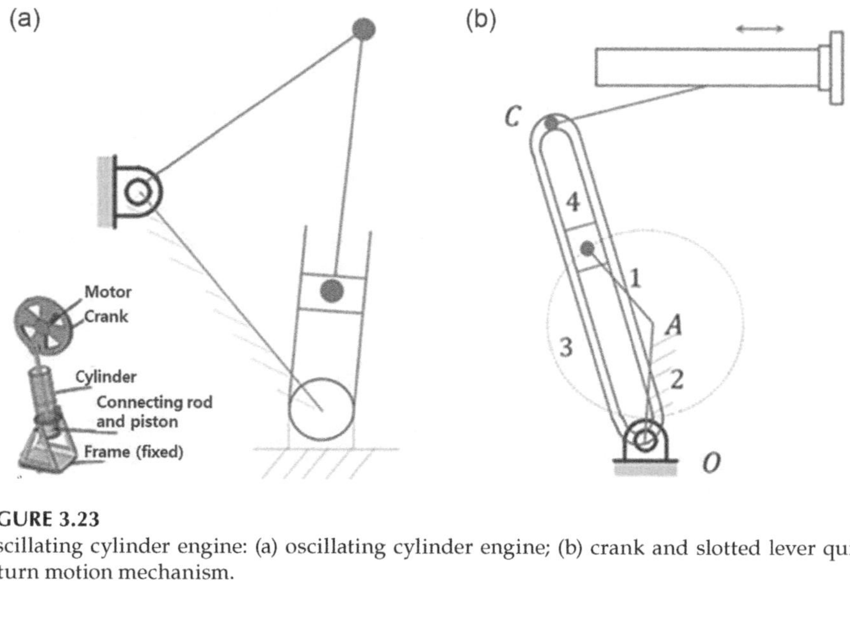

FIGURE 3.23
Oscillating cylinder engine: (a) oscillating cylinder engine; (b) crank and slotted lever quick return motion mechanism.

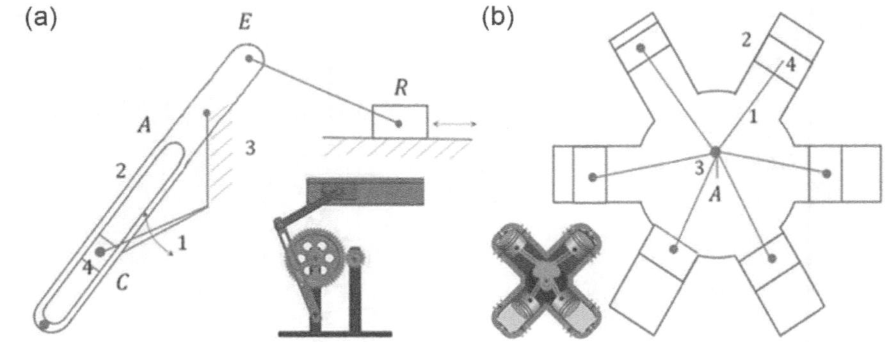

FIGURE 3.24
Inversions of slider crank chain (instance): (a) Whitworth quick return motion; (b) rotary engine.

reciprocating motion. Link 3 is fastened. Link 1 revolves about point B. The Link 4 slider is in the slotted Link 2, which oscillates about pivot point A when Link 1 rotates. The rotary engine is an untimely kind of internal combustion engine, normally designed with an odd number of cylinders per row in a radial configuration. In the rotary engine, the slider is replaced by six pistons, pivoted at point A. At this place, the crank (Link 3) is fastened. All cylinders, i.e., Link 2, revolve about the same fastened-point A. The pistons (Link 4) oscillate in these revolving cylinders' symmetrical positioning of seven or nine (Figure 3.24).

3.3.4.3 Inversions of Double Slider Crank Chain

Some main inversions of the double slider crank chain are as follows: 1) elliptical trammel, 2) Scotch yoke mechanism, and 3) Oldham's coupling. A trammel of Archimedes is a mechanism that leads to the form of an ellipse. It is made up of the two shuttles that are enclosed at right angle channels or rails and a rod that is fastened to the shuttles by pivots at fastened locations along the rod. Link 4 is the frame that is inserted and fastened. Links 1 and 3 are sliders that proceed vertically and horizontally. Link 2 is the connection for Links 1 and 3 and is extended to point C in a small amount (Figure 3.25(a)). The Scotch yoke mechanism is a reciprocating movement mechanism, changing the linear movement of a slider into the rotational movement of a crank or vice versa. It is attained when Link 1 (slider) is fastened and pivoted to the frame at point A. Link 2 is a crank that revolves entirely. The rotation of Link 2 makes Link 3 slide in the frame. So, with Link 4, the frame reciprocates. Rotary motion therefore is changed to reciprocating movement (Figure 3.25(b)). The third inversion of the double slider

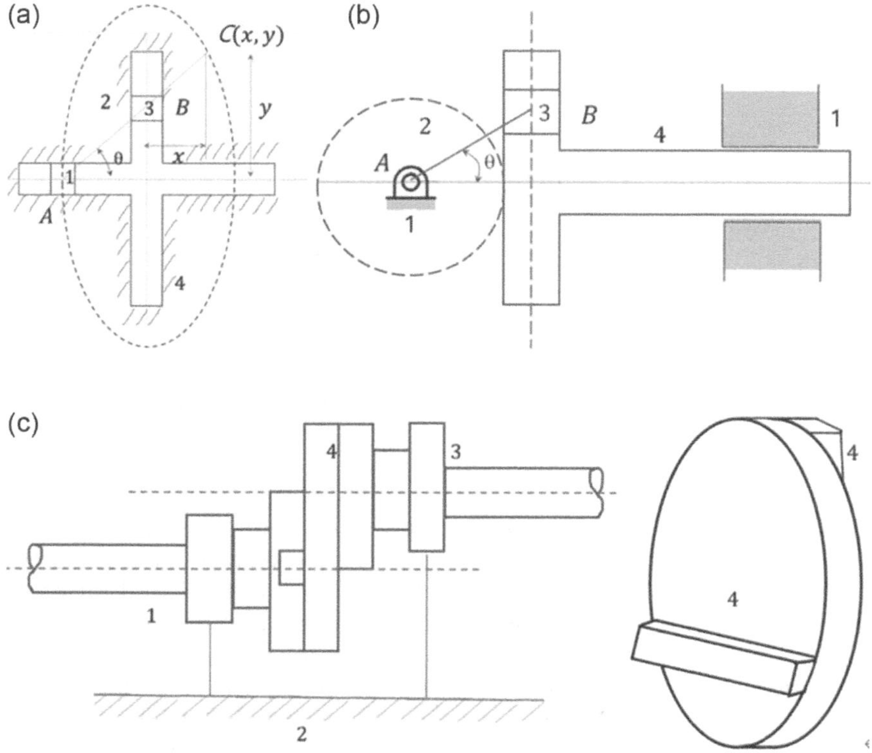

FIGURE 3.25
Inversions of double slider crank chain (example): (a) elliptical trammel; (b) Scotch yoke mechanism; (c) Oldham's coupling.

crank mechanism is Oldham's coupling. It has the shape of flexible coupling designed for implements that should be free from backlash. This is utilized to join the two parallel shafts whose axes are at a small-scale interval away from each other. In Link 2, the frame is fastened. Link 3 is made up of a fitted shaft (Figure 3.25(c)).

There are other intermittent movement mechanisms: 1) Geneva wheel, 2) ratchet and pawl, and 3) Ackermann steering. The Geneva wheel mechanism relocates a continual rotation motion into intermittent rotary movement. The revolving wheel is furnished with a pin that reaches into a slot found in the other wheel which moves it forward by one step per period. The ratchet and pawl mechanism permits the shaft to revolve in one orientation only. A ratchet is a wheel with a shape alike to a circular saw blade or a horizontal milling cutter (Figure 3.26).

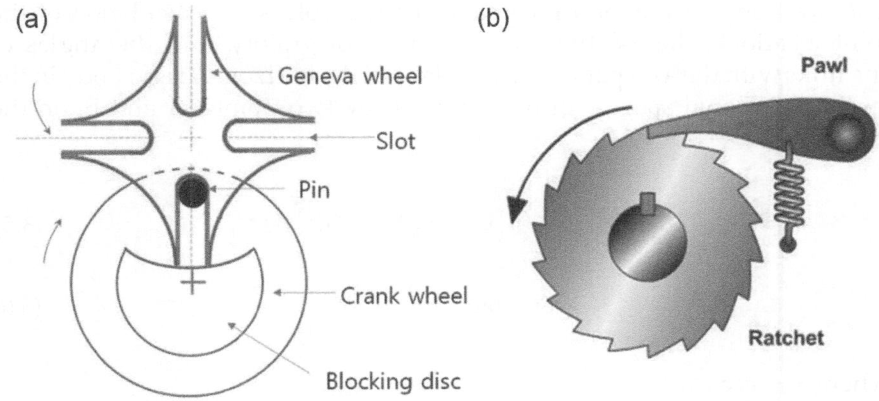

FIGURE 3.26
Intermittent movement mechanisms (example): (a) Geneva wheel; (b) ratchet and pawl.

3.3.5 Kinematic Analysis of Mechanisms

3.3.5.1 Position Analysis Utilizing the Absolute Cartesian Method

There are the two ordinary coordinate systems: Cartesian (R_x, R_y) and polar (R_A, θ. We can change between Cartesian and polar (Figure 3.27):

$$R_A = \sqrt{R_x^2 + R_y^2} \text{ or } \theta = \arctan\left(R_y / R_x\right) \tag{3.3}$$

where $R_x = R_A\cos\theta$ or $R_y = R_A\sin\theta$

Relative position may be expressed:

$$R_{B/A} = R_B - R_A \tag{3.4}$$

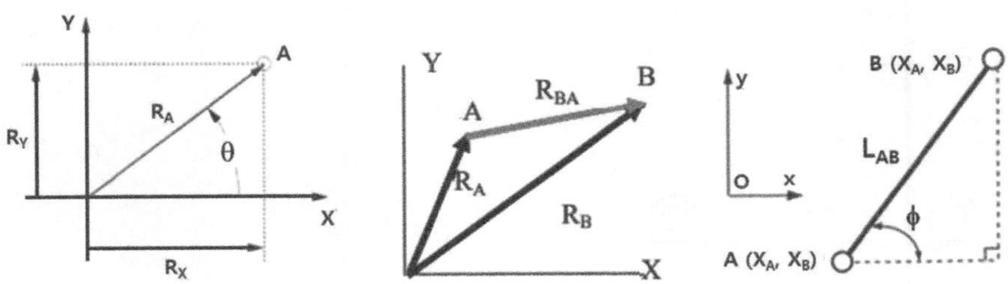

FIGURE 3.27
Coordinate systems.

The position analysis of a kinematic chain involves the calculation of the joint locations, the location of the centers of gravity, and the angles of the links with the axis parallel to the plane of the horizon. A rigid body in the two-dimensional space will be expressed by two points, A and B, on the rigid body.

$$(x_A - x_B)^2 + (y_A - y_B)^2 = AB^2 = L_{AB}^2 \tag{3.5}$$

$$m = \tan\phi = \frac{y_B - y_A}{x_B - x_A} \tag{3.6}$$

where $y = mx + n$

3.3.5.2 Graphical Method for Analyzing Velocity and Acceleration

First, we graphically examine the velocity of each of points B and C in a four-link bar (Figure 3.28).

As Link AB revolves continuously, we may state the velocity at point B:

$$V_B = ab = \omega \cdot AB \tag{3.7}$$

where AB is the length of link.

From the velocity diagram, we discover velocity V_{CB} at the intersection. Thus angular velocity is discovered:

$$\omega_C = \frac{V_{CB}}{CB} \tag{3.8}$$

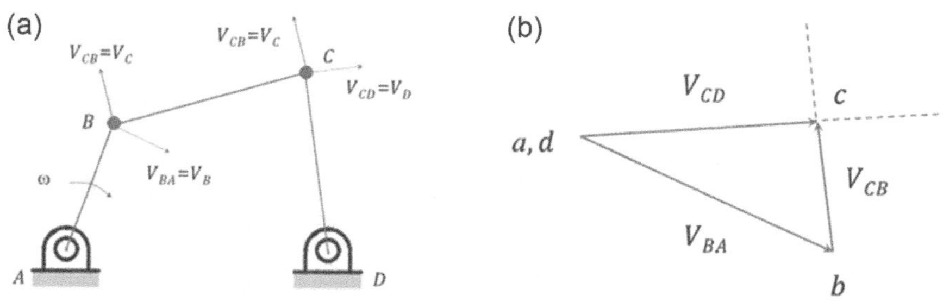

FIGURE 3.28
Velocity at four-link bar: (a) velocity in four-link bar; (b velocity diagram.

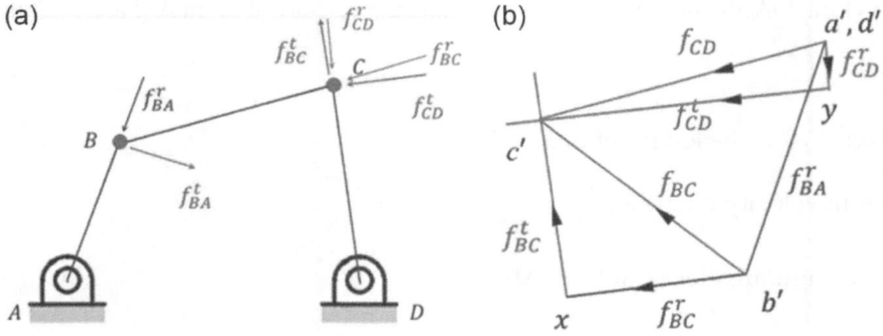

FIGURE 3.29
(a) Acceleration in four-link bar; (b) acceleration diagram.

where *CB* is the length of link. We realize that at point B, Link BC revolves counterclockwise. The acceleration of the four-link bar is computed with clear and vividly explicit details (Figure 3.29).

The tangential component of acceleration at Link AB, seen in Figure 3.29, is

$$f_{BA}^{t} = 0 \tag{3.9}$$

The radial component of acceleration at Link AB is

$$a'b' = f_{BA}^{r} = \omega^2 \cdot AB \tag{3.10}$$

The radial component of acceleration at Link BC is

$$b'x = f_{BC}^{r} = \frac{V_{BC}^2}{BC} \tag{3.11}$$

The radial component of acceleration at Link C is,

$$a'y = f_{CD}^{r} = \frac{V_{CD}^2}{CD} \tag{3.12}$$

The angular acceleration at Link BC is

$$\alpha_{BC} = \frac{f_{BC}^{t}}{BC} = \frac{x_c}{BC} \tag{3.13}$$

The angular acceleration at Link CD is

$$\alpha_{CD} = \frac{f_{CD}^{t}}{CD} = \frac{y_c}{CD} \tag{3.14}$$

As Link OA always revolves, we may state the velocity at point A (Figure 3.30):

$$oa = V_A = \omega \cdot OA \quad (3.15)$$

where OA is the length of link.

From velocity diagram, $\frac{ac}{ab} = \frac{AC}{AB}$ or $ac = ab \cdot \frac{AC}{AB}$.

The angular velocity at Link AB is

$$\omega_{AB} = \frac{V_{AB}}{AB} \text{ or } \omega_c = \frac{V_c}{AC} \quad (3.16)$$

where AB and AC are the lengths of the link.

The tangential component of acceleration at Link AO is (Figure 3.31)

$$f_{AO}^t = 0 \quad (3.17)$$

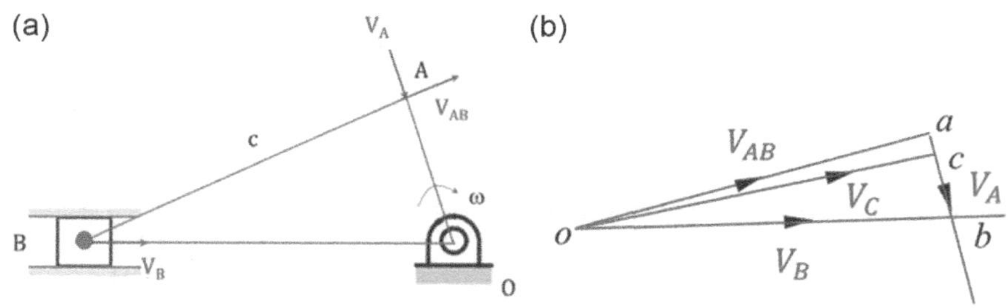

FIGURE 3.30
Velocity at the slider crank mechanism: (a) velocity in slider crank; (b) velocity diagram.

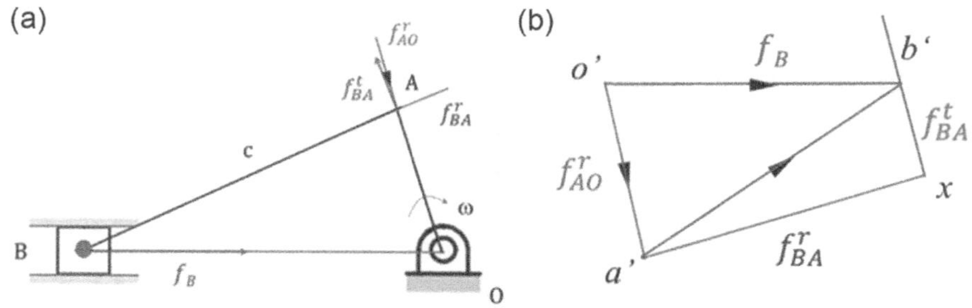

FIGURE 3.31
Acceleration in slider crank: (a) acceleration in slider crank; (b) acceleration diagram.

The radial component of acceleration at Link AO is

$$o'a' = f^r_{AO} = \omega^2 \cdot OA \tag{3.18}$$

The radial component of acceleration at Link AB is

$$a'x = f^r_{AB} = \omega^2_{AB} \cdot AB \tag{3.19}$$

Example 3.3

Because N_{BA} = 100 RPM, the velocity of Link AB is (Figure 3.32)

$$\omega_{AB} = \frac{2\pi N_{BA}}{60} = \frac{2\pi \cdot 100}{60} = 10.47 \text{ rad / sec}$$

$$V_{AB} = \omega_{AB} \cdot AB = 10.47 \cdot 30 = 314.1 \text{ mm / sec}$$

From the velocity diagram, because 100 mm/sec = 1 mm, $V_{AB} = 31.41$ mm long (vector length),
velocity V_{BC} is

$$V_{BC} = 12 \text{mm} \cdot 10 = 120.0 \text{ mm / sec}$$

$$\omega_{BC} = \frac{V_{BC}}{BC} = \frac{120}{120} = 1 \text{rad / sec}$$

On the other hand, velocity V_{CD} is

$$V_{CD} = 26 \text{mm} \cdot 10 = 260.0 \text{ mm / sec}$$

$$\omega_{CD} = \frac{V_{CD}}{CD} = \frac{260}{60} = 4.33 \text{ rad / sec}$$

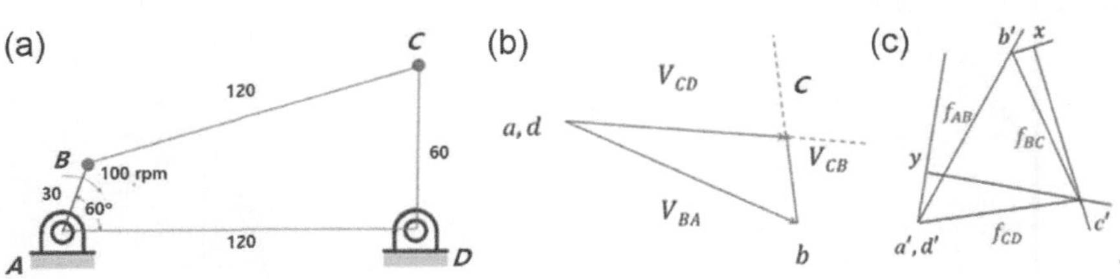

FIGURE 3.32
Four-link bar mechanism: (a) four-link bar; (b) velocity diagram; (c) acceleration diagram.

The acceleration at Link AB is $f_{BA}^t = 0$:

$$\text{vector}\, a'b' = f_{BA}^r = \omega_{AB}^2 \cdot AB = 10.47^2 \cdot 30 = 3,288.63\ \text{mm/sec}^2$$

The acceleration at Link BC is

$$\text{vector}\, b'x = f_{BC}^r = \omega_{BC}^2 \cdot BC = 1.0^2 \cdot 120 = 120\ \text{mm/sec}^2$$

Acceleration at Link C is

$$\text{vector}\, a'y = f_{CD}^r = \omega_{CD}^2 \cdot CD = 4.33^2 \cdot 60 = 1,124.93\ \text{mm/sec}^2$$

The angular accelerations at Links BC and CD are

$$\alpha_{BC} = \frac{f_{BC}^t}{BC} = \frac{x_c}{BC} = \frac{2,600}{120} = 21.66\ \text{rad/sec}^2,$$

$$\alpha_{CD} = \frac{f_{CD}^t}{CD} = \frac{y_c}{CD} = \frac{2,500}{60} = 41.66\ \text{rad/sec}^2$$

3.3.5.3 Instantaneous Center of Velocity

For anybody engaging in planar movement, there always exists a point in the flat surface of movement at which the velocity is instantly zero (if it is attached to the body in a stiff way). This point is named the instant center of zero velocity (IC). If the particular position of this point may be controlled, the velocity analysis may be clarified because the body looks to revolve about this point at that instant.

Example 3.4

A four-link bar connected by a pin is as seen in the Figure 3.33. As Link AB revolves at 100 RPM, discover the angular velocity of the Link BC utilizing an instantaneous center of rotation when the angle BAD is 60°.

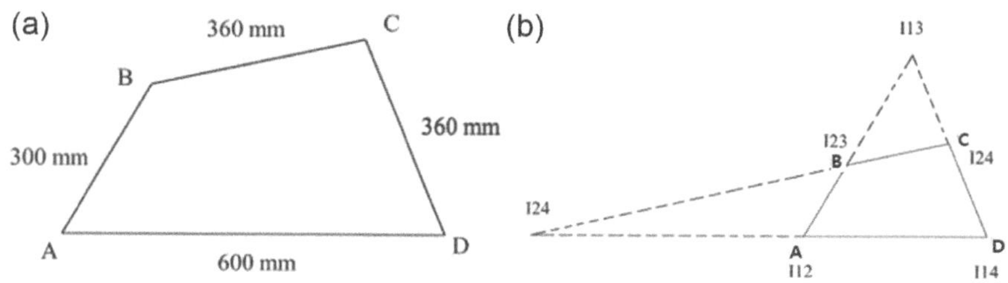

FIGURE 3.33
Four-link bar joined by pin: (a) four-link bar; (b) instantaneous center.

$$\omega_{AB} = \frac{2\pi N_{AB}}{60} = \frac{2\pi \cdot 100}{60} = 10.47 \text{ rad/sec}$$

Because AB = 300 mm = 0.30 m, the velocity at point B is

$$V_{BA} = \omega_{AB} \cdot AB = 10.47 \cdot 0.30 = 3.141 \text{ m/sec}$$

$$V_B = \omega_{BC} \cdot \text{I13B}$$

$$\omega_{AB} = \frac{V_B}{\text{I13B}} = \frac{3.141}{0.495} = 6.345 \text{ rad/sec}$$

Example 3.5

A slider crank mechanism is as seen in the Figure 3.34. The crank BC revolves at 100 RPM. Determine the angular velocity of the link BC utilizing the instantaneous center of rotation when angle BCA is 45°.

$$\omega_{BC} = \frac{2\pi N_{BC}}{60} = \frac{2\pi \cdot 100}{60} = 10.47 \text{ rad/sec}$$

As BC = 220 mm = 0.22 m, the velocity at point B is

$$V_B = \omega_{BC} \cdot BC = 10.47 \cdot 0.22 = 2.303 \text{ m/sec}$$

$$V_A = V_B \cdot \frac{\text{I13A}}{\text{I13B}} = 2.303 \cdot \frac{0.68}{0.76} = 2.061 \text{ m/sec}$$

$$\omega_{AB} = \frac{V_B}{\text{I13B}} = \frac{2.303}{0.76} = 3.030 \text{ rad/sec}$$

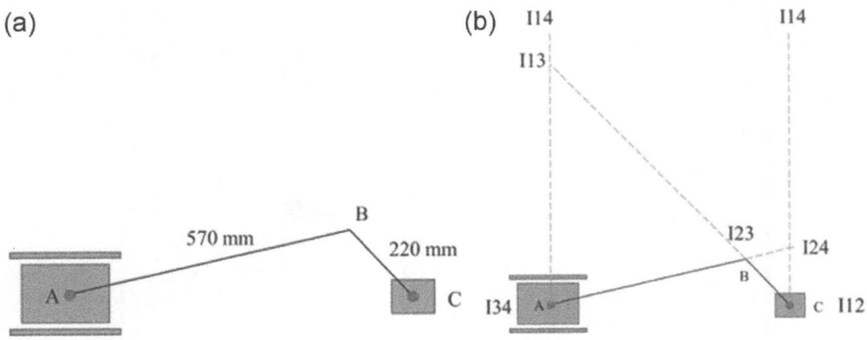

FIGURE 3.34
Angular velocity of the slider crank mechanism: (a) slider crank; (b) instantaneous center.

3.3.5.4 Klein's Construction

Klein's construction is utilized to produce a picture of the velocity and acceleration drawings for a sole slider crank mechanism. The velocity and acceleration of piston for a reciprocating engine mechanism may be decided by Figure 3.35.

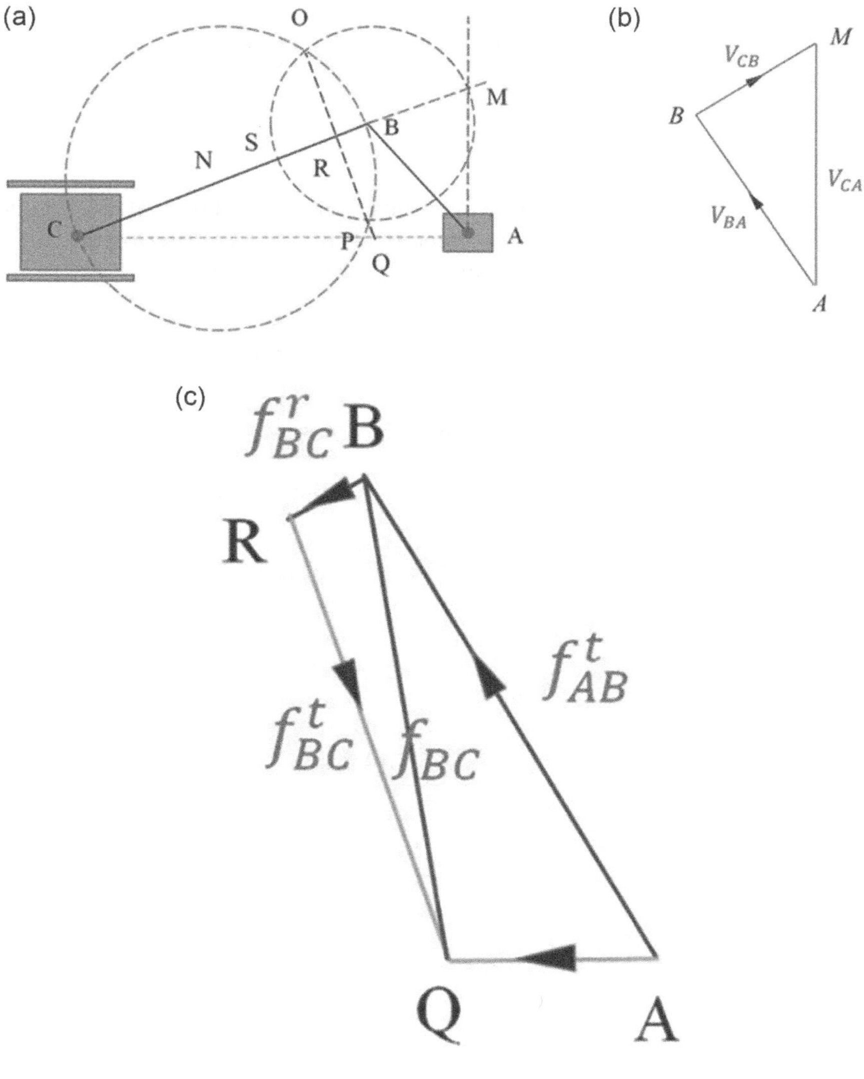

FIGURE 3.35
Klein's construction: (a) reciprocating engine mechanism; (b) velocity diagram; (c) velocity diagram.

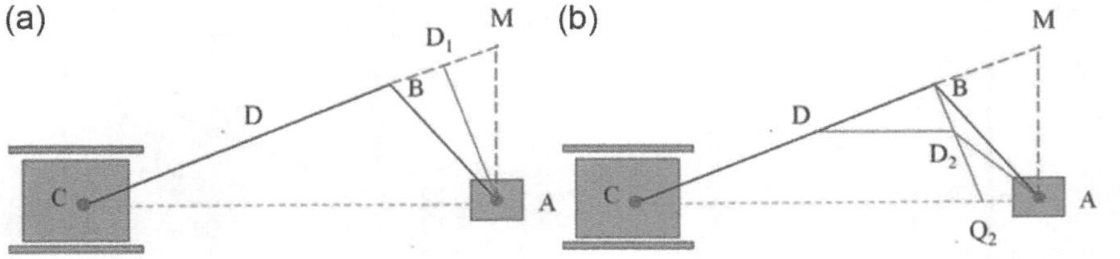

FIGURE 3.36
Klein's construction in reciprocating engine: (a) velocity; (b) acceleration.

The acceleration of the piston C with respect to crank pin B (i.e., f_{BC}) can be attained:

$$\omega^2 = \frac{f_{BC}}{BQ} \tag{3.20}$$

To discover the velocity of any point D on the attaching rod BC, split BM at f_{BC} in the same proportion as D splits BC (Figure 3.36(a)).

$$\frac{BD_1}{BM} = \frac{BD}{BC} \tag{3.21}$$

Velocity at point D is

$$V_D = \omega \times AD \tag{3.22}$$

To discover the acceleration of any point D on the attaching rod BC, draw a line from point D|| to AC that intersects BQ_2 of D_2 (Figure 3.36(b)).

$$f_D = \omega^2 \times AD_2 \tag{3.23}$$

Example 3.6

Decide the velocity and acceleration of the piston in a slider crank mechanism by Klein's construction to the subsequent specification: 1) stroke = 300 mm, 2) proportion of length of connecting rod to crank = 4, 3) speed of the engine = 300 RPM, and 4) piston of crank = 45° with inner dead center (Figure 3.37).

Solution

$$\omega_{OA} = \frac{2\pi N}{60} = \frac{2\pi 300}{60} = 31.41 \text{ rad / sec}$$

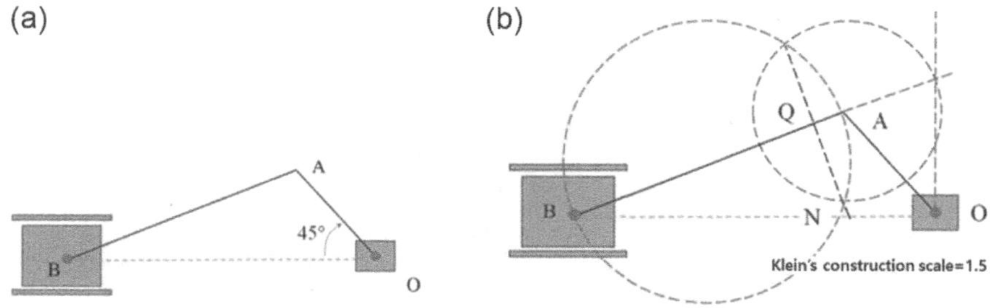

FIGURE 3.37
Slider crank mechanism by Klein's construction: (a) slider crank mechanism; (b) Klein's velocity polygon.

As engine speed is 300 RPM, ·OAM is Klein's velocity polygon:

$$V_B = OM \times \omega \times SF = 24.5 \times 31.41 \times 5 = 3{,}847.725 \text{ rad / sec}$$

To discover the acceleration of the piston,

$$f_B = \omega^2 \times ON \times SF = 31.41^2 \times 21 \times 5 = 1.035 \times 10^5 \text{ mm / sec}^2$$

Example 3.7

The extents of the crank and connecting rod of a reciprocating engine are 200 mm and 800 mm, respectively. The crank is rotating at a constant speed of 480 RPM. Utilizing Klein's construction, discover the acceleration of the piston (Figure 3.38).

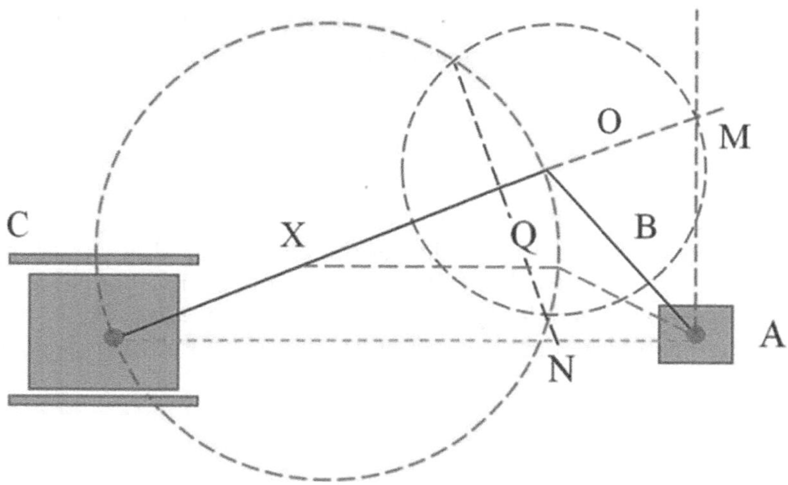

FIGURE 3.38
Klein's velocity polygon.

Solution

The acceleration of the central point of the attaching rod and angular acceleration of the connecting rod when the crank had turned through 45° from the inner dead center.

$$\omega_{\text{crank}} = \frac{2\pi N}{60} = \frac{2\pi \times 480}{60} = 50.26 \text{ rad / sec}$$

Acceleration of piston is

$$f_c = \left[AN - \omega^2\right] = \left[14 \times 10 \times \omega^2\right] = 3537 \times 10^5 \text{ mm / s}^2$$

The angular acceleration of connecting rod is

$$a_{ac} = \frac{f'BC}{BC} = \frac{QN \times \omega^2}{BC} = 14 \times 10 \times 50.26^2 / 800 = 442.06 \text{ rad / s}^2$$

3.3.5.5 Velocity and Acceleration Analysis of Mechanisms (Example)

To finish the kinematics analysis of mechanisms, we will talk about almost all the usual ways to describe the movement through velocity and acceleration. First, position and velocity analysis must be finished before acceleration analysis. All acceleration components might be defined as a one coordinate system—the inertial frame of reference of the fastened link of the mechanism. For instance, we straightforwardly think that a link at center O_2 revolves with radius R_A. Determine velocity and acceleration at point A in Figure 3.39.

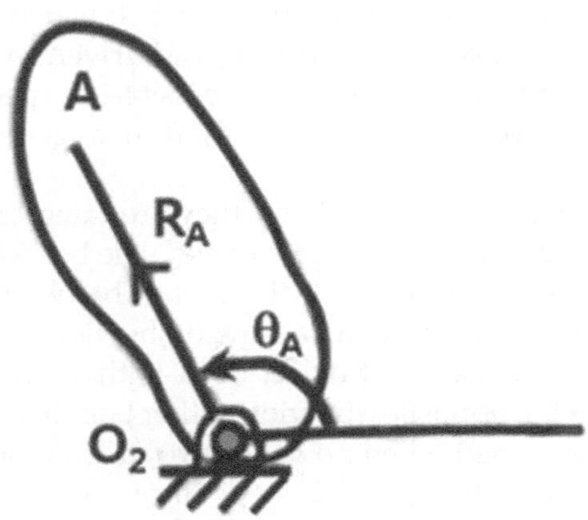

FIGURE 3.39
Analytical solution (example).

As this link is defined as a polar coordinate, we can outline the displacement of point A:

$$R_A = r_A e^{i\theta_A} \tag{3.24}$$

If Equation (3.19) is differentiated, we will find the velocity of the link:

$$V_A = \frac{dR_A}{dt} = i\omega_2 r_A e^{i\theta_A} \tag{3.25}$$

If Equation (3.20) is differentiated, we may find the acceleration of the link:

$$\begin{aligned} a_A &= \frac{dV_A}{dt} = i\frac{d\omega_2}{dt} r_A e^{i\theta_A} + i\omega_2 r_A \frac{d\left(e^{i\theta_A}\right)}{dt} \\ &= i\alpha_2 r_A e^{i\theta_A} + i\omega_2 r_A i\omega_2 e^{i\theta_A} \\ &= \underbrace{i\alpha_2 R_A}_{a_A^t} - \underbrace{\omega_2^2 R_A}_{a_A^n} = \left(-\omega_2^2 + i\alpha_2\right) R_A \end{aligned} \tag{3.26}$$

3.4 Design of the Belt Drive

3.4.1 Introduction

A belt is an easily bent mechanical component that moves power from one shaft to another. The amount of power moved by the belt depends on 1) the belt velocity, 2) the belt tension, 3) the contact arc between the smaller pulley and belt, and 4) the situation under which the belt is used. Namely, we may select the belt drive: 1) speed of the driving and driven shafts, 2) power to be moved, 3) positive drive needs, 4) space accessible, 5) speed reduction ratio, 6) center interval between the shafts, 7) shaft arrangement, and 8) service situations [4].

As stated by the material used in belts, they are categorized as 1) cotton or fabric belts, 2) leather belts, and 3) rubber belts. The belt kind is as follows: 1) flat belt, 2) V-belt, and 3) circular belt. There are the two belt drives: 1) open belt drive: employed with shafts arrayed side by side and revolving in the identical location; 2) crossed belt drive: used with shafts arranged side by side and rotating in the opposite direction. Other belt drives are 1) compound belt drives, which are used when power is moved from one shaft to another through a number of pulleys, and 2) stepped or cone pulley drives, which are used for modifying the speed of the driven shaft while the major shaft proceeds at constant speed.

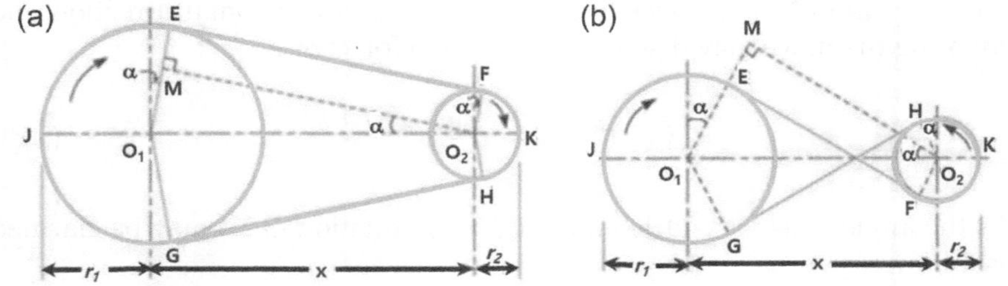

FIGURE 3.40
Schematic diagrams of (a) open and (b) closed belt.

3.4.2 Length of the Belt and Contact Angle

As the contact angle may be estimated as $\alpha \approx \frac{(r1\text{-}r2^{\circ})}{x}$, we can compute the interval of the open belt drive (Figure 3.40).

$$L=\pi(r1+r2)+2x+(r1\text{-}r2)\sin\alpha = \pi(r1+r2)+2x+\frac{(r1\text{-}r2)^2}{x} \tag{3.27}$$

where $\sin\alpha \approx \frac{(r1\text{-}r2^{\circ})}{x}$.

3.4.3 Proportion of Driving Tensions for Flat Belt

Consider a driven pulley rotating in the clockwise direction, as represented in Figure 3.41.

Consider a little bit of the belt PQ, forming an angle $\delta\theta$ at the middle point of the pulley. The belt PQ is in equilibrium under a driven pulley: 1) tension T at P; 2) Tension $(T+\delta T)$ at Q; 3) normal reaction R_N; 4) friction force, $F = \mu R_N$

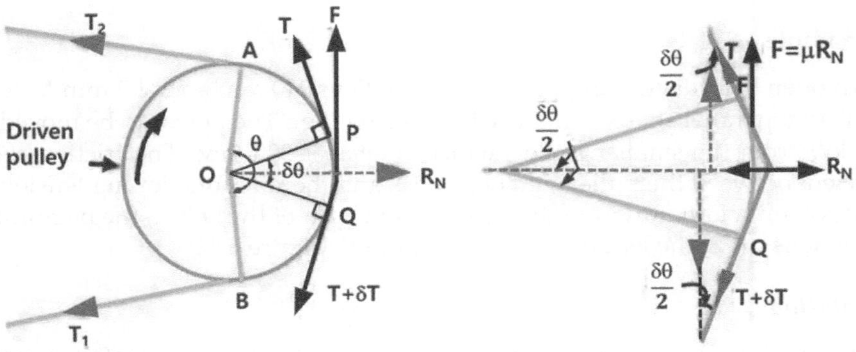

FIGURE 3.41
Free body diagram near the belt contact point.

Having found the answer to all the forces in a horizontal direction and equating them, we may discover the reaction force:

$$R_N = (T+\delta T)\sin\frac{\delta\theta}{2} + T\sin\frac{\delta\theta}{2} \tag{3.28}$$

As the angle $\delta\theta$ is very little and $\sin\frac{\delta\theta}{2} = \frac{\delta\theta}{2}$, Equation (3.28) may be clarified as follows:

$$RN = (T+T)\sin\frac{\delta\theta}{2} + T\sin\frac{\delta\theta}{2} = \frac{T\delta\theta}{2} + \frac{T\delta\theta}{2}(\approx 0) + \frac{T\delta\theta}{2} = T\delta\theta \tag{3.29}$$

Finding the forces in the vertical, we have

$$\mu RN = (T+\delta T)\cos\frac{\delta\theta}{2} - T\cos\frac{\delta\theta}{2} \tag{3.30}$$

Because the angle $\delta\theta$ is very little, $\cos\frac{\delta\theta}{2} = 1$.

$$\mu RN = T+\delta T-T=\delta T \tag{3.31}$$

Equating the values of R_N:

$$T\delta\theta = \delta T/\mu \text{ or } \delta T/T = \mu\delta\theta \tag{3.32}$$

Integrating both sides between the limits T_2 and T_1 and from 0 to θ, individually:

$$\int_{T_2}^{T_1} \frac{\delta T}{T} = \mu\int_0^\theta \delta\theta \text{ or } log\left(\frac{T_1}{T_2}\right) = \mu\theta\,, \frac{T_1}{T_2} = e^{\mu\theta} \tag{3.33}$$

Example 3.8

An open belt functioning over the two pulleys 240 mm and 600 mm fastens the two parallel shafts 3 m away from each other. The power to be moved is 4 kW from the smaller pulley, which rotates at 300 RPM. The friction coefficient between the pulley and belt is 0.3, and the safe functioning tension is 10 N/mm width. Resolve the 1) minimum width of the belt, 2) the beginning belt tension, 3) the length of the belt required (Figure 3.42).

Solution

Specified data: D_1 = 600 mm, D_2 = 240 mm, P = 4 kW, μ = 0.3, safe tension = 10 N/mm width

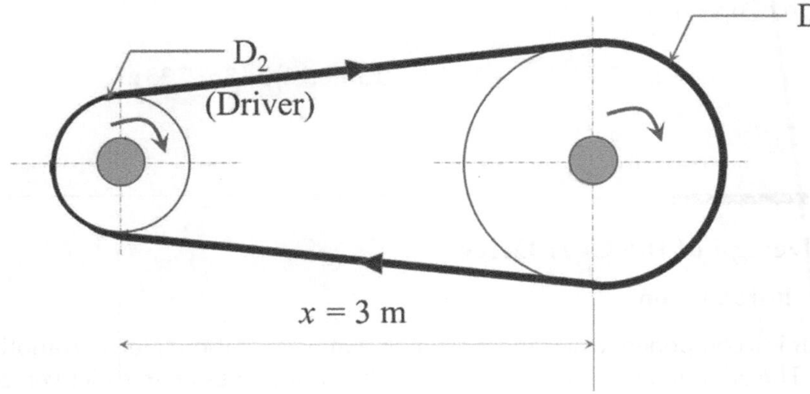

FIGURE 3.42
Open belt rotating over two pulleys.

The velocity of the belt examining a smaller pulley is defined as follows:

$$V_2 = \frac{\pi D_2 N_2}{60} = \frac{\pi \times 0.24 \times 300}{60} = 3.77 \text{ m/s}$$

At the moment, the power moved by the belt driver is

$$P = (T_1 - T_2) \cdot V_2$$

$$\text{Namely, } 4 \times 10^3 = (T_1 - T_2) \cdot 3.77 \text{ or } (T_1 - T_2) = 1{,}061$$

The lap angle for the open belt drive is presented:

$$\theta = 180^\circ - 2\alpha = 180 - 2 \times 3.44 = 173.12^\circ \times \frac{\pi}{180} = 3.022^\circ \text{ radians}$$

where $\alpha = \sin^{-1}\left(\frac{D_1 - D_2}{2x}\right) = \sin^{-1}\left(\frac{0.6-0.24}{2\times3}\right)$ =3.44 degrees

The belt tension ratio is

$$\frac{T_1}{T_2} = e^{\mu\theta} = e^{0.33.022} = 2.5$$

$(2.5T_2 - T_2) = 1{,}061$. That is,
$T_2 = 718N$, $T_1 = 1{,}779N$

Because 1mm width = 10 N,b = width of belt needed = 1,779/10 = 177.9 mm

Initial tension in the belt (T_o):

$$T_0 = (T_1 + T_2)/2 = (1779 + 718)/2 = 1248.5\text{N} = 1{,}248.5 \text{ N}$$

Length of open belt drive:

$$L = \pi(0.3+0.12)+(2\times3)+(0.3-0.12)2/3 = 7.33\,\text{m}$$

3.5 Design of the Gear Drive

3.5.1 Introduction

A gear is a component of a transmission that moves rotational force to another gear. The advantages of a gear drive are that 1) it moves at an exact velocity ratio, 2) it may be used to move large powers, 3) it has high efficiency, 4) it has fine service, and 5) it has close-packed positioning. On the other hand, the disadvantages of a gear drive are that 1) the manufacture of gears requires specific tools and apparatus, 2) the error in cutting teeth can produce vibration and noise in working [5].

3.5.2 Kinds of Gears

In a manner conforming with the position axes of the shafts, gears may be grouped as shown in Figure 3.43.

3.5.3 Nomenclature of (Spur) Gears

Some nomenclature is utilized in (spur) gears (Figure 3.44):

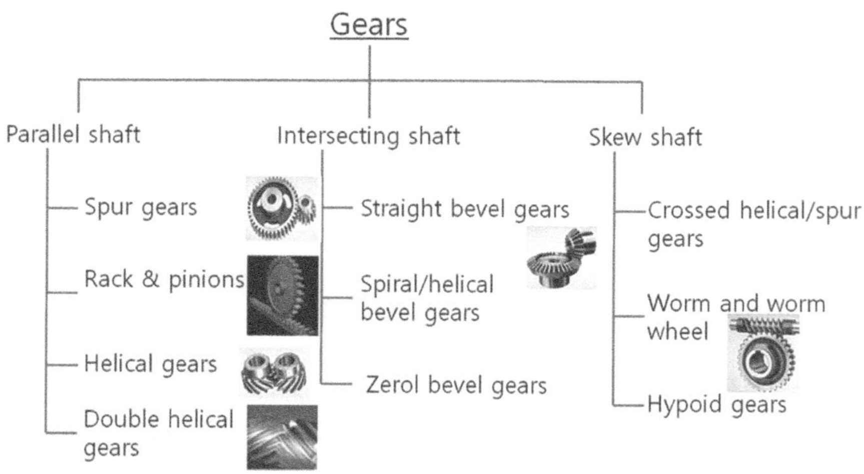

FIGURE 3.43
Gears in a manner conforming with the positions of axes of the shafts.

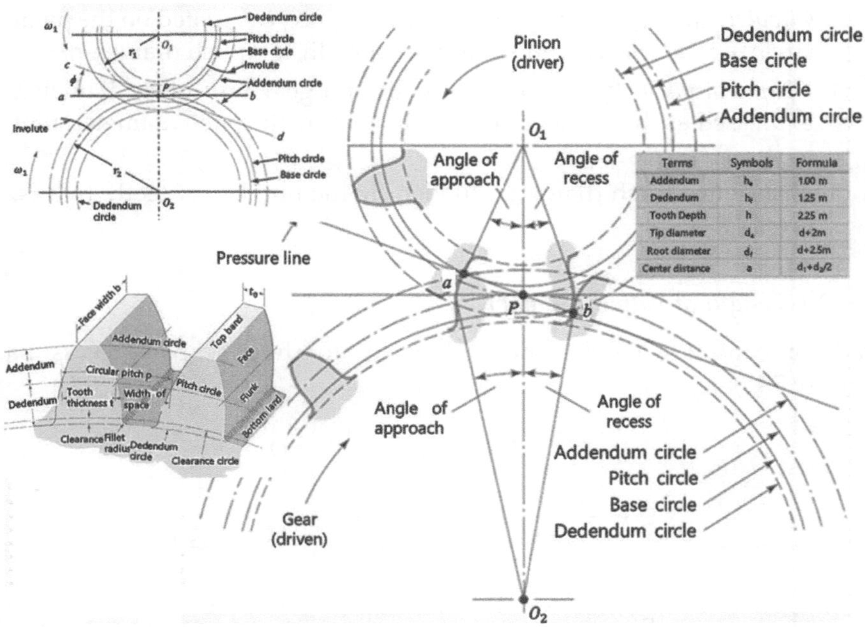

Terms	Symbols	Formula
Addendum	h_a	1.00 m
Dedendum	h_f	1.25 m
Tooth Depth	h	2.25 m
Tip diameter	d_a	d+2m
Root diameter	d_f	d+2.5m
Center distance	a	d_1+d_2/2

FIGURE 3.44
Nomenclature of (spur) gears.

1. *Pitch surface*: the outer surface of an imaginary friction wheel that's used to set gear teeth
2. *Pitch circle*: the circumference utilized as the standard of a pitch to indicate the dimensions of gear teeth
3. *Addendum circle*: a circle touching the outermost points of the teeth of a circular gear wheel
4. *Root circle*: the circle bounding the spaces between the teeth
5. *Addendum*: radial distance between the addendum circle and pitch circle
6. *Dedendum*: radial distance between the root circle and pitch circle
7. *Clearance*: difference between the addendum of the mating gear and dedendum of one gear
8. *Face of a tooth*: part of tooth surface lying outside the pitch surface
9. *Flank of a tooth*: part of tooth surface lying inside the pitch surface
10. *Circular thickness*: thickness of tooth calculated on the pitch circle
11. *Tooth space*: interval between adjacent teeth calculated on pitch circle
12. *Backlash*: interval between the circle thickness of one gear and the tooth space of the mating gear

13. *Circular pitch* (P_c): width of a tooth and a space, calculated on the pitch circle. $p = \pi d / T$ where T = number of teeth, d = pitch diameter.
14. *Diametral pitch* (P_d): number of teeth of a gear unit pitch diameter, expressed as the number of teeth divided by the pitch diameter; that is, $P = T / d$.
15. *Module* (m): pitch diameter divided by the number of teeth; that is, $m = d/T$.

3.5.3.1 Length of Path of Contact

Path of contact (*CD*) is expressed as approach path plus recess path (Figure 3.45):

$$CD = CP + PD = (FC - FP) + (ED - EP)$$

$$= \sqrt{R_a^2 - R^2\cos^2\phi} - R\sin\phi + \sqrt{r_a^2 - r^2\cos^2\phi} - r\sin\phi$$

$$= \sqrt{R_a^2 - R^2\cos^2\phi} - (R + r)\sin\phi + \sqrt{r_a^2 - r^2\cos^2\phi} \tag{3.34}$$

where $FC = \sqrt{R_a^2 - R^2\cos^2\phi}$, $FC = R\sin\phi$, $ED = \sqrt{r_a^2 - r^2\cos^2\phi}$, $EP = r\sin\phi$.

3.5.3.2 Derivation of Arc of Contact

As t_a is the period of travel across the approach arc, the approach arc PP′ is

$$PP' = [\text{tangential velocity of } P'] \times t_a = \omega_a r \times t_a = \omega_a (r\cos\phi)\frac{1}{\cos\phi} \times t_a$$

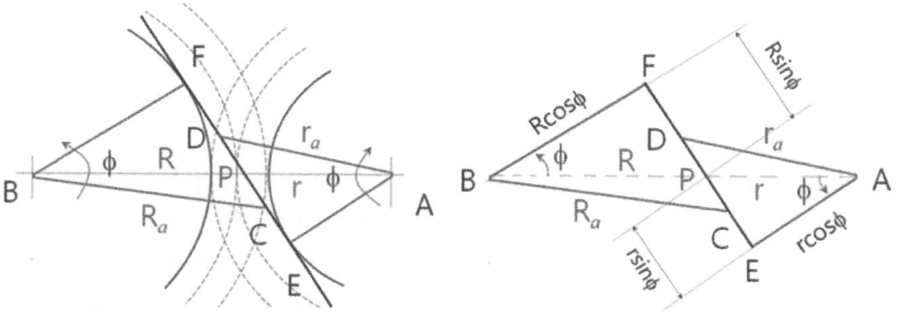

FIGURE 3.45
Contact path of (spur) gears.

$$= [\text{tangential velocity of H}] \times \frac{t_a}{\cos\phi}$$

$$= \frac{HK}{\cos\phi} = \frac{FK - FH}{\cos\phi} = \frac{FP - FC}{\cos\phi} = \frac{CP}{\cos\phi} \tag{3.35}$$

Recess arc of PP″ is

$$PP'' = [\text{tangential velocity of P}] \times t_a = \omega_a r \times t_a = \omega_a (r\cos\phi) \frac{1}{\cos\phi} \times t_a$$

$$= [\text{tangential velocity of K}] \times \frac{t_a}{\cos\phi} \tag{3.36}$$

$$= \frac{KL}{\cos\phi} = \frac{FL - FK}{\cos\phi} = \frac{FD - FP}{\cos\phi} = \frac{PD}{\cos\phi}$$

Arc of contact is expressed as the approach arc plus recess arc (Figure 3.46):

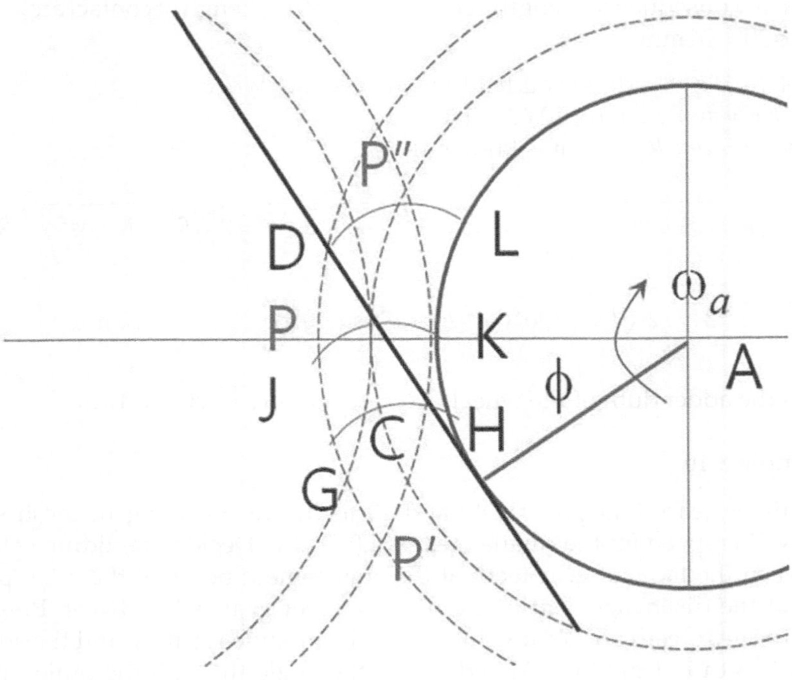

FIGURE 3.46
Arc of contact.

$$\text{Length of arc of contact} = PP' + PP'' = \frac{CP}{\cos\phi} + \frac{PD}{\cos\phi} = \frac{CD}{\cos\phi} \quad (3.37)$$
$$\left(= \frac{\text{Length of Path of contact}}{\cos\phi}\right)$$

$$\text{Contact ratio}(\text{CR}) == \frac{\text{Length of arc of contact}}{\textit{circular pitch}} = \frac{\frac{CD}{\cos\phi}}{\pi m} = \frac{CD}{\pi \cdot m \cdot \cos\phi} \quad (3.38)$$

Example 3.9

The numeric of teeth on each of the two identical spur gears in mesh are 40. The teeth have a 20° involute outline, and the module is 6 mm. As the arc of contact is 1.75 times the circular pitch, discover the addendum.

Solution

Specified: $T = t = 40, \varphi = 20°, m = 6$ mm

WKT the circular pitch,

$p = \pi m = \pi \times 6 = 18.85\text{mm}$

$\text{Lengh of contact arc} = 1.75 \times 18.85 = 33\,\text{mm}$

As a consequence, length of contact path $= \text{Lengh of contact arc} \times \cos = 33\cos 20 = 31\,\text{mm}$

Let $R_a = r_a = \text{radius of addendum circle of each wheel}$

$R = r = m \cdot T / 2 = 6 \times 40 / 2 = 120\,\text{mm}$

As $R = r\, and\, R_a = r_a$, it is known that

$$CD = \sqrt{R_a^2 - R^2\cos^2\phi} - (R + r)\sin\phi + \sqrt{r_a^2 - r^2\cos^2\phi} = 2\left[\sqrt{R_a^2 - R^2\cos^2\phi} - R\sin\phi\right]$$

$$31 = 2\left[\sqrt{R_a^2 - 120^2\cos^2 20} - 120\sin 20\right] \text{ or } R_a = 126.12 \text{ mm}$$

Thus the addendum of the wheel = $R_a - R = 126.12 - 120 = 6.12$.

Example 3.10

A pair of gears that possess 40 and 20 teeth are revolving in mesh separately. The speed for the smaller gear is 2,000 RPM. Decide the sliding velocity between the faces of gear teeth at the engagement point, at the pitch point, and at the disengagement point, if the smaller gear is the driver. Presume that the gear teeth are 20° involute form, the module is 5 mm, and the addendum length is 1 module. Also discover the angle through the pinion turns while any pairs of teeth are in contact and the contact proportion.

Solution

Given: $T = 40$, $t = 20$, $N_1 = 2{,}000$ RPM, $\phi = 20°$, addendum = 5 mm, $m = 5$ mm

Velocity of smaller gear (pinion):

$$\omega_1 = \frac{2\pi N_1}{60} = \frac{2\pi \times 2{,}000}{60} = 209.5 \text{ rad / sec}$$

Velocity of larger gear:

$$\omega_2 = \acute{E}_1 \frac{t}{T} = 209.5 \times \frac{20}{40} = 104.75 \text{ rad / sec}$$

Pitch circle radius of pinion:

$$\mathrm{r} = m \cdot \frac{t}{2} = 5 \times \frac{20}{2} = 50 \text{ mm}$$

Pitch circle radius of gear:

$$\mathrm{R} = m \cdot \frac{T}{2} = 5 \times \frac{40}{2} = 100 \text{ mm}$$

Addendum circle radius of pinion:

$$r_a = r + \text{Addendum} = 50 + 5 = 55 \text{ mm}$$

Addendum circle radius of gear:

$$R_a = R + \text{Addendum} = 100 + 5 = 105 \text{ mm}$$

Length of path of contact $CD = CP + PD$:

$$CP = \sqrt{R_a^2 - R^2 \cos^2 \phi} - R\sin\phi = \sqrt{105^2 - 100^2 \cos^2 20} - 100\sin 20 = 12.65 \text{ mm}$$

$$PD = \sqrt{r_a^2 - r^2 \cos^2 \phi} - r\sin\phi = \sqrt{55^2 - 50^2 \cos^2 20} - 50\sin 20 = 11.5 \text{ mm}$$

Velocity of sliding at point of engagement C:

$$V_{sc} = (\omega_1 + \omega_2)CP = (209.5 + 104.75)12.65 = 3{,}975 \text{ mm / sec}$$

Velocity of sliding at pitch point P:

$$V_{sp} = 0$$

Velocity of sliding at point of disengagement D:

$$V_{sc} = (\omega_1 + \omega_2)PD = (209.5 + 104.75)11.5 = 3,614 \text{ mm / sec}$$

$$\text{Length of arc of contact} = \frac{\text{Length of Path of contact}}{\cos\phi}$$

$$= \frac{CP + PD}{\cos\phi} = \frac{12.65 + 11.5}{\cos 20} = 25.69 \text{ mm}$$

$$\text{Contact ratio} = \frac{\text{Length of arc of contact}}{\cos\phi}$$

$$= \frac{25.69}{\cos 20} = 27.34, \text{ i.e., } 27 \text{ (approximately)}$$

Angle through which pinion turns:

$$\text{Angle through which pinion turns} = \text{Length of arc of contact} \times \frac{360°}{2\pi r_a}$$

$$= \frac{25.69 \times 360°}{2\pi 55} = 29.45°$$

3.5.4 Design of (Spur) Gear Trains

As seen in Figure 3.47, a gear train is two or more gears functioning jointly by meshing their teeth and turning each other in a system to construct power and speed. If electric motors are the power source, the gear systems transfer movement from one shaft to another, reduce speed, and increase torque. Established on kinematic relations such as system inputs (RPM1→ RPM2, power, space constraints, and material), we may attain the gear design—the

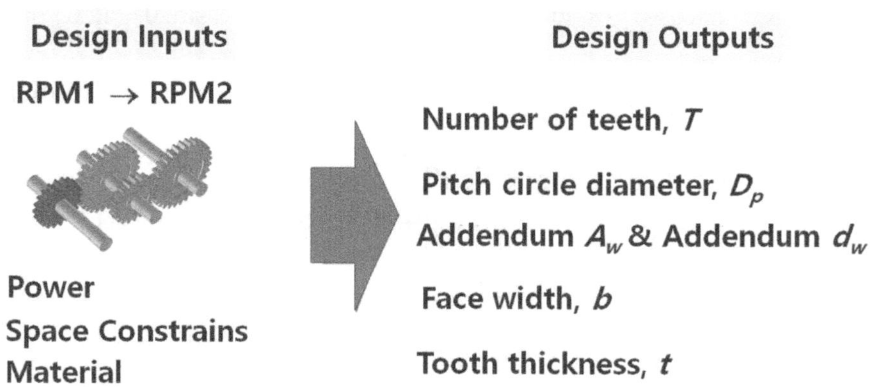

FIGURE 3.47
Basic concept of gear design.

numeric of teeth T, pitch circle diameter D_p, addendum A_w, and dedendum d_w, face width b, and tooth thickness t.

As the circular pitch of gear 1 is identical to the circular pitch of gear 2, we may decide the numeric of teeth N and pitch diameter d_p:

$$P_1 = P_2 \tag{3.39(a)}$$

$$\frac{N_1}{d_1} = \frac{N_2}{d_2} \tag{3.39(b)}$$

$$\frac{N_1}{N_2} = \frac{d_1}{d_2} \tag{3.39(c)}$$

As the elapsed path at a given time is the same $\widehat{AB} = \widehat{AC}$ or $\frac{\widehat{AB}}{t} = \frac{\widehat{AC}}{t}$, we can decide the rotational velocity. Kennedy's theorem expresses that, if three bodies have plane movements relative to one another, their instantaneous centers lie in a linear line.

$$V_1 = V_2 \tag{3.40(a)}$$

$$\omega_1 \cdot r_1 = \omega_2 \cdot r_2 \tag{3.40 (b)}$$

$$\frac{2}{2} \cdot \frac{r_1}{r_2} = \frac{\omega_2 / 2\pi}{\omega_1 / 2\pi} \tag{3.40 (c)}$$

$$\frac{d_1}{d_2} = \frac{n_2}{n_1} \tag{3.40 (d)}$$

As the power of gear 1 is identical to the power of gear 2, $H_1 = H_2$, we may decide the torque and revolutions per minute:

$$\omega_1 \cdot T_1 = \omega_2 \cdot T_2 \tag{3.41(a)}$$

$$\left[\omega 1\left(\frac{60}{2\pi}\right)\right] \cdot T_1 = \left[\omega 2\left(\frac{60}{2\pi}\right)\right] \cdot T_2 \tag{3.41(b)}$$

$$n_1 \cdot T_1 = n_2 \cdot T_2 \tag{3.41(c)}$$

$$\frac{n_1}{n_2} = \frac{T_2}{T_1} \tag{3.41(d)}$$

From Equation (3.39) ~ (3.41), we can decide the gear relationship:

$$\frac{N_1}{N_2} = \frac{d_1}{d_2} = \frac{n_2}{n_1} = \frac{T_1}{T_2} \tag{3.42}$$

Example 3.11

The mechanical power, including the gear train and BELT, moves as shown in Figure 3.48 with a specified input speed of 600 RPM). Determine the terminating output speed n_9.

Solution

$$n_9 = \frac{N_8 \cdot N_6 \cdot N_4 \cdot N_2}{N_9 \cdot N_7 \cdot N_5 \cdot N_3} \cdot n_2$$

$$\frac{N_6}{N_7} = \frac{d_6}{d_7}$$

$$n_9 = \frac{N_8 \cdot d_6 \cdot N_4 \cdot N_2}{N_9 \cdot d_7 \cdot N_5 \cdot N_3} \cdot n_2$$

$$n_9 = \frac{(3)(4)(18)(6)}{(36)(28)(36)(10)} \cdot 1{,}050 \text{ RPM}$$

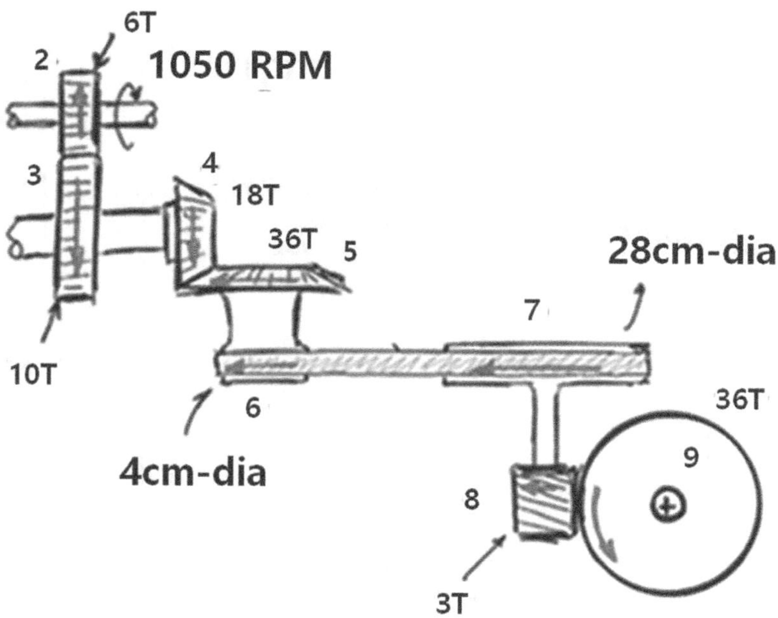

FIGURE 3.48
Design of gear trains.

$$n_9 = 3.75 \text{ RPM}$$

3.5.5 Force Components and Reaction Forces of (Spur) Gears

If the gear force is expressed as W, the tangential force W^t, radial force W^r, and axial force W^a are shown in Figure 3.49.

For the two gear forces (W_{32} and W_{23}) in power transmission, the sizes are the same, but the orientation is opposite:

$$W_{32} = W_{23} \tag{3.43(a)}$$

$$W_{23}^t = W_{23}\cos\varnothing \tag{3.43(b)}$$

$$W_{23}^r = W_{23}\sin\varnothing \tag{3.43(c)}$$

So power can be expressed:

$$H = T \cdot \omega = T \cdot \left(\frac{2\pi}{60} \cdot n\right) \tag{3.44}$$

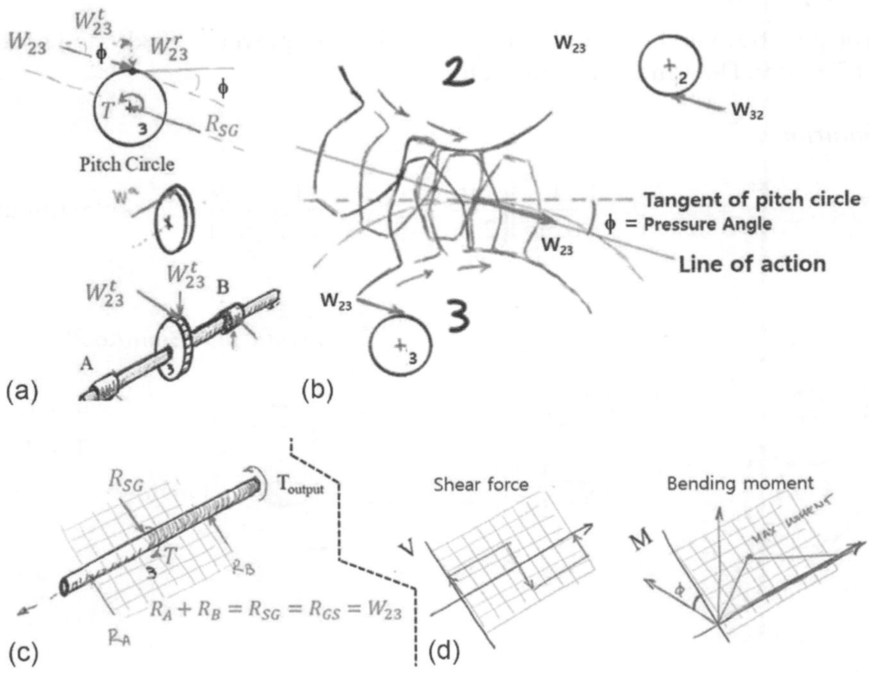

FIGURE 3.49
Force components and reaction forces of (spur) gears: (a) gear force components; (b) reaction forces; (c) reaction force; (d) shear force and bending moment.

As the power and speed are known, torque is expressed as:

$$T = \frac{H \cdot 60}{2\pi \cdot n} \tag{3.45(a)}$$

$$T = r_2 \cdot W_{23}^t \tag{3.45(b)}$$

From Equation (3.40), the tangential force can be redefined:

$$W_{23}^t = \frac{H \cdot 60}{2\pi \cdot n \cdot r} \tag{3.46}$$

As the tangential force in Equation (3.41) is known, the reaction forces can be found on the bearing.

$$W_{23} = \frac{W_{23}^t}{\cos\varnothing} \tag{3.47(a)}$$

$$R_{SG} = R_{GS} = W_{23} \tag{3.47(b)}$$

Example 3.12

The gear train is shown in Figure 3.50 with input power of 50 kW and speed of 200 RPM. Determine the output torque.

Solution

$$T_a = \frac{H \cdot 60}{2\pi \cdot n_a} = 50{,}000W \cdot \left(\frac{60s}{1\,\text{min}}\right)\left(\frac{1\text{rev}}{2\pi[\text{rad}]}\right)\left(\frac{1}{200\ \text{rev}/\text{min}}\right) = 2.39\ \text{kNm (input)}$$

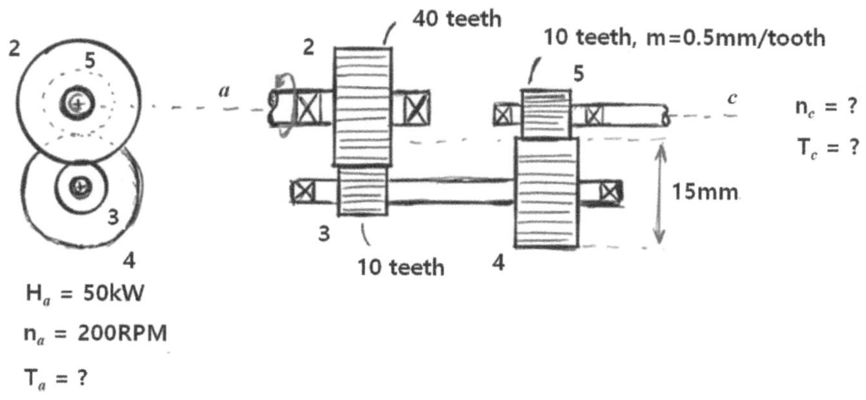

FIGURE 3.50
Gear trains.

$$n_c = n_5$$

$$n_2 = n_a = 200 \text{ RPM}$$

$$\frac{n_2}{n_3} = \frac{N_3}{N_2} \rightarrow n_3 = \frac{N_2}{N_3} \cdot n_2 = \frac{40}{10} \cdot 200 = 800 \text{ RPM}$$

$$n_4 = n_3 = 800 \text{ RPM}$$

$$\text{m} = 0.5 = \frac{d_5}{N_5} \rightarrow d_5 = 0.5 \cdot N_5 = 0.5 \cdot 10 = 5\text{mm}$$

$$\frac{d_5}{d_4} = \frac{n_4}{n_5} \rightarrow n_5 = \frac{d_4}{d_5} \cdot n_4 = \frac{15}{5} \cdot 800 = 2400 \text{ RPM}$$

$$T_c = \frac{H \cdot 60}{2\pi \cdot n_c} = 50{,}000W \cdot \left(\frac{60s}{1\text{ min}}\right)\left(\frac{1\text{rev}}{2\pi[\text{rad}]}\right)\left(\frac{1}{200\text{ rev / min}}\right) = 198.9\text{Nm}(\text{output})$$

Example 3.13

The engineer must raise the weight with the compound gear system (Figure 3.51). Presume there are no losses.

- How much torque does the motor have to exert to raise the weight?
- What orientation must the motor take to raise the weight?
- How fast must the motor turn to raise the weight?

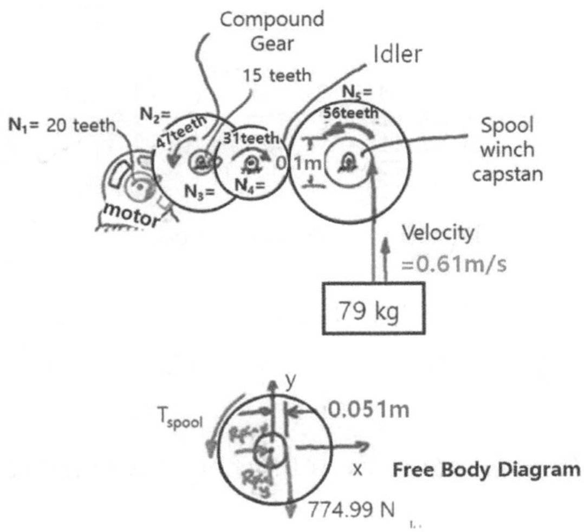

FIGURE 3.51
Compound gear system.

Solution

Free body diagram at spool winch capstan

Moment balance at the end of spool winch capstan

$$\sum M = T_{\text{spool}} - \left(79\text{ kg}\cdot\frac{9.81\text{ m}}{\text{s}^2}\right)(0.051\text{m}) = 0.\ T_{\text{spool}} = 39.37\text{ Nm}$$

$$T_{\text{spool}} = \frac{N_5\cdot N_4\cdot N_2}{N_4\cdot N_3\cdot N_1}\cdot T_{\text{motor}} = \frac{56\cdot 31\cdot 47}{31\cdot 15\cdot 20}(=8.77)\cdot T_{\text{motor}}$$

$$T_{\text{motor}} = \frac{T_{\text{spool}}}{8.77} = 4.487\text{ Nm}$$

As Speed $\cdot\,\pi\text{d} = V$, we may attain the speed of the spool winch capstan:

$$Speed_{\text{spool}} = \frac{0.61\text{ m/s}}{\pi\cdot 0.1\text{ m}} = 1.91\frac{\text{rev}}{\text{sec}}\left(\frac{60\text{ sec}}{\text{min}}\right) = \text{RPM}$$

$$Speed_{\text{motor}} = Speed_{\text{spool}}\frac{N_5\cdot N_4\cdot N_2}{N_4\cdot N_3\cdot N_1} = 114.59\frac{56\cdot 31\cdot 47}{31\cdot 15\cdot 20}(=8.77) = 1005.34\text{ RPM}$$

3.5.6 Lewis Bending Equation for Design of (Spur) Gear

As $\sigma > S_y$ (moderate use) or S_e (fatigue), a gear may begin the bending failure of the teeth. The clarified model of the cantilever beam may be defined (Figure 3.52).

$$|\sigma| = \frac{My}{I} = \frac{M(t/2)}{\frac{1}{12}\text{ft}^2} = \frac{\left(W^t l\right)(t/2)}{\frac{1}{12}\text{ft}^3} = \frac{6W^t l}{\text{ft}^2} \tag{3.48}$$

As $\frac{t/2}{X} = \frac{l}{t/2}$ or $l = \frac{t^2}{4X}$, Equation (3.48) is restated:

$$\sigma = \frac{6W^t\left(t^2/4X\right)}{\text{Ft}^2} = \frac{6W^t}{4FX} = \frac{W^t p}{F\left(\frac{2}{3}X\right)p} = \frac{W^t}{Fpy} \tag{3.49}$$

where $\text{y} = \frac{2x}{3p}$ = *Lewis form factor* as a function of N (number of teeth) and φ (pressure angle).

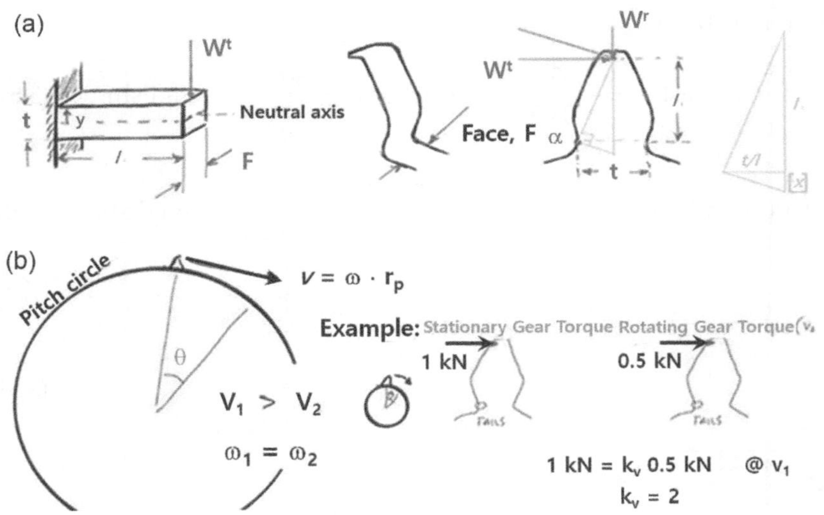

FIGURE 3.52
(a) Clarified model for Lewis form factor computation; (b) stress k_v (dynamic) factor.

Because $p = \text{circular pitch} = \frac{\pi d}{N} = \pi m$ and $P = \text{diametral pitch} = \frac{N}{d} = \frac{\pi}{p}$, Equation (3.40) can be stated as

$$\sigma = \frac{W^t}{F(m \cdot \pi)y} = \frac{W^t}{FmY} \text{ (metric)} \tag{3.50(a)}$$

$$\sigma = \frac{W^t}{F\left(\frac{\pi}{P}\right)y} = \frac{W^t P}{F[\pi y]} = \frac{W^t P}{FY} \text{ (English)} \tag{3.50(b)}$$

where $Y = y\pi = \frac{2XP}{3}$ (Lewis form factor Y, which shall be obtained from the table).

As the bending stress is computed from the Lewis bending equation for the design of a (spur) gear, we might decide the stress k_v (dynamic) factor from the exploratory data. For instance, the revolving gear force is 0.5 kN at a certain line velocity, which is expressed as $v = \omega \cdot r_p$. If failure happens at a stationary gear force of 1 kN, the stress k_v (dynamic) factor is 2 (see Figure 3.52(b)).

So k_v may be defined:

$$\text{Cast } k_v = \frac{3.05 + v}{3.05} \text{ (metric) or } k_v = \frac{600 + v}{600} \text{ (English)} \tag{3.51(a)}$$

$$\text{Cut or milled } k_v = \frac{6.1 + v}{6.1} \text{ (metric) or } k_v = \frac{1200 + v}{1200} \text{ (English)} \tag{3.51(b)}$$

$$\text{Hobbed or shaped } k_v = \frac{3.56 + \sqrt{v}}{3.56} \text{ (metric) or } k_v = \frac{50 + \sqrt{v}}{50} \text{ (English)} \tag{3.51(c)}$$

$$\text{Shaped or ground } k_v = \sqrt{\frac{5.56 + \sqrt{v}}{5.56}} \text{ (metric) or } k_v = \sqrt{\frac{78 + \sqrt{v}}{78}} \text{ (English)} \tag{3.51(d)}$$

Example 3.14

A tooth in the gear train is assigned as shown in Figure 3.53 and specification, decide the bending stress

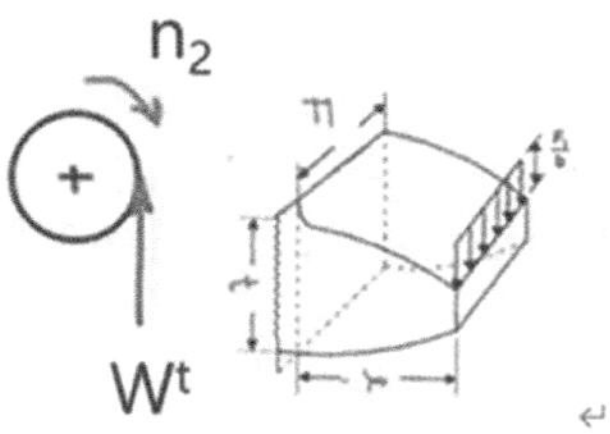

Table 3.1 Values of the Lewis Form Factor Y (These Values Are for a **Normal Pressure Angle of 20°**, Full-Depth Teeth, and a Diametral Pitch of Unity in the Plane of Rotation)

Number of Teeth	Y	Number of Teeth	Y
12	0.245	28	0.353
13	0.261	30	0.359
14	0.277	34	0.371
15	0.290	38	0.384
16	0.296	43	0.397
17	0.303	50	0.409
18	0.309	60	0.422
19	0.314	75	0.435
20	0.322	100	0.447
21	0.328	150	0.460
22	0.331	300	0.472
24	0.337	400	0.480
26	0.346	Rack	0.485

FIGURE 3.53
Tooth in the gear train for Lewis form factor.

$$\text{Module: } m = 1.25 \text{ mm/teeth } (= \frac{d}{N})$$

$$N_2 = 18 \text{ teeth}, \phi = 20°, F = 12 \text{ mm}$$

$$n_2 = 1{,}800 \text{ RPM}, H = 0.5 \text{ kW} = 500 \text{ W}$$

Solution

$$\text{Bending stress?} \rightarrow \sigma = K_v \frac{W^t}{FmY}$$

$$\sigma = K_v \frac{W^t}{FmY} = 1.348 \cdot \frac{235.8}{0.012 \cdot 0.00125 \cdot 0.309} = 68.6 \text{ MPa}$$

$$K_v = \frac{6.1 + V}{6.1} = \frac{6.1 + 2.121}{6.1} = 1.348$$

$$V = \omega \cdot r = n \cdot \frac{2\pi}{60} \cdot r = 1800 \times \frac{2\pi}{60}$$

$$V = 2{,}121 \text{ mm/s } V = 2.121 \text{ m/s}$$

$$d = \text{N} \cdot \text{m} = 18 \cdot 1.25 = 22.5 \text{ mm} \rightarrow r = 11.25$$

$$\text{H} = \text{T} \cdot = \left(W^t \cdot r\right) \cdot \left(n \frac{2\pi}{60}\right)$$

$$W^{to} = \frac{60 \cdot \text{H}}{r \cdot n \cdot 2\pi} = \frac{60 \cdot 500}{0.01125 \cdot 1800 \cdot 2\pi} = 235.8 \text{ N}$$

3.5.7 Design of Bevel Gears

Bevel gears are gears wherein the axes of the two shafts intersect and the tooth-bearing faces of the gears themselves are conically shaped. Bevel gears are placed on shafts that are 90° apart but that will be designed to work at other angles as well. The pitch surface of bevel gears is a cone. Thus the tangential, radial, and axial components of helical gears are defined (Figure 3.54):

$$\text{Tangential } W^t = W^t \tag{3.52(a)}$$

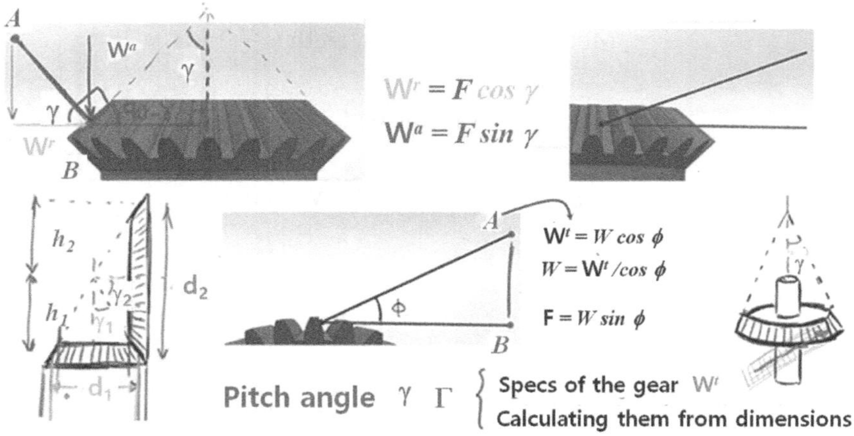

FIGURE 3.54
Force elements of bevel gears.

$$\text{Radial } W^r = F\cos\Upsilon = W\sin\phi\cos\Upsilon = \frac{W^t}{\cos\cdot}\cdot\sin\phi\cos\Upsilon = W^t\tan\phi\cos\Upsilon \quad (3.52(b))$$

$$\text{Axial (thrust) } W^a = F\sin\Upsilon = W\sin\phi\sin\Upsilon = \frac{W^t}{\cos\cdot}\cdot\sin\phi\sin\Upsilon \quad (3.52(c))$$
$$= W^t\tan\phi\sin\Upsilon$$

The pitch angle is decided:

$$h_1 = h_2 = \frac{1}{2}d_2 \quad (3.53(a))$$

$$\Upsilon_1 = \tan^{-1}\left(\frac{d_1}{d_2}\right) \quad (3.53(b))$$

$$\Upsilon_2 = \tan^{-1}\left(\frac{d_2}{d_1}\right) \quad (3.53(c))$$

Example 3.15

Gear 3 has 32 teeth. The power to be moved is 2.5 hp at 240 RPM. However, gear 3 has 16 teeth and a pressure angle of 20°. As the subsequent figure is given, discover the reaction forces at bearings A and B (Figure 3.55).

Solution

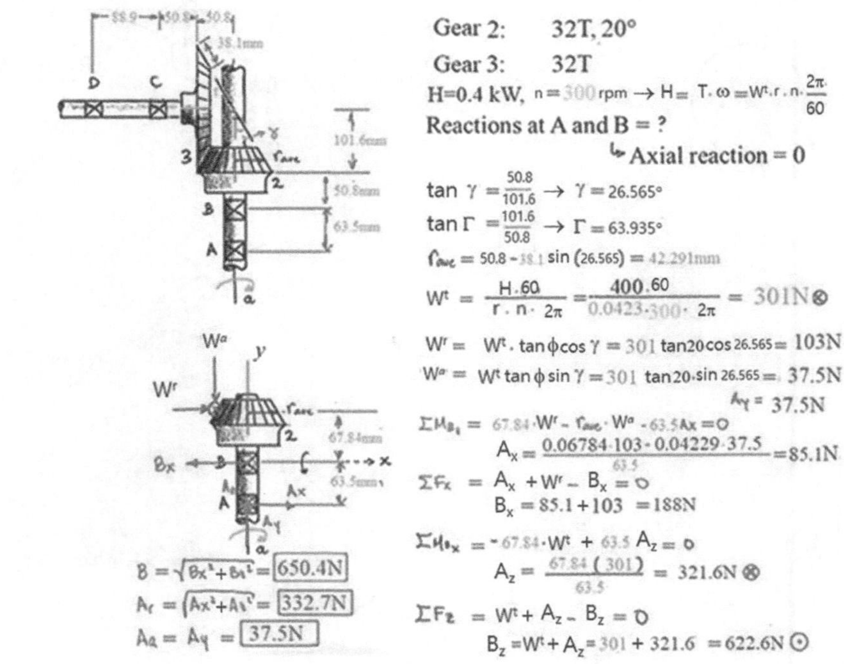

FIGURE 3.55
Design of bevel gears (example).

3.5.8 Design of Helical Gears

Helical gears are like spur gears but for the fact that their teeth are cut at an angle to the hole rather than straight and parallel to the hole like the teeth of a spur gear. If power H and speed n are given, the tangential component of helical gears for the applied force in the process of gear interaction is specified (Figure 3.56):

$$T = \frac{H}{\omega} = \frac{H \cdot 60}{2\pi \cdot n} \tag{3.54(a)}$$

$$T = W^t \cdot r \tag{3.54(b)}$$

$$W^t = \frac{H \cdot 60}{2\pi \cdot n \cdot r} \tag{3.54(c)}$$

Thus the tangential, radial, and axial elements of helical gears are defined:

$$\text{Tangential } W^t = W\cos\phi \cdot \cos\Psi \tag{3.55(a)}$$

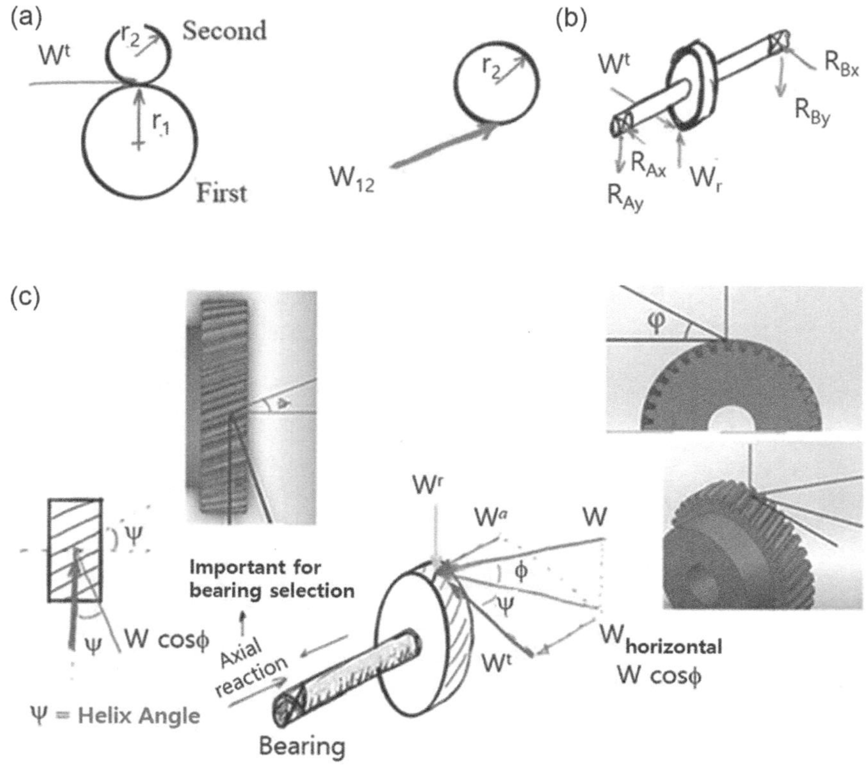

FIGURE 3.56
Force components of helical gears: (a) tangential component of helical and bevel gears; (b) reaction forces; (c) force components of helical gears.

$$\text{Radial } W^r = W\sin\phi = \frac{W^t}{\cos\phi\cos\Psi}\cdot\sin\phi = \frac{W^t\tan\phi}{\cos\Psi} \tag{3.55(b)}$$

$$\text{Axial (thrust) } W^a = W\cos\phi\cdot\sin\Psi = \frac{W^t}{\cos\cdot\cos\Psi}\cdot\cos\phi\cdot\sin\Psi = W^t\tan\Psi \tag{3.55(c)}$$

Example 3.16

When the motor with 1 HP is rotating at 1,800 RPM, power is transferred through the bevel gear (d_2 = 3.81 cm and ψ = 30°) to the shaft. Determine the reaction at each bearing A and B (Figure 3.57).

Solution

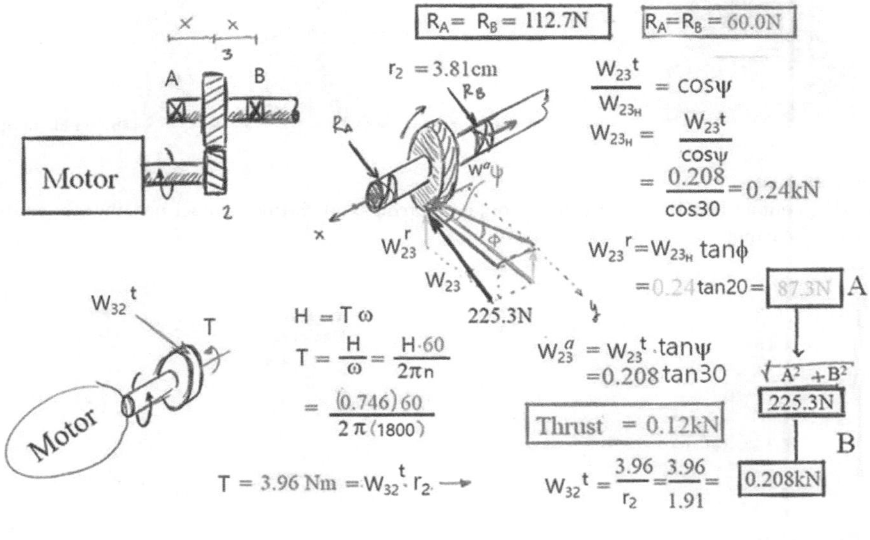

FIGURE 3.57
Design of helical gears (example).

3.6 Design of the Bearing

3.6.1 Introduction

A bearing is a machinery component that supports another moving machine component. It requires support to ensure stability and frictionless rotation (confinement of movement) and allows a relative motion between the contact surfaces of the members. As the rubbing surfaces are in straight contact, there will be quick wear. All bearings supply some lubrication positioning to lessen friction. The lubricant normally consists of vegetable oils, mineral oil, grease, silicon oils, etc.

Depending on their attributes, there are the two kinds of contact: 1) sliding contact bearings and 2) rolling contact bearings. In sliding contact bearings, sliding happens along the surface of the contact between the moving part and the fixed part. For rolling contact bearings, steel balls or rollers are inserted between the moving and fixed parts. Each ball or roller supplies rolling friction at two points (Figure 3.58).

Depending on the positioning of the load to be supported, there are two kinds: 1) radial bearings and 2) thrust bearings. In radial bearings, the load acts at right angles to the motion of the moving element. In thrust bearings, the load occurs along the axis of rotation (Figure 3.59).

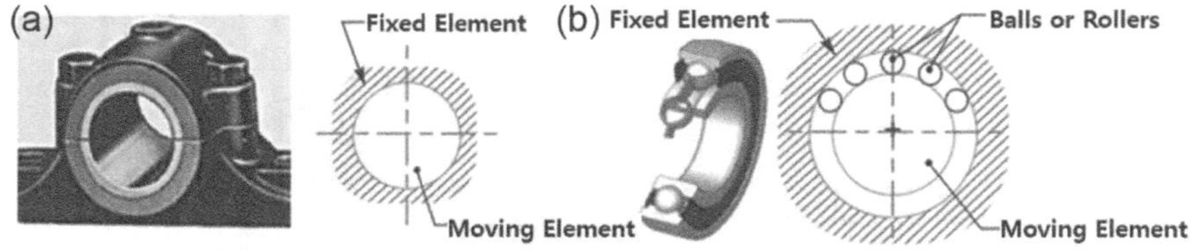

FIGURE 3.58
Sliding contact bearings and rolling contact bearings: (a) sliding contact bearing; (b) rolling contact bearing.

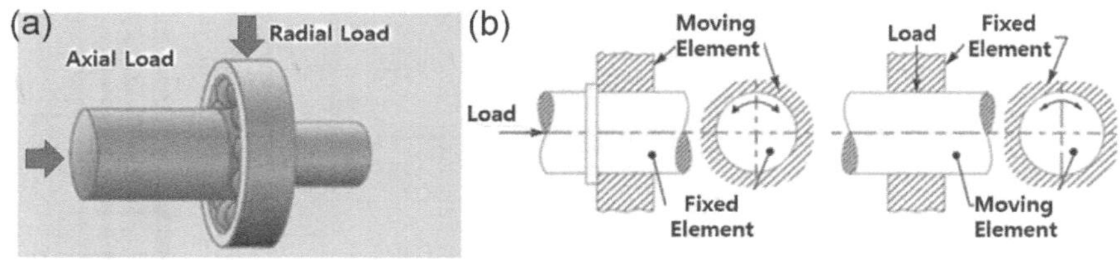

FIGURE 3.59
(a) Radial bearing and (b) thrust bearing.

3.6.2 Bearing Life and Load Ratings

Bearing failure consists of spalling or pitting of a district of 0.01 in^2 (ABMA). The life for one bearing will be indicated as the numeric of revolutions (or hours at a given speed) until failure. For a group of bearings, the rating life is one that is required for 10% of samples to be failed and is named the minimum life or L10 life. On the other hand, the median life for numerous groups of bearings is the mean life required for 50% of samples to fail, regarded as the mean life or median life. Thus median life is usually four or five times that of L10 life (Figure 3.60).

There are the four load ratings: 1) catalog load rating, C_{10}; 2) basic load rating, C; 3) static load rating, C_o; and 4) equivalent radial load, F_e.

1. *Catalog load rating,* C_{10}: This is a constant radial load that conducts 10% of a category of bearings to fail at the end of the maker's rating lifetime. It depends on the kind, geometry, accuracy of fabrication, and material of the bearing. It is also known as basic dynamic load rating and capacity.
2. *Basic load rating,* C: This reference is a catalog load rating constructed on a rating life of 10^6 revolutions of the inner ring. The radial load required to conduct failure at such a low life is incredibly high.

FIGURE 3.60
Bearing life definitions.

3. *Static load rating, C_o*: The static radial load balances a permanent deformation of rolling element and race at the greatest degree of stressing contact of 0.0001 d (= diameter of roller). It is used to inspect for permanent deformation and in integrating radial and thrust loads into an equivalent radial load.
4. *Equivalent radial load, F_e*: This is a constant stationary load applied to the bearing with a rotating inner ring, which has the same life as the real load and rotation.

3.6.3 Bearing Design

Similar groups of bearings are tested to decide the life-failure standard at some loads. We can plot load versus life on a log-log scale, which is approximately linear. Using a regression equation, it may be expressed (Figure 3.61):

$$FL^{\frac{1}{a}} = \text{constant} \tag{3.56(a)}$$

$$F_R L_R^{\frac{1}{a}} = F_D L_D^{\frac{1}{a}} \tag{3.56(b)}$$

where R = rated, D = desired, a = 3 (ball bearings), and a = 10/3 (roller bearings).

Example 3.17

As mechanical power is operated through the gear system, the reaction forces at A and B are computed as 0.89 kN and 1.33 kN, respectively. The rated life of each bearing is 106 (rating life: number of revolutions until 90% of samples have not yet failed). The desired bearing life is 8,000 hours @ 1,600 RPM. In this case, decide the dynamic load ratings at bearings A and B (Figure 3.62)

FIGURE 3.61
Load–life relationship.

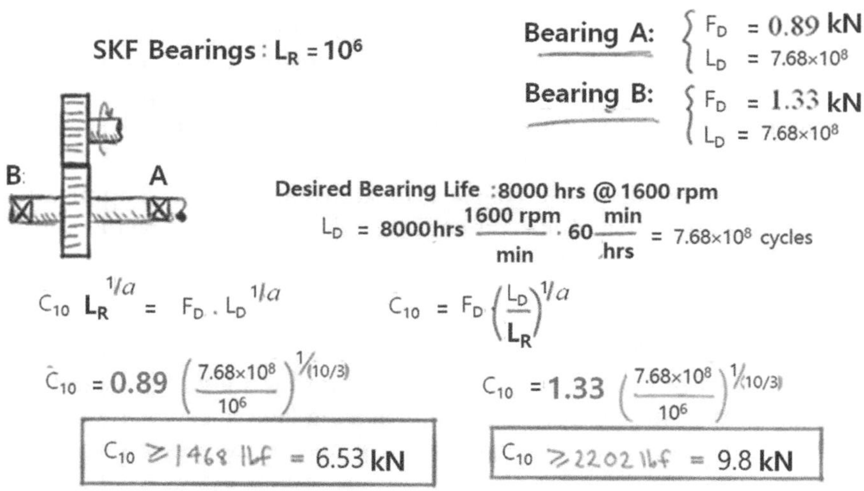

FIGURE 3.62
Load calculation in bearing.

As bearings are exposed to united radial and axial loads, we might discover the equal radial load F_e because $F_e \neq F_a + F_r$ or $F_e \neq \sqrt{F_a^2 + F_r^2}$ (Figure 3.63).

The connection of the equivalent radial load F_e on the plot may be defined:

$$\frac{F_e}{V \cdot F_r} = 1 \text{ or } F_e = V \cdot F_r \text{ for } \frac{F_a}{V \cdot F_r} \leq e \tag{3.57(a)}$$

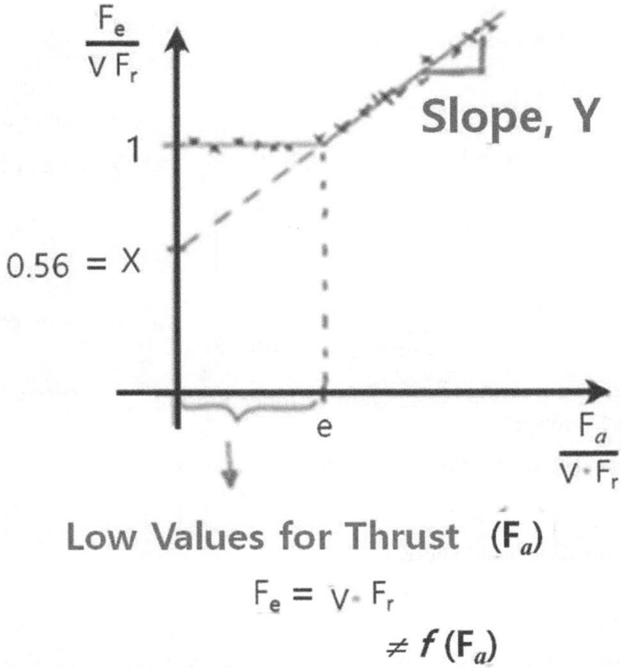

FIGURE 3.63
Connection of the equivalent radial load F_e.

$$\frac{F_e}{V \cdot F_r} = Y\left(\frac{F_a}{V \cdot F_r}\right) + X \text{ for } \frac{F_a}{V \cdot F_r} \geq e \tag{3.57(b)}$$

We know that Equation (3.57(b)) has the relationship of a straight line $y = m \cdot x + b$. Thus if we redefine Equation (3.57(b)), we obtain:

$$F_e = X_i \cdot V \cdot F_r + Y_i \cdot F_a \tag{3.58}$$

where V = 1 (inner ring rotates), V = 1.2 (outer ring rotates) attained from experimentation.

For ball bearings, $X_i = 0.56$ and $Y_i = f\left(\frac{F_a}{C_o}\right)$ may be obtained—(equivalent radial load factors). Thus we may outline the steps of bearing design exposed to combined axial and radial loads (Figure 3.64).

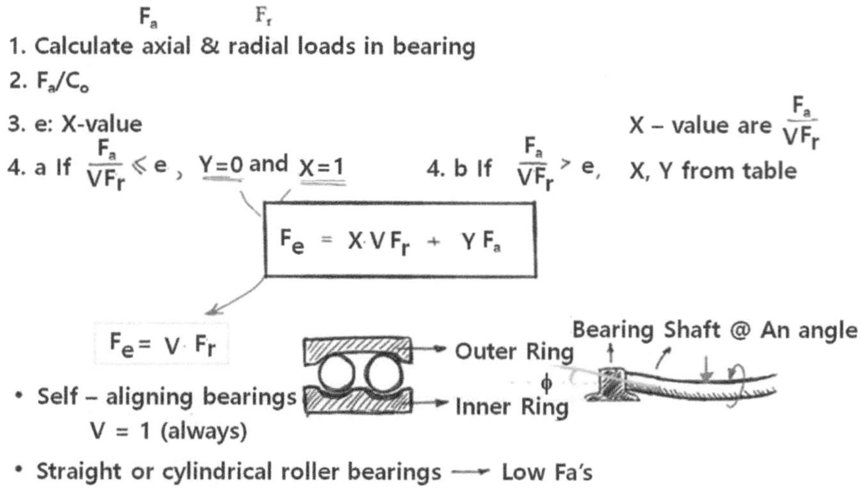

FIGURE 3.64
Procedure for equivalent loads of bearing.

Example 3.18

For the (angular contact) ball bearing, if C_o =20 kN, C_{10} =35.6 kN, and $L_{10} = 10^6$ are assigned. Determine the life for $F_a = 1.7$ kN, $F_r = 2.1$ kN, and $n = 680$ RPM.

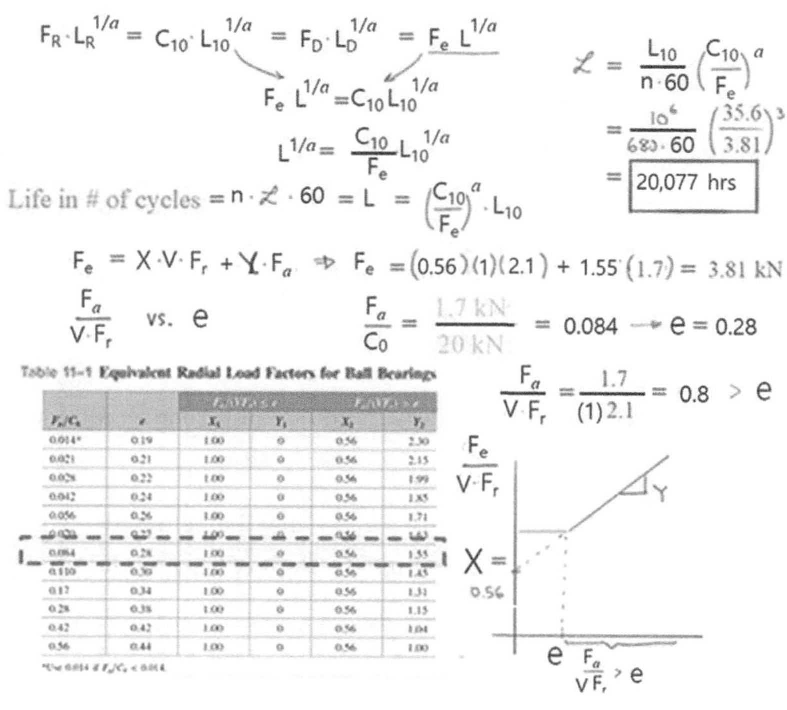

F_a/C_o	e	$F_a/(VF_r) \le e$		$F_a/(VF_r) > e$	
		X_1	Y_1	X_2	Y_2
0.014*	0.19	1.00	0	0.56	2.30
0.021	0.21	1.00	0	0.56	2.15
0.028	0.22	1.00	0	0.56	1.99
0.042	0.24	1.00	0	0.56	1.85
0.056	0.26	1.00	0	0.56	1.71
[illegible]	0.27	1.00	0	0.56	1.63
0.084	0.28	1.00	0	0.56	1.55
0.110	0.30	1.00	0	0.56	1.45
0.17	0.34	1.00	0	0.56	1.31
0.28	0.38	1.00	0	0.56	1.15
0.42	0.42	1.00	0	0.56	1.04
0.56	0.44	1.00	0	0.56	1.00

*Use 0.014 if F_a/C_o < 0.014.

3.7 Design of the Shaft

3.7.1 Introduction

A shaft is a revolving part of a round, hollow or solid cross section that transfers (or receives) power, depending on the implementation. There are the two sorts of shafts: transmission shafts and machine shafts. Transmission shafts get power from one end and bring power to another end. On the other hand, machine shafts are a critical part of a machine, or we can say that they are the chief member in a machine. For instance, there is a crankshaft in internal combustion (IC) engines. The axle is a stationary member utilized as a support for revolving components—wheels and idler gears. The spindle is a little shaft or axle—the head-stock spindle of a lathe (Figure 3.65).

Shaft materials are mostly steel or carbon steel. The elements installed on the shaft are gears, flywheels, clutches, pulleys, sprockets, etc. Press fit, dowel, keys, pins, and splines are utilized to add these machine components on the shaft. To become involved in axial loads and find the rotating elements, some kinds of keeping rings, thrust bearings, grooves, and steps in the shaft are utilized. Couplings are also utilized to transfer power from the drive–the motor—to the driven shaft—the gearbox, wheels.

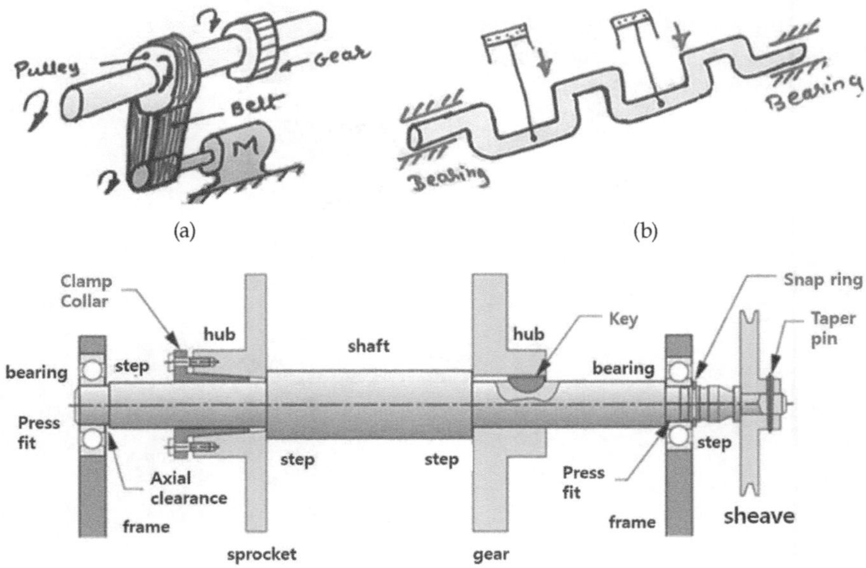

(c)

FIGURE 3.65
General shaft elements: (a) transmission shafts; (b) machine shafts; (c) shaft with various attachments.

Shaft design factors are as follows: 1) size and spacing of components (as on a general assembly drawing), tolerances; 2) material selection and material treatments; 3) defection and rigidity (bending/torsional/shear defection, slope at bearings); 4) stress strength such as static strength and fatigue; and 5) frequency response/manufacturing restrictions.

Common design rules for shafts are as follows:

1. *Retain shafts short, with bearings near to the exerted loads*: This shall lessen deflections and bending moments. On the other hand, this will encourage critical speeds
2. *Put required stress raisers away from very stressed shaft area if possible.*
3. *Utilize cheap steels for deflection-critical shafts because all steels have basically the same modulus of elasticity.*
4. *In the early design of any specified shaft, an approximation is generally made of whether strength or deflection shall be the critical factor.* A preparatory design is established on that basis.
5. *Deflection must not bring mating gear teeth to split more than approximately 1 mm.*

3.7.2 Simple Steps for Shaft Design

First, a free body diagram is created by changing the many machine elements installed on the shaft by their statistical or torque components (Figure 3.66).

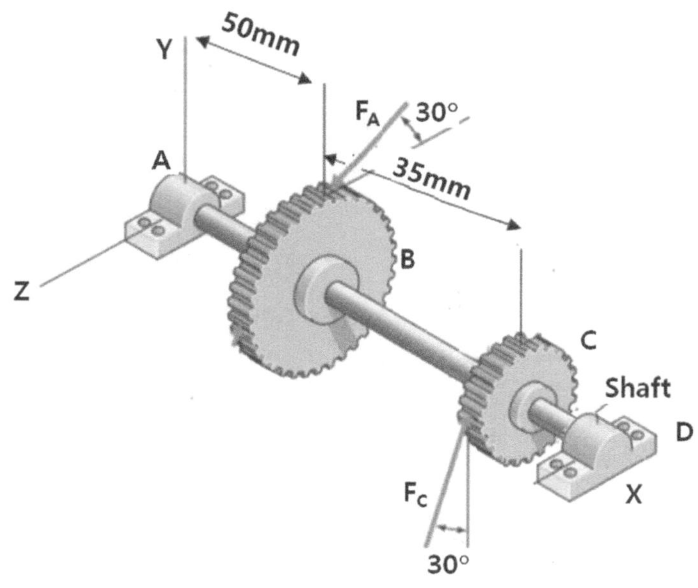

FIGURE 3.66
Free body diagram.

Second, sketch the x–y and x–z planes. Compute the resultant internal moment at several sections along the shaft (Figure 3.67).

Third, evolve a torque diagram. The torque evolved from the power-transmitting element should balance the torque from other power transmission elements (Figure 3.68).

Fourth, set up the position of the critical cross section, or the x location where the torque and moment are the largest.

Finally, for ductile materials, the distortion-energy theory or maximum-shear–stress theory is used. On the other hand, for brittle materials, the maximum normal stress is utilized.

For the plane stress state, when $\sigma_y = 0$, the principal normal stresses may be defined:

$$\sigma_1, \sigma_2 = \frac{\sigma_x}{2} \pm \sqrt{\left(\frac{\sigma_x}{2}\right)^2 + \tau_{xy}^2} \tag{3.59}$$

Because $\sigma_x = \dfrac{Mc}{I} = \dfrac{M\left(\dfrac{d}{2}\right)}{\dfrac{\pi d^4}{64}} = \dfrac{64Md}{2\pi d^4} = \dfrac{32M}{\pi d^3}$ and $\tau_{xy} = \dfrac{Tc}{J} = \dfrac{T\left(\dfrac{d}{2}\right)}{\dfrac{\pi d^4}{32}} = \dfrac{16T}{\pi d^3}$

Equation (3.59) is redefined:

$$\sigma_1, \sigma_2 = \frac{16M}{\pi d^3} \pm \sqrt{\left(\frac{16M}{\pi d^3}\right)^2 + \left(\frac{16T}{\pi d^3}\right)^2} = \frac{16}{\pi d^3}\left(M \pm \sqrt{M^2 + T^2}\right) \tag{3.60}$$

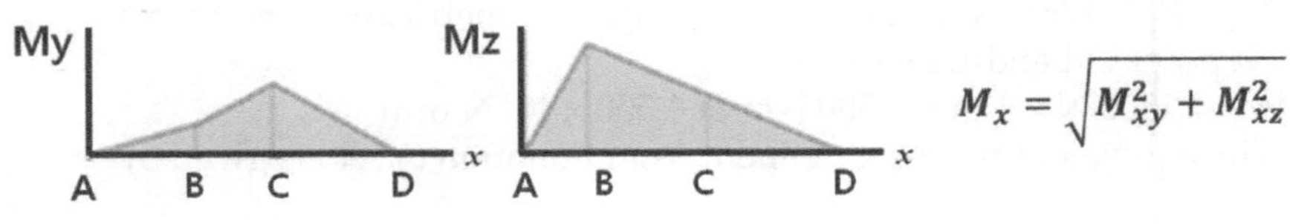

FIGURE 3.67
Bending moment diagram.

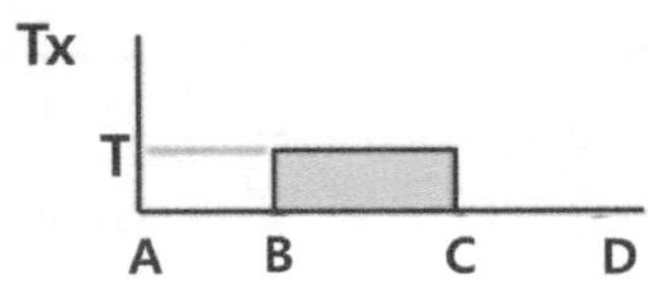

FIGURE 3.68
Torque diagram.

The principal normal stresses can be defined:

$$\tau_1, \tau_2 = \frac{\sigma_x}{2} \pm \sqrt{\left(\frac{\sigma_x}{2}\right)^2 + \tau_{xy}^2} \tag{3.61(a)}$$

$$\tau_1, \tau_2 = \pm\frac{16}{\pi d^3}\sqrt{M^2 + T^2} \tag{3.61(b)}$$

Example 3.19

A shaft consisting of mild steel is necessary to transfer 100 kW at 300 RPM. The supported length of the shaft is 3 m. It conveys two pulleys, each weighing 1,500 N bore at an interval of 1 m from the ends separately. Presuming a safe value of stress of 60 MPa, determine the diameter of the shaft (Figure 3.69).

Solution

Given data: Solid shaft (type shaft)

Power (P) = 100 kW; speed (N) = 300 RPM; weight (W) = 1,500 N; safe stress = 60 MPa = 60 N/mm^2.

The specified shaft is exposed to both bending and twisting. Here, based on the twisting moment and bending moment, the shaft will be designed.

The power transferred by the shaft is assigned by

$$P = \frac{2\pi NT}{60}$$

$T = \frac{P60}{2\pi N}$ = $10010^3 60/2\pi 300 = 3{,}183\,\text{Nm}\, or\, 318{,}310^3\,\text{N mm}$

At each supported end, reactions

$R_A = R_B$ = 1,500 N (because the beam is symmetrical)

At point C, bending moment

M_C = 1500 N × 1 m = 1,500 Nm or 1,500 × 10^3 N mm

The equivalent twisting moment from Equation (3.52b) is given by

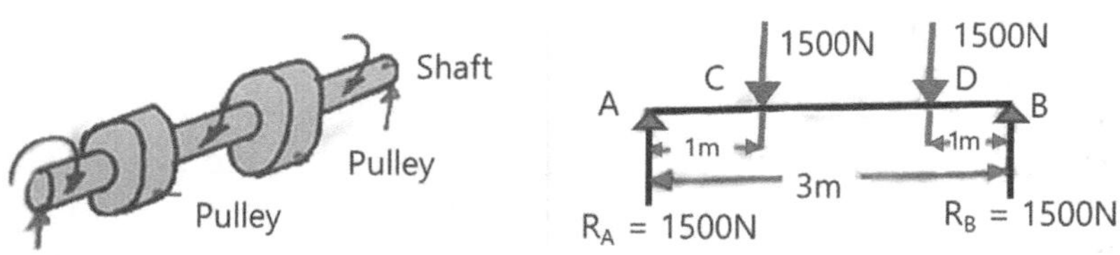

FIGURE 3.69
Pulley on shaft and free body diagram.

$$T_{eq} = \sqrt{T^2 + M^2} = \sqrt{(3183)^2 + (1500)^2} = 3519\, Nm\, or\, 351910^3\, N\,\text{mm}$$

$$T_{eq} = \frac{\pi}{16} D^3 \cdot \tau$$

$$3519 \times 10^3 = \frac{\pi}{16} D^3 \cdot 60 \; D^3 = \frac{3519 \times 10^3 \times 16}{\pi \times 60}$$

$$D = 66.8^\circ\, or\, ^\circ 70 \text{ mm}$$

References

1. Sonntag, R.E.; Borgnakke, C., *Introduction to Engineering Thermodynamics*, John Wiley & Sons, Inc.: York, PA, 2007.
2. Tsai, L.-W., *Mechanism Design: Enumeration of Kinematic Structures According to Function*, CRC Press: Boca Raton, FL, 2001.
3. Mallik. A.K., *Kinematic Analysis and Synthesis of Mechanisms*, CRC Press: Boca Raton, FL, 1994.
4. Norton, R., *Machine Design: An Integrated Approach*, Pearson: London, 2020.
5. Lynwander, P., *Gear Drive Systems: Design and Application (Mechanical Engineering)*, CRC Press: Boca Raton, FL, 1983.

4

Engineering (Dynamic) Load Analysis

4.1 Introduction

Load can be expressed as the amount of weight conveyed, particularly by a vehicle, of a construction such as a bridge or an animal. There are the two sorts of loads: static and dynamic. Static is the subdivision of traditional mechanics related to the examination of force and torque acting on a physical system that does not undergo an acceleration but rather is in equilibrium with its surroundings. Static or dynamic loads—due to tension, compression, shear, or twist—bring about displacement, strain, and stress. The mechanical product is designed to supply sufficient strength in its life to support such repetitive (dynamic) loads: 1) constructional load and deflection versus material stress and strain, 2) tension and compression loads, and 3) torsion and bending loads.

On the other hand, the dynamics branch of mechanics, which has to do with the movement of material objects in connection with the physical factors influencing them: force, mass, momentum, and energy. A dynamic load is a force exerted on a body in movement, normally in a relatively short period. As such loads are variable, we can call it the dynamic load. It brings about movements such as vibration, cumulative damage, fatigue, and fracture. Therefore, a product might be designed to have sufficient strength/stiffness in the structure over its lifetime to withstand repetitive (dynamic) loads: 1) impact, vibration, and shock loads and 2) unbalanced inertia loads.

A basic analysis for such loads may be performed: 1) express the problem; 2) draw figures; 3) express movement: numeric of degrees of freedom⇒ numeric of coordinates⇒ allocate coordinate⇒ numeric of equations of motion (EOM) ⇒ perform the kinematics (compute velocity and acceleration). Dynamic modeling for EOM is a method of describing a system's behavior over time, including the sequencing of events and states. It can be used to understand how a system works and to simulate and analyze how it might behave in the future. There are several ways for dynamic modeling—Newtonian, Lagrangian, Hamiltonian, etc.—to be included in the mechanical system [1,2].

DOI: 10.1201/9781003451471-4

4.2 Statics Principles

4.2.1 Equilibrium in Statics

At all times, forces happen in pairs acting on two dissimilar objects that act in opposite directions and have identical magnitudes. A force may be exchanged by component forces if their combination produces the same outcomes as the primary force. The resultant force is a force that has the same outcome as some simultaneous forces.

Equilibrium in physics is the state of a system when neither its condition of movement nor its internal energy condition tends to alter with period. That is, if all the forces acting on an object are balanced, the object can be said to be in a condition of equilibrium. For a rigid object, we have the following situations:

1. *Translational equilibrium* This is the condition in which balanced forces are acting on a body.

$$\sum F=0 \tag{4.1}$$

2. *Rotational equilibrium*: The circumstance in which the entire of all the clockwise moments is equal to the entirety of all the counterclockwise moments about a pivot point.

$$\sum \mathrm{M}=0 \tag{4.2}$$

Example 4.1

The system in Figure 4.1 is in equilibrium with the string in the middle point, which is exactly horizontal. Find (a) tension T_1, (b) tension T_2, (c) tension T_3 and (d) angle θ.

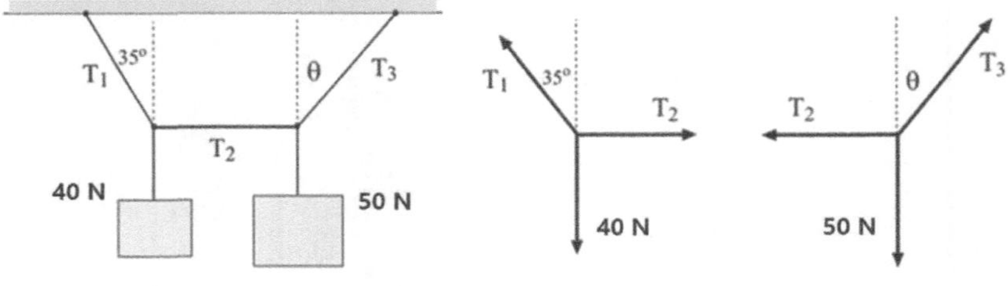

FIGURE 4.1
Forces at the left/right joint of the strings.

Solution

We know that there are the four unknowns (T_1, T_2, T_3, and θ) to solve it. This junction in the strings is equilibrium in static. Therefore, the entirety of the *y* components of the forces in the left junction of the strings is zero:

$\sum F_y = 0 : +T_1 \cos 35° - 40\text{ N} = 0 \rightarrow T_1 = 48.8\text{ N}$

and the entirety of the *x* components of the forces is zero:

$\sum F_x = 0 : -T_1 \sin 35° + T_2 = 0 \rightarrow T_2 = 28.0\text{ N}$

At the moment, the whole of the *x* or *y* components of the forces is zero. We look at the right joint of the strings.

$\sum F_x = 0 : -T_2 + T_3 \sin\theta = 0$ or $\sum F_y = 0 : +T_3 \cos\theta - 50.0\text{ N} = 0$

$T_3 \sin\theta = T_2 = 28.0\text{ N}$ or $T_3 \cos\theta = 50.0\text{ N}$

$\tan\theta = 28.0\text{ N}/50.0\text{ N} = 0.560 \Leftrightarrow \theta = \tan^{-1}(0.560) = 29.3°$. In the end, we obtain

$T_3 = (50.0\text{ N})/(\cos 29.3°) = 57.3\text{ N}$

$T_1 = 48.8\text{ N}, T_2 = 28.0\text{ N}, T_3 = 57.3\text{ N}, \theta = 29.3°$

Example 4.2

The system in Figure 4.2 is in static equilibrium. A mass of 225 kg is fastened to the end of the uniform strut, whose mass is 45.0 kg. Find (a) the cable tension *T* and the (b) horizontal and (c) vertical force components exerted on the strut by the hinge.

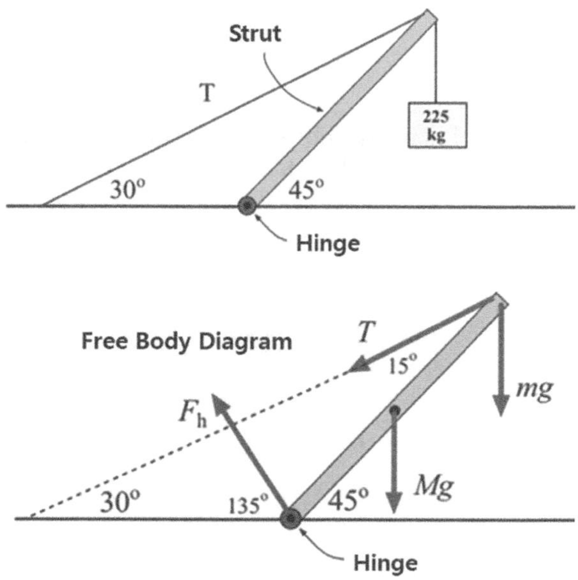

FIGURE 4.2
Forces in equilibrium.

Solution

As the swinging loosely mass is $m = 225$ kg, the string supporting it exerts a downward force of magnitude mg at the crest end of the strut.

$\sum F_x = 0$ (the sum of the F_x must give zero). Namely, $F_{h,x} - T\cos 30° = 0$

$F_{h,x} = T\cos 30° = (6{,}630\text{ N})\cos 30° = 5.74 \times 10^3$ N

$\sum F_y = 0$ (the sum of the F_y should give zero). Namely, $F_{h,y} - T\sin 30° - Mg - mg = 0$

$F_{h,y} = T\sin 30° + Mg + mg = (6{,}630\text{ N})\sin 30° + (45\text{ kg})(9.80\text{ m/s}^2) + (225\text{ kg})(9.80\text{ m/s}^2) = 5.96\times10^3$ N

$\sum \text{MB} = 0$ (the circumstance for zero net torque) $- Mg(L/2)\sin 45° - mg(L)\sin 45° + T(L)\sin 15° = 0$

$T\sin 15° = (Mg/2)\sin 45° + mg\sin 45° = (45\text{ kg}/2)(9.80\text{ m/s}^2)\sin 45° + (225\text{ kg})(9.80\text{ m/s}^2)\sin 45$

$T = (1{,}715\text{ N})/\sin 15° = 6.63 \times 10^3$ N

Example 4.3

As seen in Figure 4.3, a ladder with a mass m stops against a smooth stand with the back straight wall at an angle of 60°. The lowest part end stops on a parallel surface where the coefficient of static friction is $\mu_s = 0.40$. A student with a mass $M = 2$ m attempts to climb the ladder. What fraction of the length L of the ladder will the student have ascended when the ladder begins to slip?

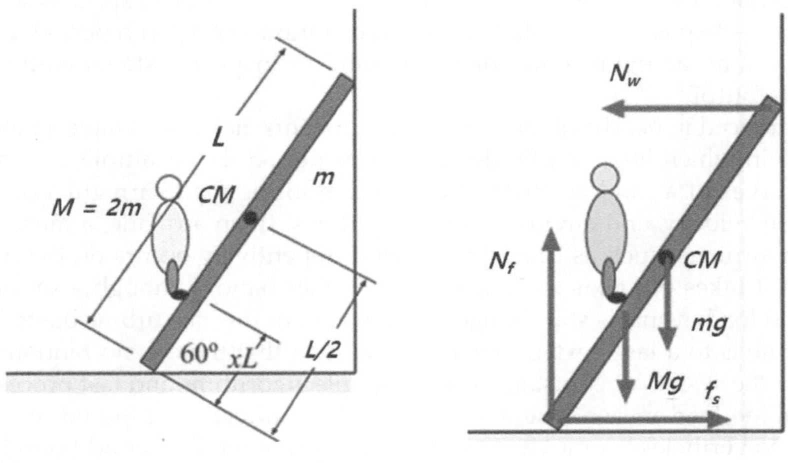

FIGURE 4.3
Ladder forces in equilibrium.

Solution

We exert the circumstances for static equilibrium:

$\sum F_x=0$ (the whole of the F_x should be zero). $f_s - Nw = 0$

$\sum F_y=0$ (the whole of the F_y should be zero). $N_f - Mg(=2\ mg) - mg = 0$

$\sum$MB=0 (the circumstance for zero net torque)

$-(x\ L)(2\ mg)\sin 30° - (L/2)(mg)\sin 30° + (L)(N_w)\sin 60° = 0$

$-x\ mg - mg/4 + \sqrt{3}/2\ N_w = 0$

As the ladder is on the verge of slipping, we have the equality that $f_s = \mu sNf \Leftrightarrow \mu_s Nf - Nw = 0 \Leftrightarrow \mu_s Nf\ (= 3\ mg) - N_w = 0 \Leftrightarrow N_w = 3\mu_s\ mg$

$-x\ mg - mg/4 + \sqrt{3}/2\ N_w\ (= 3\mu_s mg) = 0 \Leftrightarrow -x - 1/4 + 3\sqrt{3}/2\ \mu_s = 0$

$x = -1/4 + 3\sqrt{3}/2\ \mu_s\ (= 0.4) = 0.798$

The student may ascend a fraction of almost 80% of the length of the ladder before it starts to slip.

4.3 (Dynamic) Modeling of Mechanical System (Power System)

4.3.1 Introduction

The activity of making three-dimensional models of a mechanical product, involving construction and mechanism, is a mathematical representation of the dynamic structures to comprehend its characteristics under the forces at work. The methods of a common modeling used in the dynamic systems are Newton, D'Alembert's law, Lagrange, and Hamiltonian mechanics. As a product, models should report the system attributes, which will be described as random variables in a state space. The state space is seen as vectors—displacement, velocity, and acceleration—in a given period. It also provides an acceptable and dense method to inspect systems with some inputs/outputs.

As noticed in nearly all mechanical components, field loads have a random curve in which load amplitudes vary over period. For example, cars show the curves of a random stochastic load due to the road's irregular surface, vehicle velocity, and environmental conditions. In an airplane, a mean load of an airplane, such as drag due to wind, repetitively occurs on the wings when it takes off, taxis, or lands. On the other hand, although a set of the related load events is still changeable, the load of the gas turbine blade in an airplane is to a large extent deterministic such that there is no randomness under the system circumstances. With simple algorithms and fast processors, an online load measurement for parts will be directly computed in operation. Nevertheless, a measurement in function is totally period-bound and therefore impractical in deciding the whole accumulated loads in the product's lifetime. To successfully comprehend the load circumstances of a system in period, the engineer depends on mathematical modeling, utilized by the Newtonian method and response analysis.

Mechanics is a physical discipline that is a branch of applied mathematics dealing with motion and forces, which deals with the state of stopping or motion of bodies under the effect of forces. No other subjects play a more critical part in engineering analysis than mechanics. Thus engineering started with the investigation of mechanics. Contemporary research and development in the area of system reliability could rely on the basic scientific hypothesis of mechanics, involving probability and statistical procedures. Mechanics also is the oldest of the physical sciences. The untimely investigation was that of Archimedes (287–212 BCE), which was connected to the systematic theorem such as buoyancy or the lever. The early investigation of a dynamic object was attributed to Galileo (1564–1642), with his tests of vertically falling stones. The exact arrangement of the law of gravitation and the laws of motion was developed by Newton (1642–1727). A fundamental benefit for the development of mechanics was also attained by D'Alembert, Lagrange, Laplace, etc.

The idea of mechanics as a science relies on mathematics. However, the objective of engineering mechanics is the application of its findings to product design. Although the basic principles of mechanics are relatively few in number, they have wide fields of application such as vibrations, automatic control, engine performance, machines, strength of structures, fluid flow stability, rocketry, and spacecraft design. For instance, by employing only a few equations, scientists may define the motion of a projectile moving through a space and predict the eclipses of the moon. The mathematical analysis of the motion is expressed in Newtonian mechanics because almost the entire investigation is built on the processes achieved by Isaac Newton. Some rigorous principles and laws central to Newtonian mechanics are as follows:

- *Newton's first law of motion*: In the absence of forces, a particle moves with continual velocity.
- *Newton's second law of motion*: The acceleration of a body is instantly connected to the net force and connected to its mass in the opposite manner.

$$F = ma = \frac{dp}{dt} \tag{4.3}$$

- *Newton's third law of motion*: As the pair objects act reciprocally, they exert forces on one another that are equal to magnitude and opposite in direction. Namely, $f_{21} = -f_{12}$.

4.3.2 Reference Frames and Vectors—Displacement, Velocity, and Acceleration

The equation of motion (EOM) is therefore discovered: 1) selecting independent coordinate systems, 2) discovering the numeric of freedom, 3) choosing EOM, 4) producing the free body diagram by making some symbols, 5) exerting force ($\sum_{external} F = ma$) or torque ($\sum_{external} \tau = \frac{d^A h}{dt} + {}^O v_A \times {}^O P$). As the displacement vector is given with respect to the reference frame, velocity and acceleration at Cartesian coordinates may be expressed (Figure 4.4).

Displacement at point B with respect to reference frame O is

$$r_{B/O} = r_{A/O} + r_{B/A} \tag{4.4}$$

where $r_{B/O} = r_{Bx}\hat{i} + r_{By}\hat{j} + r_{Bz}\hat{k}$

If Equation (4.4) is differentiated, the velocity at point B is

$$\frac{dr_{B/O}}{dt} = v_{B/O} = v_{A/O} + v_{B/A} \tag{4.5}$$

Where $v_{B/O} = \frac{dr_{B/O}}{dt} = \dot{r}_{Bx}\hat{i} + \dot{r}_{By}\hat{j} + \dot{r}_{Bz}\hat{k}$

If Equation (4.5) is differentiated, acceleration at point B is

$$\frac{d^2 r_{B/O}}{dt^2} = a_{B/O} = a_{A/O} + a_{B/A} \tag{4.6}$$

where $a_{B/O} = \ddot{r}_{Bx}\hat{i} + \ddot{r}_{By}\hat{j} + \ddot{r}_{Bz}\hat{k}$

If $\Delta r_{B/O}$ moves in Δt, distance $r'_{B/O}$ can be defined as follows (see Figure 4.4(b)):

$$r'_{B/O} = r_{B/O} + \Delta r_{B/O} \tag{4.7}$$

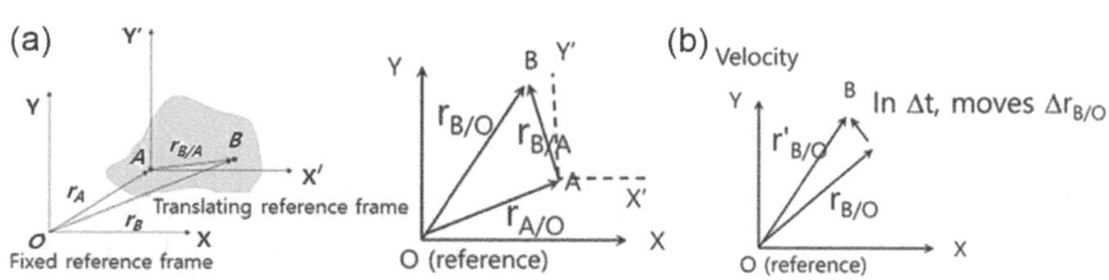

FIGURE 4.4
Displacement or velocity vector at the reference frame: (a) displacement; (b) velocity.

As Δt approaches 0, velocity $v_{B/O}$ can be obtained:

$$v_{B/O} = \lim_{\Delta t \to 0} \frac{r_{B/O}}{t} = \frac{dr_{B/O}}{dt} \tag{4.8}$$

The time derivative of a vector may be expressed:

$$\left(\frac{dA}{dt}\right)_{Oxyz} = \left(\frac{dA}{dt}\right)_{Axyz} + \omega_{/o} \times A \tag{4.9}$$

Therefore, from Equation (4.8), the velocity with respect to a rotating frame A at point O is

$$v_{B/O} = \frac{dr_{B/O}}{dt} = \left(\frac{dr_{A/O}}{dt}\right)_{/O} + \left(\frac{dr_{B/A}}{dt}\right)_{/O} = v_{A/O} + \left(\frac{dr_{B/A}}{dt} = v_{B/A}\right)_{Axyz'} + \left(v_{B/A}\right)_A \tag{4.10}$$

where $\left(\frac{dr_{B/A}}{dt}\right)_{Axyz'}$ is a dog running with $\omega_{/o} = 0$: $\left(\frac{dr_{B/A}}{dt}\right)_{\omega_{/o}=0}$

If $\omega = \omega\hat{k}$, $r_{B/A} = r_{B/A}\hat{i}$, $\left(v_{B/A}\right)_A$ is expressed:

$$\left(v_{B/A}\right)_A = \omega_{/O} \times r_{B/A} = \omega_{/O}\hat{k} \times r_{B/A}\hat{i} = \omega_{/O} r_{B/A} \tag{4.11}$$

Finally, Equation (4.11) can be redefined:

$$v_{B/O} = \frac{dr_{B/O}}{dt} = \left(\frac{dr_{A/O}}{dt}\right)_{/O} + \left(\frac{dr_{B/A}}{dt}\right)_{/O} = v_{A/O} + \left(\frac{dr_{B/A}}{dt} = v_{B/A}\right)_{Axyz'} + \omega_{/O} \times r_{B/A} \tag{4.12}$$

The acceleration from Equation (4.12) can be defined:

$$a_{B/O} = \left(\frac{dv_{B/O}}{dt}\right)_{Oxyz} = a_{A/O} + \left(a_{B/A}\right)_{Axyz'} + 2\omega_{/O} \times \left(v_{B/A}\right)_{Axyz'} + \dot{\omega}_{/O}\left(v_{B/A}\right)_A + \omega_{/O} \times \left(\omega_{/O} \times r_{B/A}\right) \tag{4.13}$$

For a cylindrical coordinate system, the velocity and acceleration from Equations (4.12) and (4.13) may be defined:

$$v_{B/O} = v_{A/O} + \left(\frac{dr_{B/A}}{dt}\right)_{Axyz'} + \omega_{/O} \times r_{B/A} = \frac{d\left(r\hat{r} + z\hat{k}\right)}{dt} = v_{A/O} + \dot{r}\hat{r} + r\dot{\theta}\hat{\theta} + \dot{z}\hat{k} \tag{4.14}$$

$$a_{B/O} = a_{A/O} + \left(\ddot{r} - r\dot{\theta}^2\right)\hat{r} + \ddot{z}\hat{k} + \left(r\ddot{\theta} + 2\dot{r}\dot{\theta}\right)\hat{\theta} \tag{4.15}$$

Example 4.4

As seen in the Figure 4.5, there is a smooth ball B in the smooth hollow cylinder, revolving with continual angular velocity $\dot{\theta}$ about an axis. Inside that cylinder is a particle of mass that can move freely along the longitudinal axis. If so, we may discover the velocity and acceleration of a ball established on the definition of Equations (4.14) and (4.15).

$$v_{B/O} = 0 + \dot{r}\hat{r} + r\dot{\theta}\hat{\theta} \tag{4.16}$$

$$a_{B/O} = 0 + \left(\ddot{r} - r\dot{\theta}^2\right)\hat{r} + \ddot{z}\hat{k} + \left(r\ddot{\theta} + 2\dot{r}\dot{\theta}\right)\hat{\theta} \tag{4.17}$$

4.3.3 Velocity at Two Rotating Frames

A revolving frame of reference is a particular case of a non-inertial reference frame, which is revolving relative to an inertial reference frame. An instance of a revolving reference frame is the outside of the Earth. By taking time derivatives, the velocity and acceleration of the particle in the two frames can be derived. Utilizing these accelerations, the fictitious forces are recognized by comparing Newton's second law as expressed in the two dissimilar frames. All non-inertial reference frames exhibit fictitious forces: 1) the centrifugal force, 2) the Coriolis force, and 3) the Euler force (non-uniformly rotating reference frames).

The two revolving frames are shown in Figure 4.6. We state velocity at the B point with respect to the A point.

$${}^{O}v_B = {}^{o}v_B + {}^{o}v_{B|\omega=0} + {}^{o}\omega_B \times {}^{A}r_B \tag{4.18}$$

where ${}^{o}\omega_B = {}^{o}\omega_1 + {}^{o}\omega_2\left(arm\,\widehat{AB}\right) = (\omega_1 + \omega_2)\hat{k}$

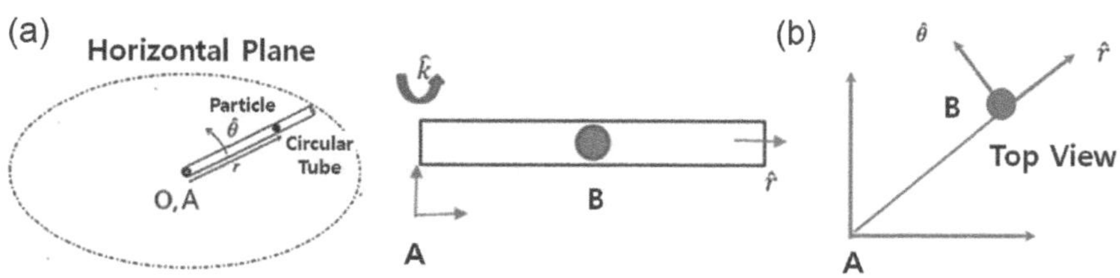

FIGURE 4.5
Equation of the movement of a particle inside a rotating tube: (a) ball motion in the rotating hollow cylinder; (b) top view.

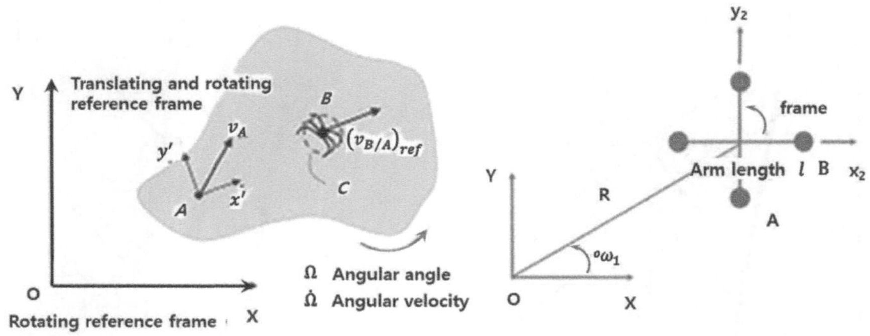

FIGURE 4.6
Two rotating frames.

If arm's length is l, Equation (4.18) may be expressed as

$$^O v_B = R\omega_1 \hat{j}_1 + (\omega_1 + \omega_2)\hat{k} \times l\hat{i}_2 = R\omega_1 \hat{j}_1 + l(\omega_1 + \omega_2)\hat{j}_2 \tag{4.19}$$

4.3.4 Linear Impulse Momentum

An object placed on a leaned surface will slide down the surface. The speed rate relies on how much the surface leans; that is, the greater the slope of the surface, the faster the rate at which the object will glide down it. In physics, a leaned surface is called a slipped plane. Objects are known to begin to quickly move more down inclined planes because of an unbalanced force. To comprehend this sort of movement, it is critical to examine the forces acting on an object on a sloped plane.

As an object with mass m is exposed to an outer force F_{ext}, the motion of the equation is

$$\sum F_{ext} = ma = m\frac{dv}{dt} \tag{4.20}$$

If Equation (4.20) is integrated for interval $[t_1, t_2]$, it may be defined:

$$\int_{t_1}^{t_2} \sum F_{ext} dt = \int_{v_1}^{v_2} m dv \text{ or } \sum \int_{t_1}^{t_2} F_{ext} dt = mv_2 - mv_1 \tag{4.21}$$

Finally, Equation (4.21) can be summarized:

$$\mathrm{m}v_1 + \sum \int_{t_1}^{t_2} F_{\mathrm{ext}} \mathrm{dt} = \mathrm{m}v_2 \tag{4.22}$$

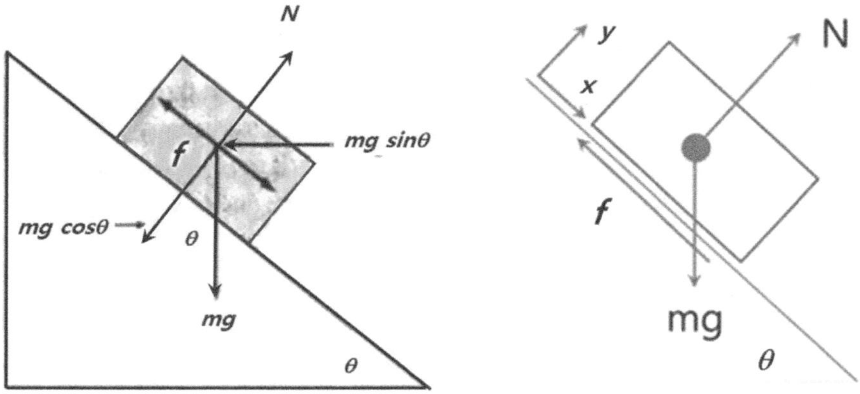

FIGURE 4.7
Moving smoothly object in the slant plane.

Example 4.5

There is a gliding object in the slant plane. In this case, we can discover the final velocity after 3 sec elapses, established in Equation (4.22) (Figure 4.7).

$$\mathrm{m}v_{1x} + \int_0^t [mg\sin\theta - \alpha mg\cos\theta]dt = \mathrm{m}v_{2x} \tag{4.23}$$

where $\sum F_x = mg\sin\theta - \alpha mg\cos\theta$.

Presume the initial velocity $v_{1x} = 0$ in Equation (4.23), we may find the final velocity at t = 3sec.

$$[mg\sin\theta - \alpha mg\cos\theta]\mathrm{t} = \mathrm{m}v_{2x} \text{ or } v_{2x} = 3[g\sin\theta - \alpha g\cos\theta] \tag{4.24}$$

4.3.5 Angular Momentum of a Particle

Angular momentum in physics has both a direction and a magnitude as a vector. It is the rotational information that can be expressed by a variational linear momentum and is a main physical amount because the whole angular momentum of a closed system is conserved. Examples are motorcycles and bicycles, rifled bullets, flying discs, and gyroscopes. Conservation of an angular momentum also explains why hurricanes are a kind of spirals and neutron stars with high rotational rates.

There is a particle with some velocity (Figure 4.8). Angular momentum of a particle (or rigid body) in the space is specified:

$$h_{B/O} = r_{B/O} \times P_{B/O} \tag{4.25}$$

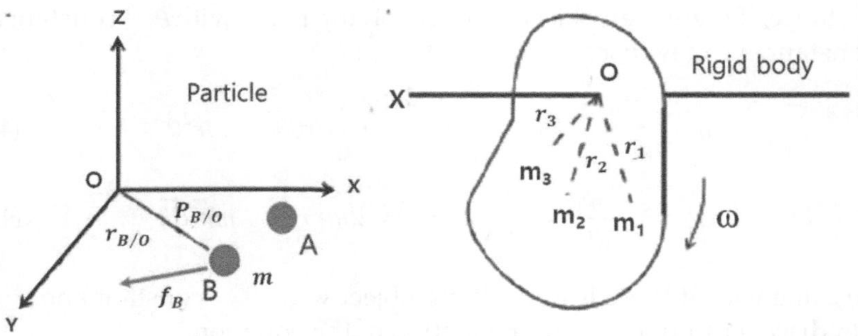

FIGURE 4.8
Angular momentum of a rigid body (or particle) in a space.

If Equation (4.25) is differentiated, it may be defined:

$$\frac{dh_{B/O}}{dt} = \sum \tau_{B/O} \tag{4.26}$$

Because $h_{B/A} = r_{B/A} \times P_{B/O}$, torque $\tau_{B/A}$ is

$$\tau_{B/A} = \frac{dh_{B/A}}{dt} + v_{B/A} \times P_{B/O} \tag{4.27}$$

The movement of the equation at B is

$$\sum\nolimits_{\text{ext}} f_B = \frac{dP_{B/O}}{dt} = m\frac{dv_{B/O}}{dt} \tag{4.28}$$

Torque B with respect to A is

$$\tau_{B/A} = r_{B/A} \times f_B = r_{B/A} \times \frac{dP_{B/O}}{dt} = \frac{d\left(r_{B/A} \times P_{B/O}\right)}{dt} - \frac{dr_{B/A}}{dt} \times P_{B/O} \tag{4.29}$$

Because $r_{B/A} = r_{B/O} - r_{A/O}$ and $\frac{dr_{B/A}}{dt} = v_{B/O} - v_{A/O}$, Equation (4.29) is

$$\tau_{B/A} = \frac{dh_{B/A}}{dt} - \left(v_{B/O} - v_{A/O}\right) \times P_{B/O} \tag{4.30}$$

where $P_{B/O} = mv_{B/O}$

Case I: If $v_{A/O} = 0$, $\tau_{B/A} = \frac{dh_{B/A}}{dt}$ from Equation (4.30)

Case II: If $v_{A/O} \parallel$ to $P_{B/O}$ (correct when A = center of mass), $\tau_{/A} = \frac{dh_{/A}}{dt}$ from Equation (4.30)

For instance, if there is an object in the revolving plane with $\dot{\theta}$ = constant and $\dot{r}$ = constant, we may discover $h_{B/O}$ and $\tau_{B/O}$:

$$h_{B/O} = r_{B/O} \times P_{B/O} = r\hat{r} \times m\left(\dot{r}\hat{r} + r\dot{\theta}\hat{\theta}\right) = mr^2\dot{\theta} \tag{4.31}$$

$$\tau_{B/o} = \frac{dh_{B/O}}{dt} = \frac{d}{dt}\left(mr^2\dot{\theta}\hat{k}\right) = 2mr\,\dot{r}\,\dot{\theta}\hat{k} + mr^2\ddot{\theta}\hat{k} \tag{4.32}$$

In the same way, if there is a revolving object with $\dot{\theta}$ = constant and $\dot{r}$ = 0, we can discover torque T from the motion of the equation.

$$v_{A/O} = r_1\dot{\theta}\hat{\theta} \tag{4.33}$$

$$P_{A/O} = mr_1\dot{\theta}\hat{\theta} \tag{4.34}$$

$$\sum\nolimits_{ext} F = \frac{dP_{A/O}}{dt} = -mr_1\dot{\theta}^2\hat{r} = -T\hat{r} \tag{4.35}$$

$$T = mr_1\dot{\theta}^2 \tag{4.36}$$

Example 4.6

As seen in Figure 4.9, a small-scale object of mass m is rotating in an almost horizontal circle of radius r at a constant speed. That is, if $\dot{r}$ = constant, find the motion of equation.

Solution

$$v_{A/O} = \dot{r}\hat{r} + r\dot{\theta}\hat{\theta} \tag{4.37}$$

$$P_{A/O} = mv_{A/O} = m\left(\dot{r}\hat{r} + r\dot{\theta}\hat{\theta}\right) \tag{4.38}$$

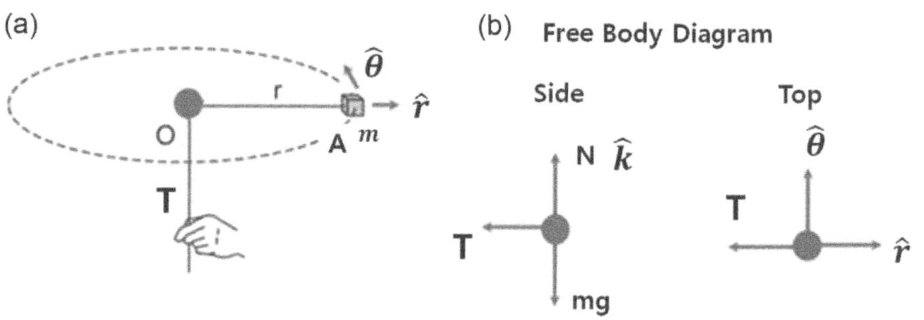

FIGURE 4.9
Revolving object with continual speed and radius: (a) revolving object; (b) free body diagram.

$$\sum_{ext} F = \frac{dP_{A/O}}{dt} = m\left[\left(-r\dot{\theta}^2\right)\hat{r} + \left(r\ddot{\theta} + 2\dot{r}\dot{\theta}\right)\hat{\theta}\right] \tag{4.39}$$

Because a small object of mass m is spinning at constant speed, then the $\hat{\theta}$ direction : $\sum_{ext} F = 0$. That is,

$$\hat{\theta}\text{ direction: } \frac{dP_{A/O}}{dt} = 0 = m\left(r\ddot{\theta} + 2\dot{r}\dot{\theta}\right) \tag{4.40}$$

$$r\ddot{\theta} = -2\dot{r}\dot{\theta} \tag{4.41}$$

$$\hat{r}\text{ direction: } \sum_{ext} F = -mr\dot{\theta}^2 = -T \text{ or } T = mr\dot{\theta}^2 \tag{4.42}$$

$$h_{A/O} = r_{A/O} \times P_{A/O} = r\hat{r} \times m\left(\dot{r}\hat{r} + r\dot{\theta}\hat{\theta}\right) = mr^2\dot{\theta}. \tag{4.43}$$

$$\hat{k}\text{ direction: } \sum_{ext} \tau_{A/O} = \frac{dh_{A/O}}{dt} = m\left(2r\dot{r}\dot{\theta} + r^2\ddot{\theta}\right) = 0 \text{ or } 2\dot{r}\dot{\theta} = -r\ddot{\theta} \tag{4.44}$$

As $r = r_1$ at t_1, $h_{A/O} = mr_1^2\dot{\theta}_1$. On the other hand, as $r = r_2$ at t_2, $h_{A/O} = mr_2^2\dot{\theta}_2$
Because $h_{A/O} =$ constant, we know that $mr_1^2\dot{\theta}_1 = mr_2^2\dot{\theta}_2$.

Example 4.7

Suppose we have a mass conically rotating about the *z*-axis in Figure 4.10. Determine the motion of the equation.

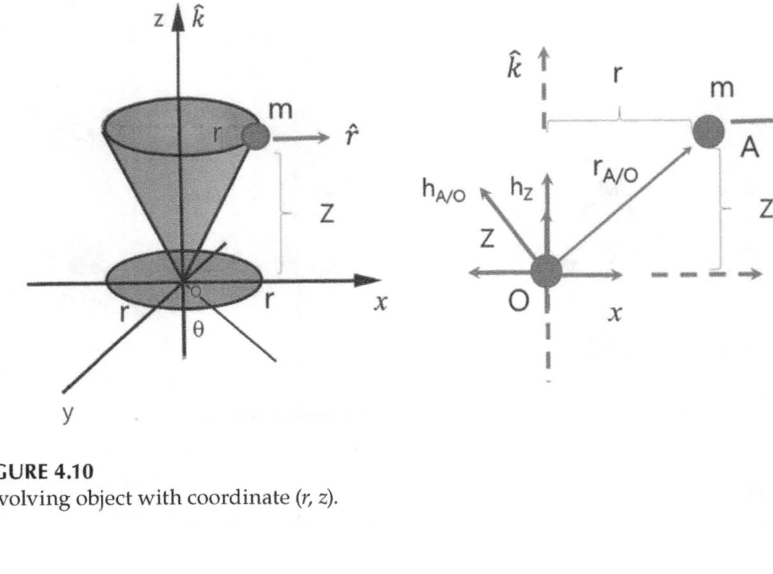

FIGURE 4.10
Revolving object with coordinate (*r*, *z*).

Solution

$$h_{A/O} = r_{A/O} \times P_{A/O} = \left(r\hat{r} + z\hat{k}\right) \times mr\dot{\theta}\hat{\theta} = mr^2\dot{\theta}\hat{k} - mrz\dot{\theta}\hat{r} \tag{4.45}$$

$$\begin{aligned}\sum\nolimits_{ext} \tau_{A/O} &= \frac{dh_{A/O}}{dt} = mr^2\,\ddot{\theta}\,\hat{k} - mrz\,\ddot{\theta}\,\hat{r} - mrz\dot{\theta}\dot{\theta}\hat{\theta} \\ &= mr^2\,\ddot{\theta}\,\hat{k} - mrz\,\ddot{\theta}\,\hat{r} - mrz\dot{\theta}^2\hat{\theta}\end{aligned} \tag{4.46}$$

Think about bodies revolving about the middle point of mass or fastened points; in such a case, the reference frames are fastened to bodies (Figure 4.11).

$$h_{i/A} = r_{i/A} \times P_{1/O} = \left(x_1\hat{i} + z_1\hat{k}\right) \times m_1 x_1 \omega_z \hat{j} = m_1 x_1^2 \omega_z \hat{k} - m_1 x_1 z_1 \omega_z \hat{i} \tag{4.47}$$

If that is the situation, the external torque τ^{ext} is

$$\frac{dh_{i/A}}{dt} = \tau^{ext} \tag{4.48}$$

The angular momentum of the rigid body is

$$H_{/A} = \sum\nolimits_i r_{i/A} \times P_{i/A} = \sum\nolimits_i m_i r_{i/A} \times \left(\omega_{/o} \times r_{i/A}\right) = H_x\hat{i} + H_y\hat{j} + H_z\hat{k} \tag{4.49}$$

For the matrix form, Equation (4.49) is expressed:

$$H_{/A} = \begin{Bmatrix} H_x \\ H_y \\ H_z \end{Bmatrix} = \begin{bmatrix} I_{xx} & I_{xy} & I_{xz} \\ I_{yx} & I_{yy} & I_{yz} \\ I_{zx} & I_{yz} & I_{zz} \end{bmatrix} \begin{Bmatrix} \omega_x \\ \omega_y \\ \omega_z \end{Bmatrix} \tag{4.50}$$

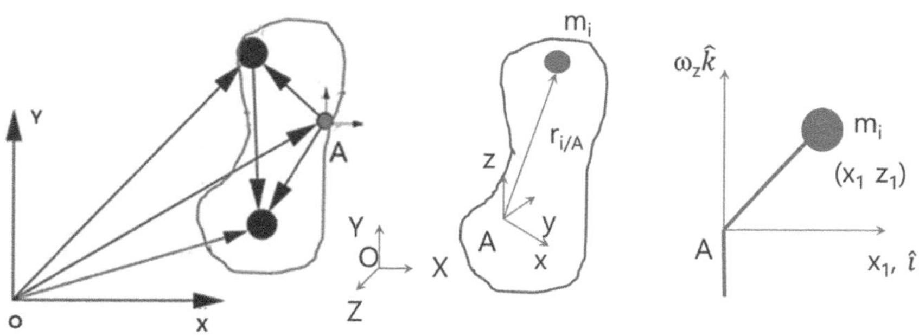

FIGURE 4.11
Bodies that revolve about center of mass with respect to fastened points.

where $I_{xz} = -\sum_i m_i x_i z_i, I_{zz} = \sum_i m_i \left(x_i^2 + y_i^2\right)$

In which case, $\omega = \omega_z \hat{k}$

$$H_{/A} = \begin{Bmatrix} H_x \\ H_y \\ H_z \end{Bmatrix} = \begin{bmatrix} I_{xx} & I_{xy} & I_{xz} \\ I_{yx} & I_{yy} & I_{yz} \\ I_{zx} & I_{yz} & I_{zz} \end{bmatrix} \begin{Bmatrix} 0 \\ 0 \\ \omega_z \end{Bmatrix} = \begin{Bmatrix} I_{xz}\omega_z \\ I_{yz}\omega_z \\ I_{zz}\omega_z \end{Bmatrix} \tag{4.50}$$

In other words,

$$H_{/A} = H_x\hat{i} + H_y\hat{j} + H_k\hat{k} = I_{xz}\omega_z\hat{i} + I_{yz}\omega_z\hat{j} + I_{zz}\omega_z\hat{k} \tag{4.51}$$

Example 4.8

Two rotating balls are mounted on rods in Figure 4.12. Find the angular momentum ${}^{A}H$ at A.

Solution

$${}^{o}P_1 = m_1\, {}^{o}v_1 = m_1 \Omega x_1 \hat{j} \tag{4.52}$$

$${}^{o}P_2 = m_2\, {}^{o}v_1 = -m_2 \Omega x_1 \hat{j} \tag{4.53}$$

$${}^{A}h_1 = {}^{A}r_1 \times {}^{o}P_1 = \left(x_1\hat{i} + z_1\hat{k}\right) \times m_1\Omega x_1\hat{j} = m_1 x_1^2 \Omega \hat{k} - m_1 x_1 z_1 \Omega \hat{i} \tag{4.54}$$

$${}^{A}h_2 = {}^{A}r_2 \times {}^{o}P_2 = \left(-x_1\hat{i} + z_1\hat{k}\right) \times -m_2\Omega x_1\hat{j} = m_2 x_1^2 \Omega \hat{k} + m_2 x_1 z_1 \Omega \hat{i} \tag{4.55}$$

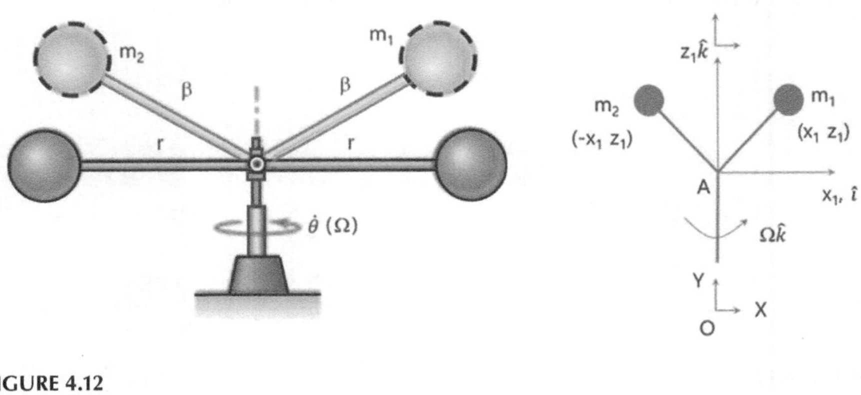

FIGURE 4.12
Two rotating balls.

If $m_1 = m_2 = m$, $\omega = \{0 \quad 0 \quad \Omega\}^T$, the angular momentum is

$$^{A}H = {}^{A}h_1 + {}^{A}h_2 = 2mx_1^2\Omega\hat{k} \tag{4.56}$$

If that is so, then the external torque τ^{ext} is

$$\frac{d\,^{A}H}{dt} = \tau^{\text{ext}} = 2mx_1^2\dot{\Omega}\hat{k} \tag{4.57}$$

The principal axis can be expressed as a group of orthogonal axes such that

$$[I] = \begin{bmatrix} I_{xx} & 0 & 0 \\ 0 & I_{yy} & 0 \\ 0 & 0 & I_{zz} \end{bmatrix}$$

The laws of symmetry for rigid bodies are 1) as there is an axis of symmetry, the axis is a principal axis; 2) as there is one plane of symmetry, a principal axis at right angles to it go through the middle point of mass, 3) as there are the two orthogonal flat surfaces of symmetry, their intersection is a principal axis.

The basic rules for rigid bodies are

$$\sum\nolimits_{ext} F = \frac{dP_{/O}}{dt} = Ma \tag{4.58}$$

$$\sum\nolimits_{ext} \tau = \frac{dH_{/A}}{dt} + v_{A/O} \times P_{/O} \text{ where } H_{/A} = \left[{}^{G}I\right]\{\omega\} \tag{4.59}$$

Example 4.9

As seen in the following Figure 4.13, there is a rotating rod with 0, 45° angle. Determine the angular momentum H of a revolving rod.

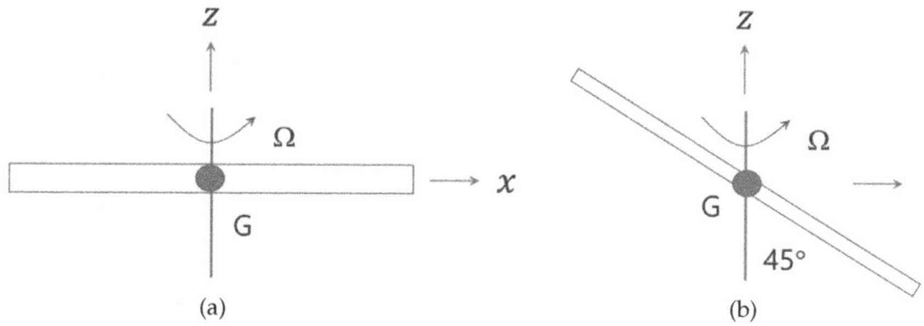

FIGURE 4.13
Rotating rod with some angle.

For case (a), because $L>>R$, the angular momentum H is

$$H = \left[{}^{G}I\right]\{\omega\} = \begin{bmatrix} I_{xx} & 0 & 0 \\ 0 & I_{yy} & 0 \\ 0 & 0 & I_{zz} \end{bmatrix} \begin{Bmatrix} \omega_x \\ \omega_y \\ \omega_z \end{Bmatrix} = \begin{bmatrix} \frac{MR^2}{2} \approx 0 & 0 & 0 \\ 0 & \frac{ML^2}{12} & 0 \\ 0 & 0 & \frac{ML^2}{12} \end{bmatrix} \begin{Bmatrix} 0 \\ 0 \\ \Omega \end{Bmatrix} \tag{4.60}$$

$$= \begin{Bmatrix} 0 \\ 0 \\ \frac{ML^2}{12}\Omega \end{Bmatrix}$$

If Equation (4.60) is differentiated,

$$\frac{dH}{dt} = \frac{ML^2}{12}\dot{\Omega}\hat{k} \tag{4.61}$$

For case (b), angular momentum, angular momentum H is

$$H = \begin{bmatrix} 0 & 0 & 0 \\ 0 & \frac{ML^2}{12} & 0 \\ 0 & 0 & \frac{ML^2}{12} \end{bmatrix} \begin{Bmatrix} -\frac{\sqrt{2}}{2}\Omega \\ 0 \\ \frac{\sqrt{2}}{2}\Omega \end{Bmatrix} = \begin{Bmatrix} 0 \\ 0 \\ \frac{ML^2}{12}\frac{\sqrt{2}}{2}\Omega \end{Bmatrix} \tag{4.62}$$

If Equation (4.62) is differentiated,

$$\frac{dH}{dt} = \frac{ML^2}{12}\frac{\sqrt{2}}{2}\dot{\Omega}\hat{k} \tag{4.63}$$

Example 4.10

An object is on the moving cart. Find the required maximum force F of an object on the cart (Figure 4.14).

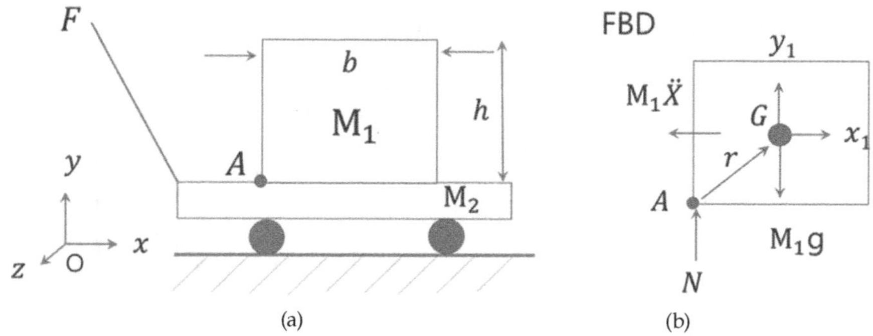

FIGURE 4.14
Object on the moving cart.

Solution

$$\sum\nolimits_{ext} F_{x,\,\text{system}\,(M_1+M_2)} = \left(M_1 + M_2\right)\ddot{x} \tag{4.64}$$

$$\sum\nolimits_{\text{ext}} \tau_{A/O} = -M_1 g\frac{b}{2}\hat{k} = \frac{dH_{/A}}{dt} + v_{A/O} \times \left(M_1 v_{G/O}\right) \tag{4.65}$$

where $v_{A/O} = \dot{X}\hat{i}$, $v_{G/O} = v_{A/O}$.

Only rotation is considered:

$$\omega = \{0 \quad 0 \quad \omega_z\}^T \tag{4.66}$$

$$H_{G/O} = \left[I_{G/O}\right]\begin{Bmatrix} 0 \\ 0 \\ \omega_z \end{Bmatrix} = I_{zz/G}\omega_z\hat{k} \tag{4.67}$$

where $H_x = H_y = 0, H_z = I_{zz/G}\omega_z\hat{k}$.

$$\begin{aligned} H_{A/O} &= H_{G/O} + r_{G/A} \times P_{/O} = I_{zz/G}\omega_z\hat{k} + \left(\frac{b}{2}\hat{i} + \frac{h}{2}\hat{j}\right) \times M_1\dot{X}\hat{i} \\ &= \left(I_{zz/G}\omega_z - M_1\frac{h}{2}\dot{X}\right)\hat{k} \end{aligned} \tag{4.68}$$

$$\sum_{ext} \tau_{/A} = -M_1 g \frac{b}{2}\hat{k} = \frac{dH_{/A}}{dt} = \left(I_{zz/G}\dot{\omega}_z - M_1 \frac{h}{2}\ddot{X}\right)\hat{k} \tag{4.69}$$

If $\dot{\omega}_z = 0$, we know:

$$M_1 g \frac{b}{2} = M_1 \frac{h}{2}\ddot{X} \tag{4.70}$$

$$\ddot{X}_{\max} = g\frac{b}{h} \tag{4.71}$$

$$F_{\max} = (M_1 + M_2)\ddot{X}_{\max} = (M_1 + M_2)g\frac{b}{h} \tag{4.72}$$

$$\sum_{ext} \tau_{A/O} = M_1 \ddot{X}\frac{h}{2} - M_1 g\frac{b}{2} \tag{4.73}$$

If $r_{G/A} = \left(a\hat{i} + c\hat{k}\right), H_{A/O}$, then

$$\begin{aligned} H_{\frac{A}{O}} &= H_{\frac{G}{O}} + r_{\frac{G}{A}} \times P_{\overline{O}} = \left[I_{/G}\right]\begin{Bmatrix} 0 \\ 0 \\ \omega_z \end{Bmatrix} + \left(a\hat{i} + c\hat{k}\right) \times Ma\omega_z\hat{j} \\ &= \left[I_{zz/G} + Ma^2\right]\omega_z\hat{k} - Mac\omega_z\hat{i} \end{aligned} \tag{4.74}$$

4.3.6 Four Classes for Rigid Rotational Motion

A rigid body is an object with a mass that keeps a stiff form, such as a phonograph turntable, when compared to the sun, which is a ball of gas. Rotational movement is everywhere in the universe; for example, 1) the movement of electrons in an atom and 2) the movement of the Moon around the Earth. Objects may not be regarded as particles when showing rotational movement since dissimilar parts of the object move with dissimilar velocities and accelerations. Thus it is required to regard the object as a system of particles. The rotational movement of a rigid body is more complex than linear movement, and only the movement of rigid bodies may be contemplated here. Many of the equations for the mechanics of revolving objects are indistinguishable to the movement equations for linear motion.

As all components of a rigid body proceed parallel to a fastened plane, the movement of the object is referred to as plane movement. There are the two sorts of plane movement: 1) the rigid body in such a movement revolves about a fastened axis, which is perpendicular to a fastened plane (the pure rotational motion); 2) the movement may be regarded as a combination of pure translational movement parallel to a fixed plane in addition to a pure rotational movement on an axis that is perpendicular to that plane (general plane motion). So there are four categories for rigid rotational movement: 1) *Class 1*: pure rotation about a fastened axis through the middle point of mass G; 2) *Class 2*: pure rotation about a fastened axis at *A*, which is not similar to the center of mass G ($A \neq G$); 3) *Class 3*: no outer constraint; 4) *Class 4*: bodies with moving points of constraint.

4.3.6.1 Class 1 (Pure Rotation)

Rotors of all sorts (Figure 4.15)

$$H_{G/O} = \left[I_{G/O}\right]\begin{Bmatrix}\omega_x \\ \omega_y \\ \omega_z\end{Bmatrix} \tag{4.75}$$

For 2D planar motion, Equation (4.75) is expressed:

$$H_{G/O} = I_{zz_G}\omega_z \text{ where } \omega = \{0 \quad 0 \quad \omega_z\}^T \tag{4.76}$$

If Equation (4.76) is differentiated:

$$\frac{dH_{G/O}}{dt} = I_{zz_G}\dot{\omega}_z = I_{zz_G}\ddot{\theta} \tag{4.77}$$

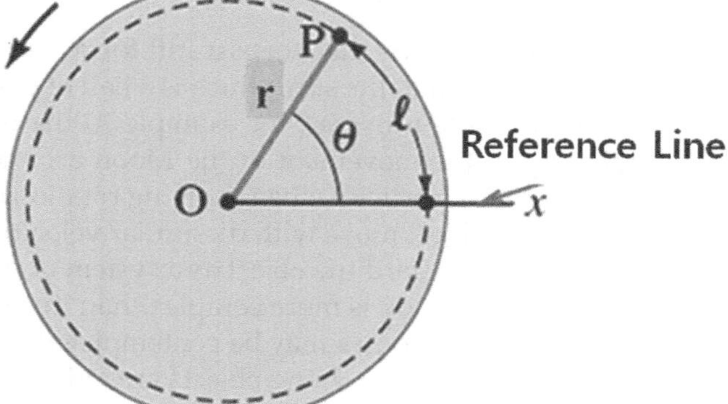

FIGURE 4.15
Pure rotation.

4.3.6.2 Class 2 (Pure Rotation on a Fastened Axis at A ≠ G)

$$\sum\nolimits_{ext} \tau_{A/O} = \frac{dH_{A/O}}{dt} = \frac{d}{dt}\left([I_{/A}] \begin{Bmatrix} \omega_x \\ \omega_y \\ \omega_z \end{Bmatrix} \right) \tag{4.78}$$

For 2D planar movement $\omega = \begin{Bmatrix} 0 \\ 0 \\ \omega_z \end{Bmatrix}$, if $[I_{G/O}] = [\because]$ (the principle axis is selected and the aligned axis theorem utilized to attain $I_{A/O}$. Because $I_{A/O} = I_{G/O} + Md^2$:

$$H_{G/O} = [I_{G/O}] \begin{Bmatrix} \omega_x \\ \omega_y \\ \omega_z \end{Bmatrix} \tag{4.79}$$

4.3.6.3 Class 3 Problem (No External Constraint)

To describe the case with no outer constraint, an instance is a wheel with mass = 75 kg, exposed to end force F = 150 N, as shown in Figure 4.16.

The degree of freedom (DOF) is

$$DOF = 6 - 3 = 3$$

where K = 0.15 m (≡ radius of gyration)

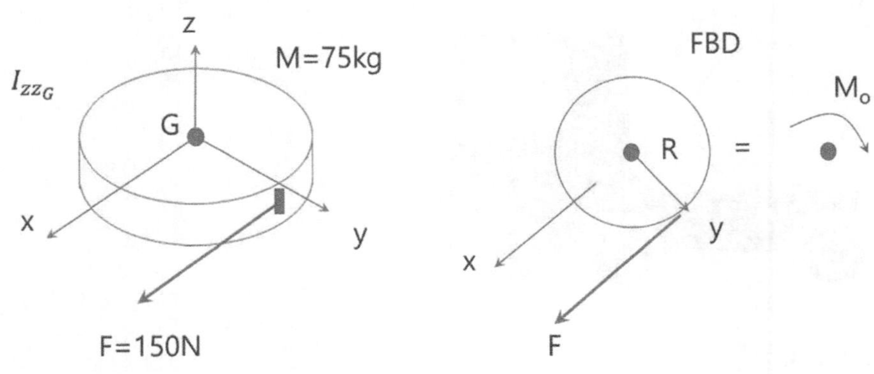

FIGURE 4.16
Class 3 problem (no external constraint).

$$I_{zz_G} = MK^2 \text{ or } K = \sqrt{\frac{I_{zz_G}}{M}} \tag{4.80}$$

As the two trivial equations show, the motion of equation at x orientation is

$$\sum F_x = F = M\ddot{x} \text{ or } \ddot{x} = \frac{F}{M} = \frac{150\,\text{N}}{75\,\text{kg}} = 2\,\text{m}/\text{s}^2$$

Class 4 problem (bodies with moving points of constraint):

Angular momentum of a rigid body (Figure 4.17) is

$$\begin{aligned} H_{A/O} &\equiv \sum r_{i/A} \times P_{i/O} = \sum_i \left(r_{G/A} + r_{i/G}\right) \times P_{i/O} \\ &= r_{G/A} \times \sum_i m_i v_{i/O} + \sum_i r_{i/G} \times m_i v_{i/O} \end{aligned} \tag{4.81}$$

$$x_2 = x_1 - R\theta \tag{4.82}$$

$$\ddot{x}_2 = \ddot{x}_1 - R\ddot{\theta} \tag{4.83}$$

Because $\tau_A = \dfrac{dH_{A/O}}{dt} + v_{A/O} \times P_{/O}$, angular momentum is

$$\begin{aligned} H_{A/O} &= I_{zz_G}\theta + r_{G/A} \times P_{/O} = I_{zz_G}\dot{\theta}\hat{k} + R\hat{j} \times M\dot{x}_2\hat{i} = I_{zz_G}\dot{\theta}\hat{k} - RM\dot{x}_2\hat{k} \\ &= I_{zz_G}\dot{\theta}\hat{k} - RM\dot{x}_2\hat{k} \end{aligned} \tag{4.84}$$

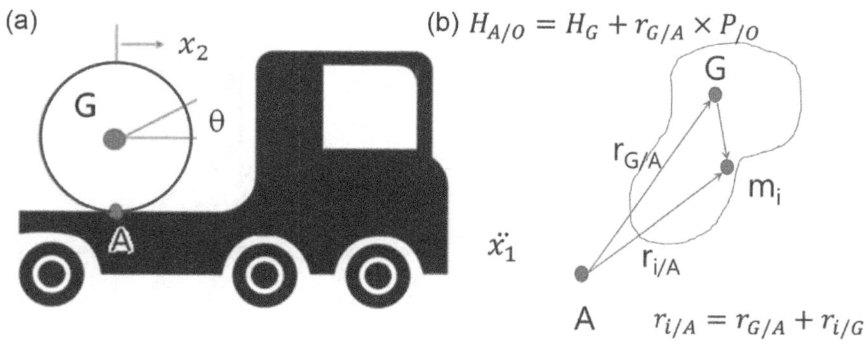

FIGURE 4.17
Class 4 problem (moving points of contact): (a) object moving points of contact; (b) angular momentum of a rigid body.

$$\frac{dH_{A/O}}{dt} = I_{zz_G}\ddot{\theta}\hat{k} - RM\ddot{x}_2\hat{k} = I_{zz_G}\ddot{\theta}\hat{k} - RM\left(\ddot{x}_1 - R\ddot{\theta}\right)\hat{k} = 0 \tag{4.85}$$

$$\ddot{\theta} = \left[MR/\left(I_{zz_G} + MR^2\right)\right]\ddot{x}_1 \tag{4.86}$$

Example 4.11

A revolving rigid object with rectangular form is shown in Figure 4.18. Determine the motion of equation. At this place, L = 32.1 cm, a = 32.1 cm, b = 1.25 cm, d = 10.2 cm.

We know that $I_{zz/G} = M\left(\frac{L^2 + a^2}{12}\right)$, $I_{yy/G} = M\left(\frac{L^2 + b^2}{12}\right)$, $I_{xx/G} = M\left(\frac{a^2 + b^2}{12}\right)$.

Because $(a/L)^2 = 0.022$ and $(b/L)^2 = 0.002$

$$I_{zz/G} = \frac{ML^2}{12} \tag{4.87}$$

If $d = L/2$, then

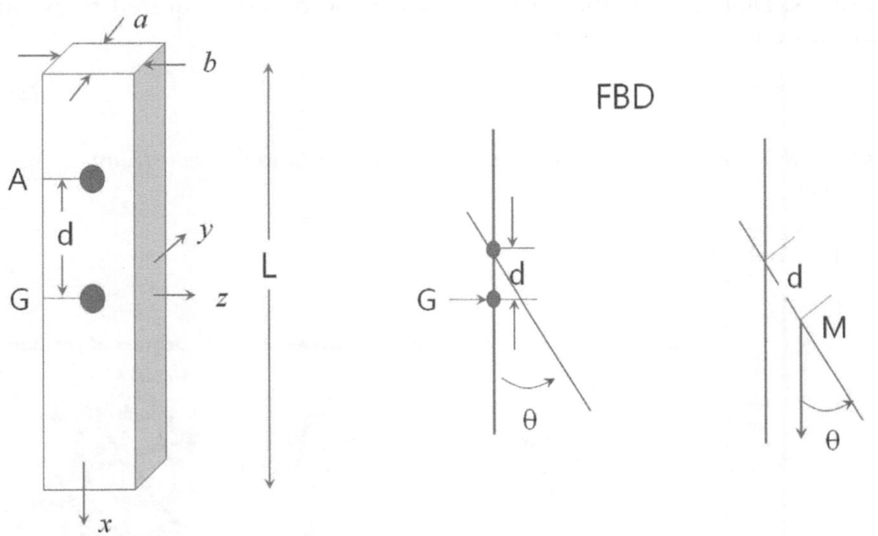

FIGURE 4.18
Revolving object with continual speed and radius.

$$I_{zz/A} = I_{zz/G} + Md^2 = \frac{ML^2}{12} + \frac{ML^2}{4} = \frac{ML^2}{3} \tag{4.88}$$

$$\begin{aligned}\sum\nolimits_{ext} \tau_{A/O} &= \frac{dH_{A/O}}{dt} = \frac{dI_{zz/A}\omega_z}{dt} = I_{zz/A}\,\ddot{\theta}_z \\ &= -Mgd\sin\theta \; or \; I_{zz/A}\,\ddot{\theta}_z + Mgd\sin\theta = 0\end{aligned} \tag{4.89}$$

If θ is small, $\sin\theta \approx \theta$. Equation (4.89) is linearized. That is,

$$I_{zz/A}\ddot{\theta}_z + Mgd\theta = 0 \tag{4.90}$$

Assume $\theta(t) = \theta_0 \sin\omega t$, we can find ω_n:

$$\left(-\omega^2 I_{zz/A} + +Mgd\right)\theta_0 \sin\omega t = 0 \tag{4.91}$$

$$\omega = \sqrt{\frac{Mgd}{I_{zz/A}}} \text{ (generic solution for pendulum with 1 DOF)} \tag{4.92}$$

4.3.7 Degree of Freedom (DOF), Coordinate System, Free Body Diagram, and Equation of Motion

There is an object in a slope plane. To explain it as motion of equation (MOE), we should explain the degree of freedom (DOF), established on the constraints. DOF is a number of independent coordinates required to explain the movement (Figure 4.19):

$$DOF = 6n + 3m - C \tag{4.93}$$

where $n = number\ of\ rigid$, $m = number\ of\ particle$, and $C = constraints$.
C is 5:

1. $y = \dot{y} = \ddot{y} = 0$ is 1.

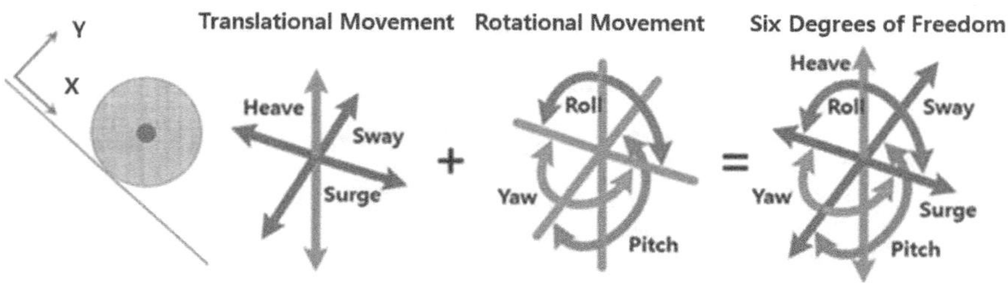

FIGURE 4.19
Object in a slope plane.

2. Presume $z = \dot{z} = \ddot{z} = 0$ is 1. That is, $\sum F_z = 0 = ma_z$.
3. Presume no rotation on x- or y-axis is 2.
4. No slip $X = -R\theta$ is 1.

From Equation (4.93), we can discover the DOF of an object in Figure 4.19. Namely, $DOF = 6 - 5 = 1$. As slip is permitted, $DOF = 6 - 4 = 2$.

Example 4.12

As shown in the subsequent Figure 4.20, a sliding ladder is leaned against the wall. In this case, determine DOE, free body diagram (FBD), and motion of equation (MOE).

Solution

C is 5:

1. Constrained in translation: x and y are 2.
2. Presume $z = \dot{z} = \ddot{z} = 0$ is 1.
3. No rotation on x and y are 2.

$$DOF = 6 - 5 = 1 \tag{4.94}$$

$$MOE \sum\nolimits_{ext} \tau_C = \frac{dh_{/C}}{dt} = +Mg\frac{L}{2}\sin\theta \tag{4.95}$$

Example 4.13

As seen in the following Figure 4.21, two bodies are connected jointly. Determine DOE, free body diagram (FBD), and motion of equation (MOE).

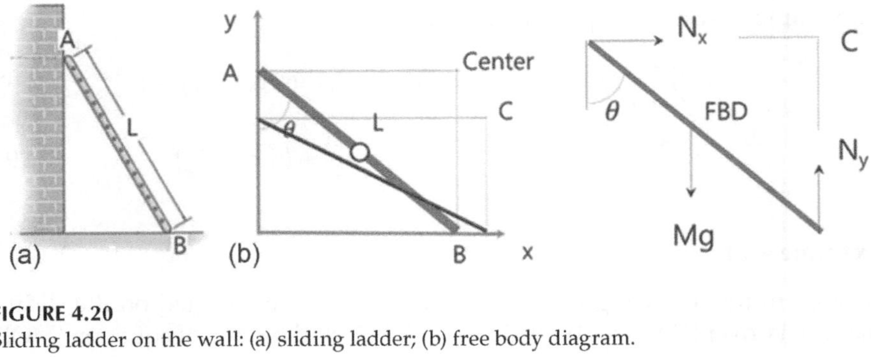

FIGURE 4.20
Sliding ladder on the wall: (a) sliding ladder; (b) free body diagram.

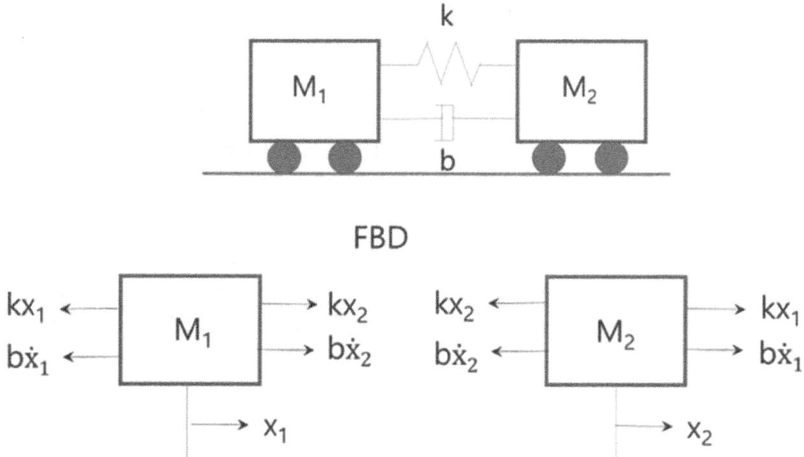

FIGURE 4.21
Two bodies connected on the plane.

Solution

In this problem, only two independent coordinates are required to explain the movement. Explain the ten constraints that one should presume to be acting on this device, so as to lessen the number of degrees of freedom to two. Namely,

C is 10:

$$DOF = 6 \times 2 - 10 = 2 \tag{4.96}$$

Produce the free body diagram by making some signs, allocating the two suitable coordinates, and determining the two equations of motion that distinguish the system. To obtain the accurate signs on the spring and dashpot forces for a period, presume positive displacements and velocities of each coordinate and conclude the following as the consequence of orientations of the spring and damper forces:

Presume x_1 and x_2 are position and MOE.

$$\sum_{ext} F_{1,x_1} = M_1\ddot{x}_1 = k(x_2 - x_1) + b(\dot{x}_2 - \dot{x}_1) \tag{4.97}$$

$$\sum_{ext} F_{2,x_2} = M_2\ddot{x}_2 = k(x_1 - x_2) + b(\dot{x}_1 - \dot{x}_2) \tag{4.98}$$

Example 4.14

As seen in the following Figure 4.22, two bodies are attached on the sliding plane. Discover DOE, free body diagram (FBD), and motion of equation (MOE).

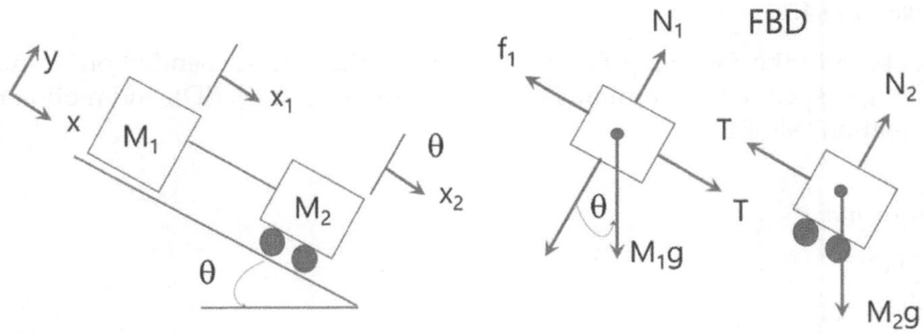

FIGURE 4.22
Two bodies attached on the sliding plane.

Solution

C is 9:

1. Moving together, M_1 & M_2 are 1.
2. No acceleration is 2.
3. No rotation is

$$DOF = 6\times 2 - C = 12 - 9 = 3 \tag{4.99}$$

There are two trivial equations:

$$\sum\nolimits_{ext} F_{Z_1} = M_1\ddot{z}_1 = 0 \text{ or } \sum\nolimits_{ext} F_{Z_2} = M_2\ddot{z}_2 = 0 \tag{4.100}$$

So it has 1 notable MOE.

$$\sum\nolimits_{ext} F_x = M\ddot{x} \text{ where } x_1 = x_2 = x \text{ or } \ddot{x}_1 = \ddot{x}_2 = \ddot{x} \tag{4.101}$$

There are 5 unknowns: $N_1, N_2, f_1, T, \ddot{x}$.

$$\sum\nolimits_{ext} F_{y_1} = M_1\ddot{y}_1 \Rightarrow N_1 \Rightarrow f_1 = \mu N_1 \text{ or } \sum\nolimits_{ext} F_{y_2} = M_2\ddot{y}_2 \Rightarrow N_2 \tag{4.102}$$

That leaves T & $\ddot{x}$.

$$\sum\nolimits_{ext} F_x = M_1\ddot{x} \text{ or } \sum\nolimits_{ext} F_x = M_2\ddot{x} \tag{4.103}$$

After removing *T*, $\ddot{x}$ may be resolved:

$$(M_1 + M_2)\ddot{x} = (M_1 + M_2)g\sin\theta - \propto M_1 g\cos\theta \tag{4.104}$$

Example 4.15

As seen in the following Figure 4.23, two bodies are suspended on a pully being suspended. Determine DOE, free body diagram (FBD), and motion of equation (MOE).

Solution

C is 1:

1. $x_1 = x_2$ is 1.
2. 4 trivial equations
3. 1 notable EOM

$$DOF = 6m + 3n - C = 6 \cdot 0 + 3 \cdot 2 - C = 6 - 1 = 5 \tag{4.105}$$

$x_1 = x_2 = x$ or θ orientation: motion of equation (MOE) is

$$\sum\nolimits_{ext} \tau_A = \frac{dh_{/A}}{dt} + v_{A/O} \times P_{/O} = (M_1 gR - M_2 gR)k \tag{4.106}$$

Because $h_{/A} = (M_1 + M_2) R\dot{x}\hat{k}$, $\frac{dh_{/A}}{dt} = (M_1 + M_2) R\ddot{x}\,\hat{k}$, Equation (4.106) is

$$(M_1 gR - M_2 gR)\hat{k} = (M_1 + M_2) R\ddot{x}\,\hat{k}$$

So acceleration is

$$\ddot{x} = \left(\frac{M_1 - M_2}{M_1 + M_2}\right) g \tag{4.107}$$

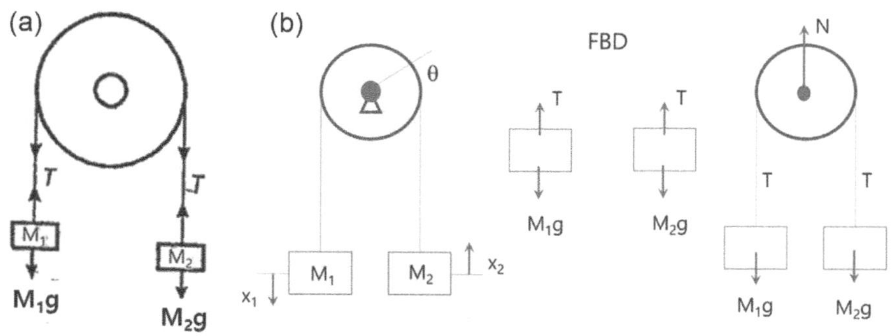

FIGURE 4.23
Two bodies in pully hanging: (a) pulley system; (b) free body diagram.

4.4 Energy Method: D'Alembert's Principle and Lagrangian Equations

4.4.1 D'Alembert's Principle for Mechanical Systems

D'Alembert's principle, also known as the Lagrange–d'Alembert principle, is an understandable statement of the fundamental classical laws of motion [3]. The theorem of virtual work describes that the whole of the relating to an increase of virtual works carried out by all outer forces F_i, acting jointly with virtual displacements δs_i of the point on which the related force is acting, is zero:

$$\delta W = \sum_i F_i \cdot \delta r_i = 0 \tag{4.108}$$

As seen in Figure 4.24, using D'Alembert's principle and free body diagram, the engineer can model a mechanical product with mass, damping, and spring. For instance, if there is an automobile in transport, we may model an uncomplicated dynamic system with a mass that is split from a wall by a dashpot and a spring.

The free body diagram is a graphical method of showing all external forces exerted on an object. Only one position in this system is defined by the variable x that is greater than zero to the right. It is assumed that $x = 0$ when the spring is in its unfastened condition. There are the four forces to develop a model from the free body diagram: 1) an external force (F_e), such as the air resistance force; 2) a spring force, which will be a force from the spring, $k \cdot x$,

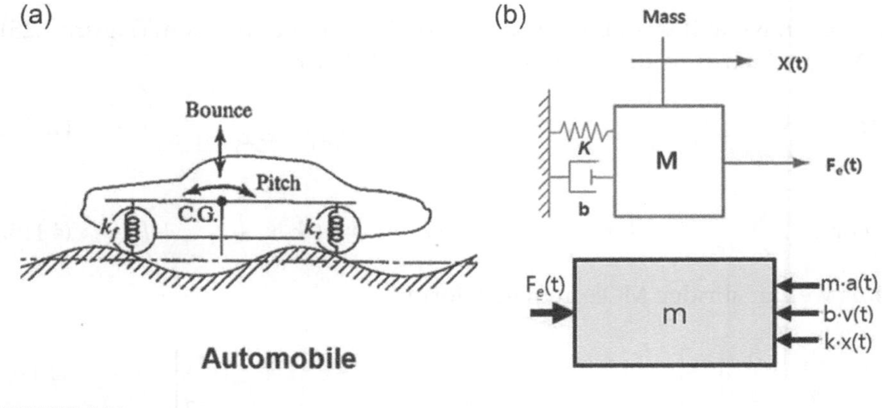

FIGURE 4.24
Free body diagram for automobile modeling: (a) car in transit; (b) modeling with free body diagram.

to the left; 3) a dashpot force, which will be a force from the dashpot, $b \cdot v$, to the left; and finally 4) there is an inertial force, which is defined to be against the stated direction of motion. This is defined by $m \cdot a$ to the left, which is called a fictitious force. It is when one multiplies true acceleration times m and puts it on the $\sum_{\text{all external}} F$ side of the equation.

Newton's second law describes that a body accelerates in the direction of an acted force and that this acceleration is proportional to the force in a way that is the opposite of removing the right-hand side results in D'Alembert's principle:

$$\sum_{\text{all external}} F = m \cdot a \text{ or } \sum_{\text{all external}} F - m \cdot a = 0 \tag{4.109}$$

As we consider the $m \cdot a$ statement to be a (fictitious) inertia force (or D'Alembert's force), D'Alembert's principle will be left:

$$\sum_{\text{all}} F \cdot \delta r = 0 \tag{4.110}$$

To assume this, consider pushing with your hand against a mass in the absence of friction in the positive direction. Your hand experiences a force in the direction opposite to that of the orientation of the force (this is the $-ma$ term). The inertial force is always in the orientation opposite to the stated positive orientation. We sum all of these forces to zero and obtain the movement of equation:

$$F_e(t) - ma(t) - bv(t) - k \cdot x(t) = 0 \text{ or } m\frac{d^2x}{dt^2} + b\frac{dx}{dt} + kx(t) = F_e(t) \tag{4.111}$$

Moreover, we will talk about objects with 2 degrees of freedom (Figure 4.25).

We can determine the motion of equation (MOE):

$$\text{For } M_1, \sum_{ext} F = M_1\ddot{x}_1 = -(k_1 + k_2)x_1 - (c_1 + c_2)\dot{x}_1 + k_2x_2 + c_2\dot{x}_2 + F_1 \tag{4.112}$$

$$\text{For } M_2, \sum_{ext} F = M_2\ddot{x}_2 = -(k_2 + k_3)x_2 - c_2\dot{x}_2 + k_2x_1 + c_2\dot{x}_1 + F_2 \tag{4.113}$$

Thus we can abridge MOE in matrix form:

$$\begin{bmatrix} M_1 & 0 \\ 0 & M_2 \end{bmatrix} \begin{Bmatrix} \ddot{x}_1 \\ \ddot{x}_2 \end{Bmatrix} + \begin{bmatrix} c_1 + c_2 & -c_2 \\ -c_2 & c_2 \end{bmatrix} \begin{Bmatrix} \dot{x}_1 \\ \dot{x}_2 \end{Bmatrix} + \begin{bmatrix} k_1 + k_2 & -k_2 \\ -k_2 & k_2 + k_3 \end{bmatrix} \begin{Bmatrix} x_1 \\ x_2 \end{Bmatrix} \tag{4.114}$$

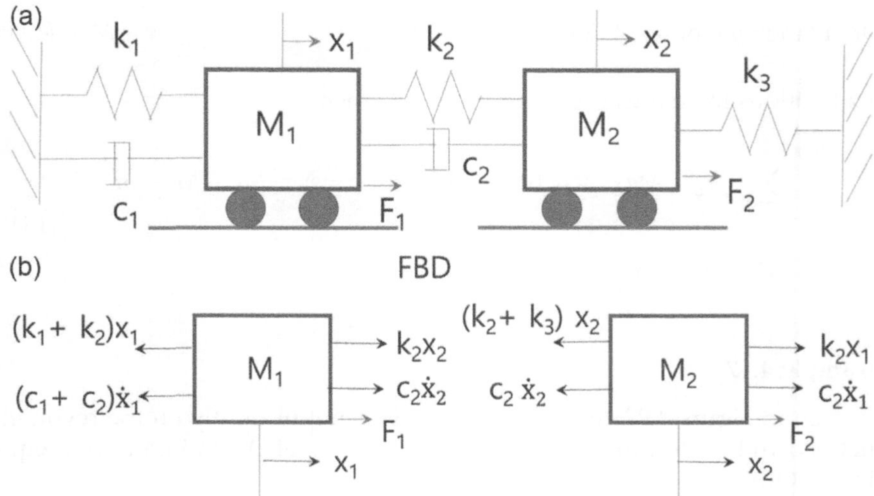

FIGURE 4.25
(a) Objects with 2 degrees of freedom; (b) free body diagram.

Example 4.16

As seen in Figure 4.26, there is a person in the elevator, lifting with some acceleration g/4. Determine the DOE, free body diagram (FBD), and motion of equation (MOE).

Solution

$$DOF = 1 \tag{4.115}$$

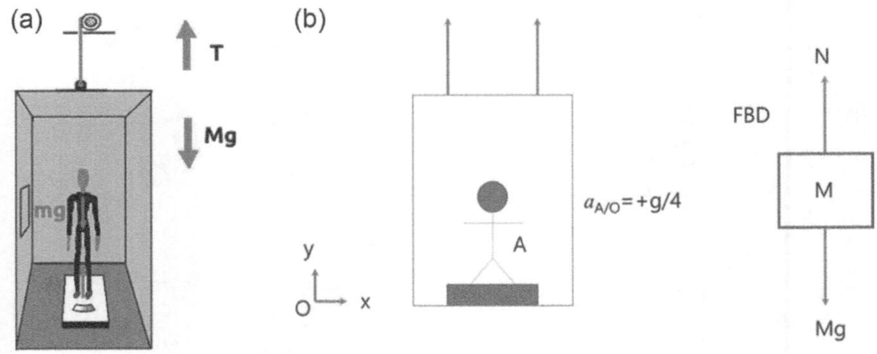

FIGURE 4.26
Lifting movement in the elevator: (a) elevator; (b) free body diagram.

Orientation y motion of Equation (MOE): $\sum_{ext} F_y = Ma_{A/O} = N - Mg$ (4.116)

If a fictitious force is exerted and $a_{A/O} = \frac{g}{4}$, then

$$\sum_{ext} F_y - Ma_{A/O}(\text{fictitious force}) = 0 = N - Mg - M\frac{g}{4} \text{ or}$$
$$N = Mg + M\frac{g}{4} = \frac{5}{4}Mg \tag{4.117}$$

Example 4.17

As seen in Figure 4.27, the m kg spool S does not fit tightly on the revolving inclined rod. Determine the free body diagram (FBD), and motion of equation (MOE).

Solution

$$\sum_{ext} F_y = Ma_{B/O} = M\left(a_{A/O} + \left(\ddot{r} - r\dot{\theta}^2\right)\hat{r} + \ddot{z}\hat{k} + \left(r\ddot{\theta} + 2\dot{r}\dot{\theta}\right)\hat{\theta}\right)$$
$$= M\left(-r\dot{\theta}^2\hat{r} + r\ddot{\theta}\,\hat{\theta}\right) \tag{4.118}$$

r direction: $\sum_{ext} F_{\hat{r}} = Ma_{B/O} = -Mr\dot{\theta}^2 = N_r \Leftrightarrow$

$$\sum_{ext} F_{\hat{r}} = N_r + Mr\dot{\theta}^2(\text{fictitious force}) = 0$$

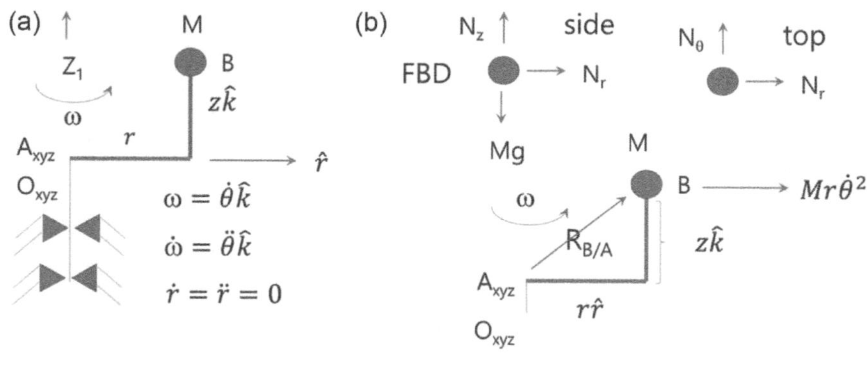

FIGURE 4.27
Rotating spool: (a) revolving spool; (b) free body diagram.

θ direction: $\sum_{ext} F_{\hat{\theta}} = Mr\ddot{\theta}\,\hat{\theta}$

Angular momentum is

$$h_{A/O} = r_{B/A} \times P_{A/O} = \left(r\hat{r} + z\hat{k}\right) \times P_{A/O} = \left(r\hat{r} + z\hat{k}\right) \times Mr\dot{\theta}\hat{\theta} \\ = Mr^2\dot{\theta}\hat{k} - Mrz\dot{\theta}\hat{r} \tag{4.119}$$

If Equation (4.119) is differentiated, then

$$\sum_{ext} \tau_{A/O} = \frac{dh_{A/O}}{dt} = Mr^2\,\ddot{\theta}\hat{k} - Mrz\,\ddot{\theta}\hat{r} - Mrz\dot{\theta}^2\hat{\theta} \tag{4.120}$$

Example 4.18

As seen in Figure 4.28, an object is compelled in a plane, revolving with an imbalanced mass. Determine the motion of equation (MOE) established on the free body diagram (FBD).

Motion of equation is

$$\sum_{ext} F_{M_b} = M_b\,\ddot{x} \tag{4.121}$$

$$\sum_{ext} F_m = ma_{B/O} = ma_{A/O} + ma_{B/A} = m\ddot{x} + m\left[\left(\ddot{r} - r\dot{\theta}^2\right)\hat{r} + \left(r\ddot{\theta} + 2\dot{r}\dot{\theta}\right)\hat{\theta}\right] \tag{4.122}$$

$$\sum_{ext} F_m = m\ddot{x}\hat{i} - me\omega^2\left(\cos\omega t\hat{i} + \sin\omega t\hat{j}\right) = F_{m,x}\hat{i} + F_{m,y}\hat{j} - mg\hat{j} \tag{4.123}$$

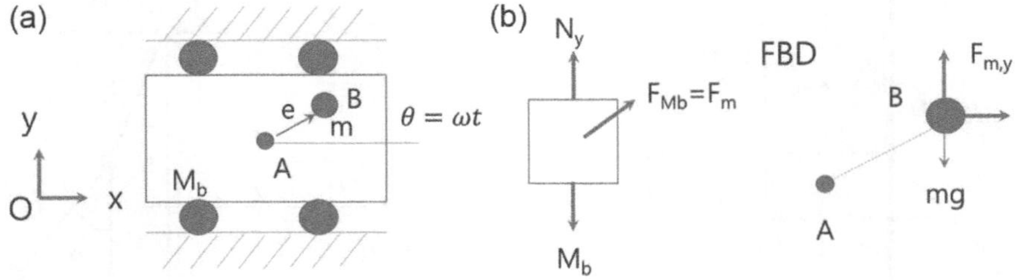

FIGURE 4.28
Object compelled in a plane, revolving with an imbalanced mass: (a) object constrained; (b) free body diagram.

$$x \text{ direction: } \sum\nolimits_{ext} F_{m,x} = m\ddot{x} - me\omega^2 \cos\omega t = F_{m,x} = - F_{M_b}, x \qquad (4.124)$$

$$y \text{ direction: } F_{m,y} = mg - me\omega^2 \sin\omega t \qquad (4.125)$$

4.4.2 Derivation of Lagrange's Equations from D'Alembert's Principle

The motion of particles in space is frequently enclosed by some constraints. Engineers must function only with independent DOFs (coordinates) with no constraints (Figure 4.29) [4].

The motions of N particles, $n = 3N$ DOFs, are exposed to k equations connecting coordinates. In classical mechanics, as all constraints of the system are holonomic, a system can be named as holonomic. A holonomic constraint should be expressed as follows:

$$f_j(x_1, \ldots, x_n, t) = c_j \quad j = 1, 2, \ldots, k \qquad (4.126)$$

Namely, a holonomic constraint depends only on the coordinates x_j and period t. The generalized coordinates explain the movement of the system relative to Cartesian coordinates exposed to k constraints:

$$x_1 = x_1(q_1, q_2, \ldots, q_{n-k}, t)$$

$$x_n = x_n(q_1, q_2, \ldots, q_{n-k}, t) \qquad (4.127)$$

where any set of $n - k = 3N - k$ independent coordinates, which completely express the system.

The change of a Cartesian coordinate may be concluded from changes in generalized coordinates:

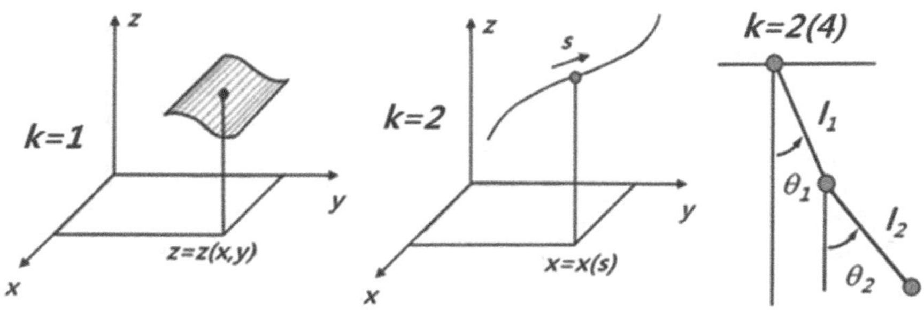

FIGURE 4.29
Movement of particles confined by constraints.

$$dx_1 = \frac{\partial x_1}{\partial q_1} dq_1 + \ldots + \frac{\partial x_1}{\partial q_{n-k}} dq_{n-k} + \frac{\partial x_1}{\partial t} dt \quad (4.128)$$

In a compact manner, we can explain:

$$dx_i = \sum_{\sigma=1}^{n-k} \frac{\partial x_i}{\partial q_\sigma} dq_\sigma + \frac{\partial x_1}{\partial t} dt \quad i = 1,\ldots,n \quad (4.129)$$

Thus the virtual displacements are named as infinitesimal, instantaneous displacements of the coordinates consistent with the constraints:

$$\delta x_i = \sum_{j=1}^{n-k} \frac{\partial x_i}{\partial q_j} \delta q_\sigma \quad i = 1,\ldots,n \quad (4.130)$$

Because the forces of the constraint are at right angles to the orientation of movement, the forces of the constraint do not function under a virtual displacement (Figure 4.30).

$$\delta W = \tau \cdot \delta s_1 - \tau \cdot \delta s_2 = -\tau \cdot \delta s_r = 0 \quad (4.131)$$

We can write again Newton's second law:

$$\dot{p}_i = F_i^{(a)} + R_i \quad i = 1,\ldots,n \text{ or } \sum_i \left(F_i^{(a)} + R_i - \dot{p}_i\right) \delta x_i = 0 \quad (4.132)$$

Because $\sum_i R_i \delta x_i = 0$, the virtual work by D'Alembert's law is restated:

$$\sum_i \left(F_i^{(a)} - \dot{p}_i\right) \delta x_i = 0 \quad (4.133)$$

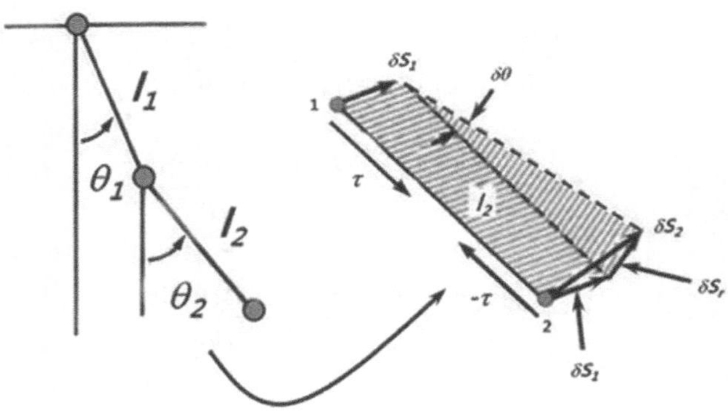

FIGURE 4.30
Forces of constraint perpendicular to the direction of motion.

We can also redefine D'Alembert's principle with respect to the generalized coordinates. The virtual work done by exerted forces under virtual displacement is expressed:

$$\delta W\left(=\sum_i\left(F_i^{(a)}-\dot{p}_i\right)\delta x_i\right)=\sum_i F_i\delta x_i=\sum_{\sigma=1}^{n-k}\left(\sum_{i=1}^{n}F_i\frac{\partial x_i}{\partial q_j}\right)dq_j$$
$$=\sum_{j=1}^{n-k}Q_j\delta q_j \tag{4.134}$$

where $\delta x_i=\sum_{j=1}^{n-k}\left(\frac{\partial x_i}{\partial q_j}\right)dq_j$ and generalized forces $Q_j=\sum_{i=1}^{n}F_i\frac{\partial x_i}{\partial q_j}$.

Let Cartesian coordinates be subjected to k constraints $x_i=x_i\left(q_1,\ldots,q_{n-k},t\right)$. The change of a Cartesian coordinate induced from changes in generalized coordinates in dt is defined:

$$dx_i=\sum_{j=1}^{n-k}\left(\frac{\partial x_i}{\partial q_j}\right)dq_j+\frac{\partial x_i}{\partial t}dt \quad i=1,\ldots,n \tag{4.135}$$

Taking the time derivatives from Equation (4.135):

$$\frac{dx_i}{dt}\equiv\dot{x}_i=\sum_{j=1}^{n-k}\left(\frac{\partial x_i}{\partial q_j}\right)\dot{q}_j+\left(\frac{\partial x_i}{\partial t}\right) \quad i=1,\ldots,n \tag{4.136}$$

where $\dot{x}_i=x_i\left(q_1,\ldots,q_{n-k};\dot{q}_1,\ldots,\dot{q}_{n-k};t\right)$.

If other variables are kept constant in taking partial derivatives,

$$\frac{\partial \dot{x}_i}{\partial \dot{q}_j}=\frac{\partial x_i}{\partial q_j}\frac{d}{dt}\left(\frac{\partial x_i}{\partial q_j}\right)=\frac{\partial}{\partial q_j}\left(\frac{dx_i}{dt}\right) \tag{4.137(a)}$$

$$\text{LHS}=\sum_\lambda\left(\frac{\partial}{\partial q_\lambda}\frac{\partial x_i}{\partial q_j}\right)\dot{q}_\lambda+\frac{\partial}{\partial t}\frac{\partial x_i}{\partial q_j} \tag{4.137(b)}$$

$$\text{RHS}=\sum_\lambda\left(\frac{\partial}{\partial q_j}\frac{\partial x_i}{\partial q_\lambda}\right)\dot{q}_\lambda+\frac{\partial}{\partial q_j}\frac{\partial x_i}{\partial t} \tag{4.137(c)}$$

Here, the order of partial derivatives may be exchanged. The momentum equations can be expressed:

$$\sum_i \dot{p}_i \delta x_i = \sum_i m_i \ddot{x}_i \delta x_i = \sum_\sigma \left(\sum_i m_i \ddot{x}_i \frac{\partial x_i}{\partial q_j} \right) \delta q_j \tag{4.138}$$

Because $\dot{p}_i = m_i \ddot{x}_i$, and $\delta x_i = \sum_{\sigma=1}^{n-k} \frac{\partial x_i}{\partial q_j} \delta q_j$, Equation (4.138) might be explained:

$$\sum_i m_i \ddot{x}_i \frac{\partial x_i}{\partial q_j} \cong \sum_i m_i \frac{d\dot{x}_i}{dt} \frac{\partial x_i}{\partial q_j} = \sum_i m_i \left[\frac{d}{dt} \left(\dot{x}_i \frac{\partial x_i}{\partial q_j} \right) - \dot{x}_i \frac{d}{dt} \frac{\partial x_i}{\partial q_j} \right] \tag{4.139}$$

Because $\frac{\partial \dot{x}_i}{\partial \dot{q}_j} = \frac{\partial x_i}{\partial q_j}$ and $\frac{d}{dt}\left(\frac{\partial x_i}{\partial q_j} \right) = \frac{\partial}{\partial q_j}\left(\frac{dx_i}{dt} \right)$, Equation (4.139) can be expressed:

$$\sum_i m_i \frac{d}{dt} \left(\dot{x}_i \frac{\partial \dot{x}_i}{\partial \dot{q}_j} \right) = \frac{d}{dt} \left[\frac{\partial}{\partial \dot{q}_j} \left(\frac{1}{2} \sum_i m_i \dot{x}_i^2 \right) \right] \tag{4.140(a)}$$

$$\sum_i m_i \dot{x}_i \frac{\partial}{\partial q_j} \frac{dx_i}{dt} = \frac{\partial}{\partial q_j} \left(\frac{1}{2} \sum_i m_i \dot{x}_i^2 \right) \tag{4.140(b)}$$

From Equation (4.138), the momentum equations may be defined:

$$\sum_i \dot{p}_i \delta x_i = \sum_j \left(\frac{d}{dt} \frac{\partial T}{\partial \dot{q}_j} - \frac{\partial T}{\partial q_j} \right) \delta q_j \tag{4.141}$$

where kinetic energy $T = \frac{1}{2} \sum_i m_i \dot{x}_i^2 = T(q_1, \ldots, q_{n-k}; \dot{q}_1, \ldots, \dot{q}_{n-k}; t)$.

Thus we can write again D'Alembert's principle with regard to the generalized coordinates:

$$\sum_j \left(\frac{d}{dt} \frac{\partial T}{\partial \dot{q}_j} - \frac{\partial T}{\partial q_j} - Q_j \right) \delta q_j = 0 \tag{4.142}$$

As all q are independent and arbitrary, Lagrange's equations are defined:

$$\frac{d}{dt} \frac{\partial T}{\partial \dot{q}_j} - \frac{\partial T}{\partial q_j} = Q_\sigma \quad \sigma = 1, \ldots, n-k \tag{4.143}$$

The $n - k$ equations for $n - k$ independent generalized coordinates are identical to Newton's second law. For conservative forces, the potential energy depends only on the location:

$$V(x_1,\ldots,x_n) = V(q_1,\ldots,q_{n-k},t) \qquad \frac{\partial V}{\partial \dot{q}_\alpha} = 0 \tag{4.144}$$

Generalized forces, Q_j, are given by negative gradients with regard to the equivalent generalized coordinate:

$$Q_j \equiv \sum_i F_i \frac{\partial x_i}{\partial q_j} = -\sum_i \left[\frac{\partial}{\partial x_i} V(x_i,\ldots,x_n)\right]\frac{\partial x_i}{\partial q_j} = -\frac{\partial}{\partial q_j} V(q_1,\ldots,q_{n-k},t) \tag{4.145}$$

Lagrange's equations may be explained:

$$\frac{d}{dt}\frac{\partial L}{\partial \dot{q}_j} - \frac{\partial L}{\partial q_j} = Q_j \quad j = 1,\ldots,n-k \tag{4.146}$$

If $L = T - V$ is plugged into Equation (4.146), then

$$\underbrace{\frac{d}{dt}\frac{\partial T}{\partial \dot{q}_j}}_{①} - \underbrace{\frac{d}{dt}\frac{\partial V}{\partial \dot{q}_j}}_{0} - \underbrace{\frac{\partial T}{\partial q_j}}_{②} + \underbrace{\frac{\partial V}{\partial q_j}}_{③} = \underbrace{Q_j}_{④} \quad j = 1,\ldots,n-k \tag{4.147}$$

Namely, it is known that ①+②+③=④.

The generalized coordinates q_σ, have certain attributes: 1) they don't have to be Cartesian, 2) not even inertial, 3) but they should be independent and absolute, 4) and the system should be holonomic. At this point, independence means that when you fix all but one coordinate, it still has a continuous range of motion in the free coordinate. All the components can be located at all times.

To explain the movement by using Lagrange's Equations, think about an object with damping and spring or with a double pendulum, as in Figure 4.31.

An equation structured as a Lagrangian equation may be abridged from Equation (4.147):

- *Left-hand side (LHS)*: 1) Decide the numeric of DOF, and select generalized coordinates, q_σ; 2) justify the absolute, independent, and holonomic; 3) calculate the Lagrangian $L = T + V$; and 4) calculate ①, ②, ③ for each generalized coordinate, q_σ.

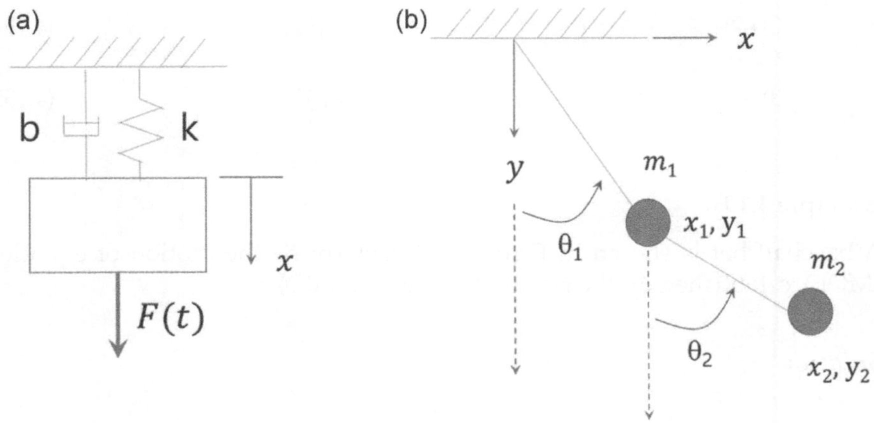

FIGURE 4.31
Object (a) with damping and spring or (b) with double pendulum.

- *Right-hand side (RHS)*: 1) for generalized coordinates, q_σ, discover the generalized forces Q_σ that go with it; 2) calculate the virtual work δW_j, linked with the virtual displacement δq_σ:

$$\delta W_j = Q_j \delta q_j$$

So we may discover the motion of equation (MOE) from Equation (4.146):

Kinetic energy: $$T = \frac{1}{2} M\dot{X}^2 \tag{4.148}$$

Potential energy: $$V = \frac{1}{2} kX^2 - MgX \tag{4.149}$$

Generalized coordinate: $$q_j = q_1 = X \tag{4.150}$$

$$① \ \frac{d}{dt}\frac{\partial T}{\partial \dot{q}_\sigma} = \frac{d}{dt}\frac{\partial}{\partial \dot{X}}\left(\frac{1}{2}M\dot{X}^2\right) = \frac{d}{dt}\left(M\dot{X}\right) = M\ddot{X} \tag{4.151}$$

$$② \ -\frac{\partial T}{\partial q_\sigma} = -\frac{\partial T}{\partial X} = 0 \tag{4.152}$$

$$③ \ \frac{\partial V}{\partial q_\sigma} = \frac{\partial V}{\partial X} = kX - Mg \tag{4.153}$$

$$①+②+③= M\ddot{X}+kX-Mg=④=Q_i=F(t)-b\dot{X} \tag{4.154}$$

$$④\ \delta W_j = Q_j\delta q_j = \left(F(t)\hat{i}-b\dot{X}\hat{i}\right)\cdot\delta X = Q_j\delta X \tag{4.155}$$

Example 4.19

A baseball bat is shown in Figure 4.32. Determine the motion of equation (MOE), established on the free body diagram (FBD).

Solution

$$\begin{aligned} a_{G/O} &= \underbrace{a_{A/O}}_{0} + \underbrace{\left(a_{G/A}\right)\Big|_{\omega=0}}_{0} + \underbrace{\dot{\omega}_{/o}\times r_{G/A}}_{\ddot{\theta}\hat{k}\times r_{G/A}\hat{i}} + \underbrace{\omega_{/o}\times\left(\omega_{/o}\times r_{G/A}\right)}_{-\omega_o^2 r_{G/A}\hat{i}} \\ &+ \underbrace{2\omega_{/o}\times\left(v_{G/A}\right)\Big|_{\omega=0}}_{0} \end{aligned} \tag{4.156}$$

$$①\sum\nolimits_{ext} F_y = R_y\hat{j} - f\hat{j} = Ma_{G/O\,y}\hat{j} = Mr_{G/A}\ddot{\theta}\hat{j} \tag{4.157}$$

If $R_y = 0$, then $f = R_y - Mr_{G/A}\ddot{\theta} = -Mr_{G/A}\ddot{\theta}$ into ③

$$②\sum\nolimits_{ext} F_x = R_x\hat{i} = Ma_{G/O\,x}\hat{i} \tag{4.158}$$

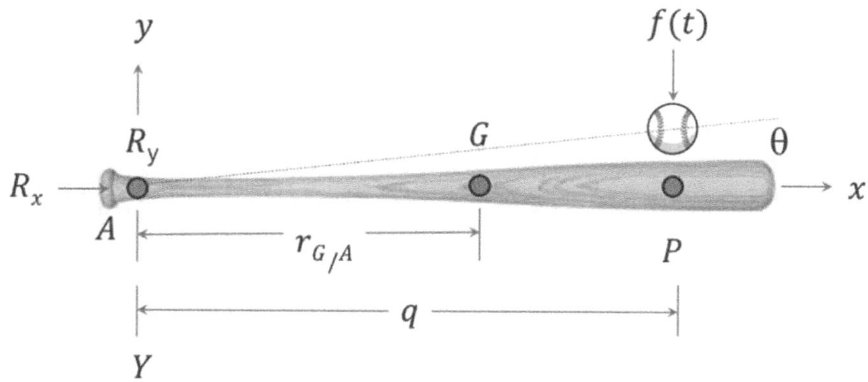

FIGURE 4.32
Baseball bat.

$$③\sum_{ext}\tau_{/A} = \frac{dH_{/A}}{dt} = I_{zz/A}\ddot{\theta}\hat{k} = -fq\hat{k} = Mr_{G/A}\ddot{\theta}q \qquad 4.159(a))$$

$$\ddot{\theta}MK_A^2 = Mr_{G/A}\ddot{\theta}q \text{ or } q = {K_A^2}/{r_{G/A}} \quad \text{where } I_{zz/A} = MK_A^2 \qquad (4.159(b))$$

Example 4.20

A cart and a pendulum are seen in Figure 4.33. Determine the motion of equation (MOE) by using the Newton or Lagrange method, established on the Free Body Diagram (FBD).

Solution

Newtonian approach:
C is 10:

1. No y, z translation on M_1 is 2.
2. No x, y, z rotation on M_1 is 3.
3. No x, y rotation on M_2 is 2.
4. No x, y, z translation on M_2 is 3.

$$DOF = 6m + 3n - C = 6\cdot 2 - 10 = 2 \qquad (4.160)$$

$$①\sum_{ext}F_{X,M_1} = M_1\ddot{X}\hat{I} = \left(-kX - b\dot{X} + F_2\cos\theta + F_1\sin\theta\right)\hat{I} \Leftarrow F_1 + F_2 \qquad (4.161)$$

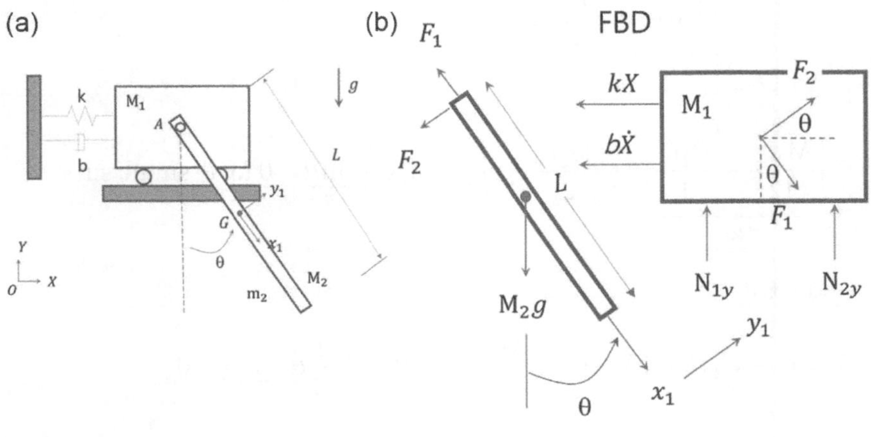

FIGURE 4.33
(a) Cart and pendulum; (b) free body diagram.

$$② \sum_{ext} F_{x_1, M_2} = M_2 \underbrace{a_{G/O}}_{A} \hat{i} = -F_1 \hat{i} + M_2 g \cos\theta \hat{i} \Rightarrow F_1 = \tag{4.162}$$

$$③ \sum_{ext} F_{y_1, M_2} = M_2 \underbrace{a_{G/O}}_{B} \hat{j} = -F_2 \hat{j} - M_2 g \sin\theta \hat{j} \Rightarrow F_2 = \tag{4.163}$$

$$④ \sum_{ext} \tau_{/G} = \frac{dH_{/G}}{dt} = F_2 \frac{L}{2} \hat{k} = \frac{d}{dt}\left(\left[I_{/G} \right] \begin{Bmatrix} 0 \\ 0 \\ \omega_z \end{Bmatrix} \right) = I_{zz/G} \ddot{\theta} \hat{k} = \frac{M_2 L^2}{12} \ddot{\theta} \hat{k}$$

$$\Leftarrow F_2 \tag{4.164}$$

We need to calculate velocity $v_{G/O}$ and acceleration $a_{G/O}$:

$$\begin{aligned} v_{G/O} &= v_{A/O} + v_{G/A} = \dot{X}\hat{I} + \left(v_{G/A}\right)\Big|_{\omega_{/o}=0} + \omega_{/o} \times r_{G/A} = \dot{X}\hat{I} + \omega_z \hat{k} \times \frac{L}{2} \\ &= \dot{X}\hat{I} + \dot{\theta}\frac{L}{2}\hat{j} \end{aligned} \tag{4.165}$$

$$\begin{aligned} a_{G/O} &= a_{A/O} + \left(a_{G/A}\right)\Big|_{\omega_{/o}=0} + \dot{\omega}_{/o} \times r_{G/A} + \omega_{/o} \times \left(\omega_{/o} \times r_{G/A}\right) \\ &\quad + 2\omega_{/o} \times \left(v_{G/A}\right)\Big|_{\omega_{/o}=0} \end{aligned}$$

$$\begin{aligned} a_{G/O} &= \frac{d}{dt}\left(v_{G/O}\right) = \ddot{X}\hat{I} + 0 + \ddot{\theta}\frac{L}{2}\hat{j} - \dot{\theta}^2 \frac{L}{2}\hat{i} + 0 = \underbrace{\left(\ddot{X}\sin\theta - \frac{L}{2}\dot{\theta}^2 \right)}_{A} \hat{i} \\ &\quad + \underbrace{\left(\ddot{X}\cos\theta + \frac{L}{2}\ddot{\theta} \right)}_{B} \hat{j} \end{aligned} \tag{4.166}$$

$$④ \underbrace{\left(\frac{M_2 L^2}{12} + \frac{M_2 L^2}{4} \right)}_{\frac{M_2 L^2}{3}} \ddot{\theta} + \frac{M_2 L}{2} \ddot{X} \underbrace{\cos\theta}_{1} + M_2 g \frac{L}{2} \underbrace{\sin\theta}_{\theta} = 0 \text{ for } \theta \text{ small, } \sin\theta \approx \theta$$

and $\cos\theta \approx 1$

$$① \left(M_1 + M_2\right)\ddot{X} + b\dot{X} + kX + \frac{M_2 L}{2} \ddot{\theta} \underbrace{\cos\theta}_{1} + M_2 \frac{L}{2} \dot{\theta}^2 \underbrace{\sin\theta}_{\theta} = 0$$

Lagrange approach: From Equation (4.146),

$$\frac{d}{dt}\frac{\partial L}{\partial \dot{q}_\sigma} - \frac{\partial L}{\partial q_\sigma} = Q_\sigma \quad \sigma = 1,\ldots,n-k$$

$$\underbrace{\frac{d}{dt}\frac{\partial T}{\partial \dot{q}_\sigma}}_{①} - \underbrace{\frac{d}{dt}\frac{\partial V}{\partial \dot{q}_\sigma}}_{0} - \underbrace{\frac{\partial T}{\partial q_\sigma}}_{②} + \underbrace{\frac{\partial V}{\partial q_\sigma}}_{③} = \underbrace{Q_\sigma}_{④} \quad \sigma = 1,\ldots,n-k$$

$$T = \frac{1}{2}M_1\dot{x}^2 + \frac{1}{2}M_2 v_{G/O}\cdot v_{G/O} + \frac{1}{2}\omega^T H_G \left(= \frac{1}{2}I_{ZZ/G_{M_2}}\dot{\theta}^2 = \frac{1}{2}\frac{M_2L^2}{12}\dot{\theta}^2 \right) \tag{4.167}$$

where

$$v_{G/O} = \left(\dot{x} + \frac{L}{2}\dot{\theta}\cos\theta\right)\hat{i} + \frac{L}{2}\dot{\theta}\sin\theta\hat{j}, v_{G/O}\cdot v_{G/O} = \left(\dot{x} + \frac{L}{2}\dot{\theta}\cos\theta\right)^2 + \left(\frac{L}{2}\dot{\theta}\sin\theta\right)^2,$$

$$\frac{\partial v_{G/O}\cdot v_{G/O}}{\partial \dot{\theta}} = 2\left(\dot{x} + \frac{L}{2}\dot{\theta}\cos\theta\right)\left(\frac{L}{2}\cos\theta\right) + 2\left(\frac{L}{2}\dot{\theta}\sin\theta\right)\left(\frac{L}{2}\sin\theta\right)\left(= \frac{L^2}{2}\dot{\theta}\sin^2\theta\right) \tag{4.168}$$

$$V = \frac{1}{2}kx^2 + M_2 g\frac{1}{2}l(1-\cos\theta) \tag{4.169}$$

θ direction of M_2:

$$\frac{\partial T}{\partial \dot{\theta}} = \frac{1}{2}I_{ZZ/G_{M_2}}2\dot{\theta} + \frac{1}{2}M_2\left[2\left(\dot{x} + \frac{L}{2}\dot{\theta}\cos\theta\right)\left(\frac{L}{2}\cos\theta\right) + \frac{L^2}{2}\dot{\theta}\sin^2\theta\right] \tag{4.170}$$

$$\frac{d}{dt}\frac{\partial T}{\partial \dot{\theta}} = I_{ZZ/G_{M_2}}\ddot{\theta} + \frac{M_2L^2}{12}\ddot{\theta} + M_2\frac{L}{2}\cos\theta\,\ddot{x} \tag{4.171}$$

$$\frac{\partial V}{\partial \theta} = \frac{M_2 g}{2}L\sin\theta \tag{4.172}$$

$$\begin{aligned} ① + ② + ③ &= I_{ZZ/G_{M_2}}\ddot{\theta} + \frac{M_2L^2}{12}\ddot{\theta} + M_2\frac{L}{2}cos\theta\,\ddot{x} + \frac{M_2 g}{2}Lsin\theta \\ &= ④ = Q_\theta \end{aligned} \tag{4.173}$$

x direction of M_1:

$$\frac{d}{dt}\frac{\partial T}{\partial \dot{x}} = M_1\ddot{x} + M_2\left(\ddot{x} + \frac{L}{2}\ddot{\theta}\cos\theta - \frac{L}{2}\dot{\theta}^2\sin\theta\right) \tag{4.174}$$

$$\frac{\partial V}{\partial x} = kx \tag{4.175}$$

$$①+②+③ = M_1\ddot{x} + M_2\left(\ddot{x} + \frac{L}{2}\ddot{\theta}\cos\theta - \frac{L}{2}\dot{\theta}^2\sin\theta\right) + kx = ④ = Q_x$$

$$\delta W = Q_x\delta x + Q\ \delta\theta = \underbrace{\left(F_1(t) - b\dot{x}\right)}_{Q_x}\delta x + \underbrace{F_2 L\cos\theta}_{Q_\theta}\delta\theta \tag{4.176}$$

Example 4.21

Figure 4.34 shows a sleeve sliding with a rod. Determine the motion of equation (MOE) by using the Newton or Lagrange method, based on the free body diagram (FBD).

Solution

Rod: $M_1, {}^{A}I_{zz_1}, L_1, G_1$

Sleeve: $M_2, {}^{G}I_{zz_2}, L_2, G_2$

Where L_o = unstretched spring length

If statically checking,

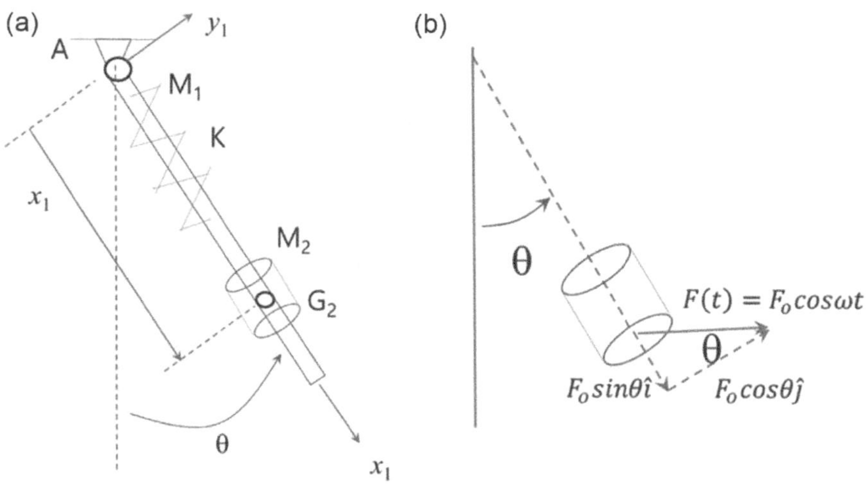

FIGURE 4.34.
(a) Sleeve sliding with rod; (b) free body diagram.

$$K\underbrace{\left(x_1 - L_o - \frac{L_2}{2}\right)}_{x_{static}} = M_2 g \tag{4.177}$$

$$T = \left(\frac{1}{2}\,{}^{A}I_{zz_1} + \frac{1}{2}\,{}^{G}I_{zz_1}\right)\dot{\theta}^2 + \underbrace{\frac{1}{2}M_2\,{}^{G}v_{G_2}\cdot{}^{G}v_{G_2}}_{\frac{1}{2}M_2\left(\dot{x}_1^2 + x_1^2\dot{\theta}^2\right)} \tag{4.178}$$

$$V = \frac{1}{2}K\left(x_1 - L_o - \frac{L_2}{2}\right)^2 + M_1 g\underbrace{\frac{L_1}{2}(1-\cos\theta)}_{\Delta h} + M_2 g\left(L_o + \frac{L_2}{2}\right) - M_2 g x_1 \cos\theta \tag{4.179}$$

If Lagrange is applied from Equation (4.178), it is in the x_1 direction:

$$① \ \frac{d}{dt}\frac{\partial T}{\partial \dot{x}_1} = \frac{d}{dt}(M_2\dot{x}_1) = M_2\,\ddot{x}_1 \tag{4.180}$$

$$② \ -\frac{\partial T}{\partial x_1} = -M_2 x_1 \dot{\theta}^2 \tag{4.181}$$

$$③ \ \frac{\partial V}{\partial x_1} = K\left(x_1 - L_o - \frac{L_2}{2}\right) - M_2 g\cos\theta \tag{4.182}$$

$$① + ② + ③ = M_2\ddot{x}_1 - M_2 x_1\dot{\theta}^2 + K\left(x_1 - L_o - \frac{L_2}{2}\right) - M_2 g\cos\theta = ④ = Q_{x_1} \tag{4.183}$$

$$\delta W_{x_1} = Q_x \cdot \delta x_1 = F(t)\cdot dr = F_o\cos\omega t\left[\sin\theta\hat{i} + \cos\theta\hat{j}\right]\cdot dx\hat{i} = \underbrace{F_o\cos\omega t\sin\theta}_{Q_{x_1}}\,dx \tag{4.184}$$

θ direction:

$$① \ \frac{d}{dt}\frac{\partial T}{\partial \dot{\theta}} = \left({}^{A}I_{zz_1} + {}^{G}I_{zz_1}\right)\ddot{\theta} + M_2 x_1^2\ddot{\theta} \tag{4.185}$$

$$② \ -\frac{\partial T}{\partial \theta} = 0 \tag{4.186}$$

$$③ \ \frac{\partial V}{\partial \theta} = M_2 g x_1 \sin\theta + M_1 g\frac{L_1}{2}\sin\theta \tag{4.187}$$

$$\textcircled{1}+\textcircled{2}+\textcircled{3}=\left({}^{A}I_{zz_1}+{}^{G}I_{zz_1}\right)\ddot{\theta}+M_2x_1^2\ddot{\theta}+M_2gx_1\sin\theta \\ +M_1g\frac{L_1}{2}\sin\theta=\textcircled{4}=Q_\theta \tag{4.188}$$

$$\delta W=F(t)\cdot dr=F_o\cos\omega t\left[\sin\theta\hat{i}+\cos\theta\hat{j}\right]\underbrace{\left(x_1+\frac{L_2}{2}\right)\delta\theta\,\hat{j}}_{dr}=\underbrace{F_o\cos\omega t\cos\theta\left(x_1+\frac{L_2}{2}\right)}_{Q_\theta}\delta\theta$$

4.5 Bond Graph Modeling

4.5.1 Introduction

A mechanical product such as an airplane or automobile produces generated power from an engine and moves it to the transmission via a drive system through the mechanism. As dissimilar engineering fields—magnetic, hydraulic, mechanical, electric, etc.—are included in a mechanical product, one of the important tools is bond graphs, which may model these physical systems. It is a straight graphical implement for modeling dynamic systems including some dissimilar engineering fields. That is, designing a new dynamic system is a potential manner to utilize in the formation of a graphical expression for sharing information of other engineers on its dynamic modeling.

In 1959, Professor Henry Payner and his previous students at MIT proposed the bond graph method, which gave the absolute notions for representing multiport systems with regard to power bonds, fastening the parts of an engineering system to the assumed junction structures that were demonstrated to be the constraints. This power exchange representation of a system is expressed as a bond graph. It is worthwhile in modeling a mechanical system because a multiport system is devised, and the power created from the engine is transferred to other systems such as the transmission, drive system, etc.

In 1961, Paynter issued a book entitled *Analysis and Simulation of Simulation of Multiport Systems*. In 2006, he published the fourth edition, *System Dynamics—Modeling and Simulation of Mechatronic Systems*. Now, the method has gone global, and many disciplines have adopted bond graphs as their modeling method [5].

4.5.2 Basic Elements, Energy Relations, and Causality of a Bond Graph

A bond graph is a graphical expression of a multiport system. Behavior with respect to energy is field independent, irrespective of the mechanical or electric product. The application of bond graphs is the same in each of the physical fields, whether dealing with the RLC circuit for an electrical system or the mass/spring/damp in a mechanical product. Established on the mutual energy exchange between dissimilar systems, the symbols in a bond graph designate a bidirectional interchange of physical energy. Thus bond graphs can be employed in multiple energy areas.

The analysis of dynamic systems could be relatively uncomplicated when it has few DOFs or a steady-state way of behaving. On the other hand, obtaining the governing equation of a model is compounded in the case of a multiport physical system. In nearly all of the outlines, the main issue for engineers is to identify a mathematical model that defines the dynamic behavior of the system and how the dissimilar parameters influence the system characteristics. That is because the dynamic equations are usually partial differential equations, whose solutions require large-scale mathematical comprehension.

As crucial as it is to understand bond graph theory, the energy flow in a system is a basic component. It is produced by one or more sources, is temporarily retained in system components or is partially consumed in resistances such as heat, and extends to "loads," where it produces some of the desired effects. Power is thus the rate of energy flow without direction.

The bond graph signifies the power flow between two systems. For instance, consider an electric-hydraulic system. The created electric power operates a motor that works the pump. In this case, the energy flow is expressed by an arrow (bond), as in Figure 4.35. Each bond directly represents the energy flow or power. The flow in each bond is nominated by two variables named called power-conjugated variables, whose product is the on-the-spot power of the bond. Because (produced) power in a mechanical system is hard to immediately calculate, engineers may utilize the two transient variables—flow and effort. All possible system domains have these two variables. For instance, in a mechanical product, flow means "velocity," and effort is the "force." In an electrical system, flow means "current," and the effort is "voltage." The multiplication of both transient variables—power—is defined (Table 4.1):

$$P = e(t) \cdot f(t) \tag{4.189}$$

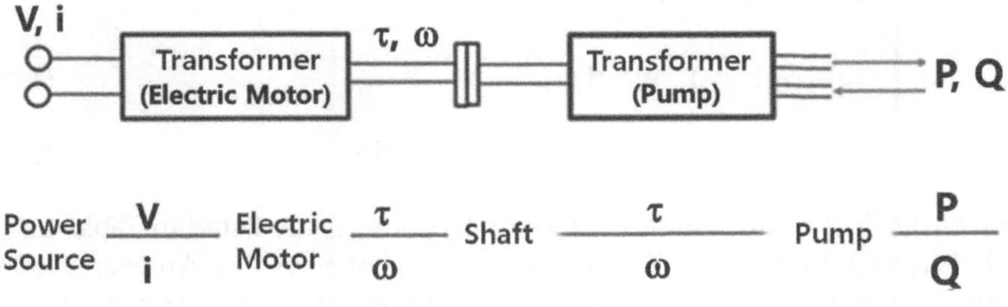

FIGURE 4.35
Power flow in the bond graph for the electric-hydraulic system.

TABLE 4.1

Energy Flow in the Multiport Engineering System

Modules	Effort, $e(t)$	Flow, $f(t)$
Mechanical translation	Force, $F(t)$	Velocity, $V(t)$
Mechanical rotation	Torque, $\tau(t)$	Angular velocity, $\omega(t)$
Compressor, pump	Pressure difference, $\otimes P(t)$	Volume flow rate, $Q(t)$
Electric	Voltage, $V(t)$	Current, $i(t)$

The procedure thus can produce a computer model of the multiple physical fields, such as thermal, mechanical, electrical, and hydraulic. Flows and efforts can be identified with a particular variable for each specific functioning physical field. The bond graph is created on the "bonds," which connect jointly with the "1-port," "2-port," and "3-port" elements. Whether the power in the bond graph is continual, every element is described in a multiport way. Ports are fastened by bonds.

For 1-ports, there are flow sources, effort sources, *I*-type elements, *C*-type elements, and *R*-type elements that will attach power in a way that has interruptions or breaks. For 2-ports, there are transformers and gyrators that may join power without interruption. For 3-ports, there are 0-junction and 1-junction that will comprise the network. Power bonds can be attached at one of the two classes of junctions: 0 and 1 junctions. In these joints, no energy may be built or dissipated. This is expressed as power continuity. In a 0 junction, Equations (4.190) and (4.191) show the flows and efforts:

$$\sum \text{flow}_{\text{input}} = \sum \text{flow}_{\text{output}} \tag{4.190}$$

$$\text{effort}_1 = \text{effort}_2 = \ldots = \text{effort}_n \tag{4.191}$$

This goes with a node in an electrical circuit where Kirchhoff's current rule applies. However, in a 1 junction, the flow and the efforts are obtained with Equations (4.192) and (4.193):

$$\sum \text{effort}_{\text{input}} = \sum \text{effort}_{\text{output}} \tag{4.192}$$

$$\text{flow}_1 = \text{flow}_2 = \ldots = \text{flow}_n \tag{4.193}$$

The basic blocs of the usual bond graph hypothesis are listed in Table 4.2.

This goes with the force balance at the mass of a system. An example of a 1-junction is a resistor in series. In the junction, the energy conservation is assumed, but no loss is allowed. There are two additional variables that are notable in the description of dynamic systems.

TABLE 4.2
Basic Components of the Bond Graph

Elements		Symbol	Relation Equations
1-port elements (sources)	Effort	S_e______	$S_e = e(t)$
	Flow	S_f______	$S_f = f(t)$
1-port elements	C-type elements	C______	$\frac{de(t)}{dt} = \frac{1}{C} f(t)$
	I-type elements	I______	$\frac{df(t)}{dt} = \frac{1}{I} e(t)$
	R-type elements	R______	$e(t) = R \cdot f(t)$
2-port elements	Transformer	___TF___	$e_2(t) = TF \cdot e_1(t)$ $f_2(t) = \frac{1}{TF} \cdot f_1(t)$
	Gyrator	___GY___	$e_2(t) = GY \cdot f_1(t)$ $f_2(t) = \frac{1}{GY} \cdot e_1(t)$
3-port junction elements	0-junction	___0___	$e_2(t) = e_1(t)$
	1-junction	___1___	$f_2(t) = f_1(t)$

For any element with a bond with power variables effort and flow—the energy variation from t_0 to t can be defined:

$$H(t) - H(t_0) = \int_{t_0}^{t} e(\tau) f(\tau) d\tau \tag{4.194}$$

For C-type storage elements, e (effort) is a function of q (displacement) as in a capacitor or spring (Figure 4.36).

As displacement is altered into its derivative, flow is defined as

$$q(t) = \int f(t) dt \Rightarrow \dot{q} = \frac{dq}{dt} = f(t) \tag{4.195}$$

which is expressed as a balance equation and establishes a part of the constitutive equations of the storage element. As Equation (4.195) is changing variables from t to q, the linear case is expressed:

$$H(q) - H(q_0) = \frac{1}{2C}\left(q^2 - q_0^2\right) \tag{4.196}$$

For I-type storage elements, f (flow) is a function of p (momentum) as an inductor or mass (Figure 4.37).

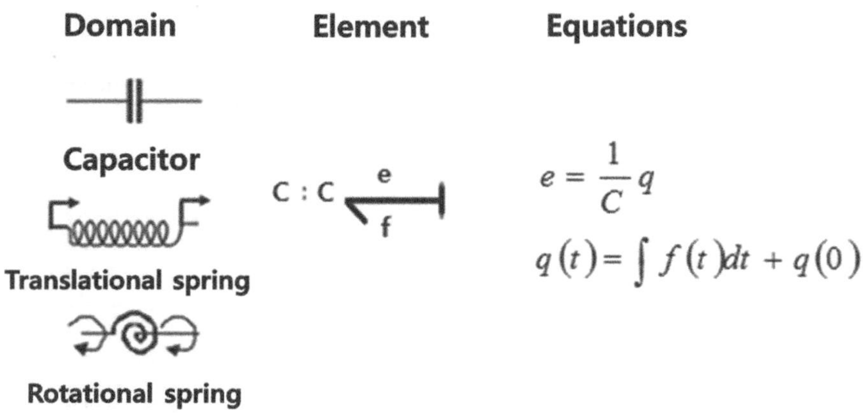

FIGURE 4.36
Instances of *C* elements.

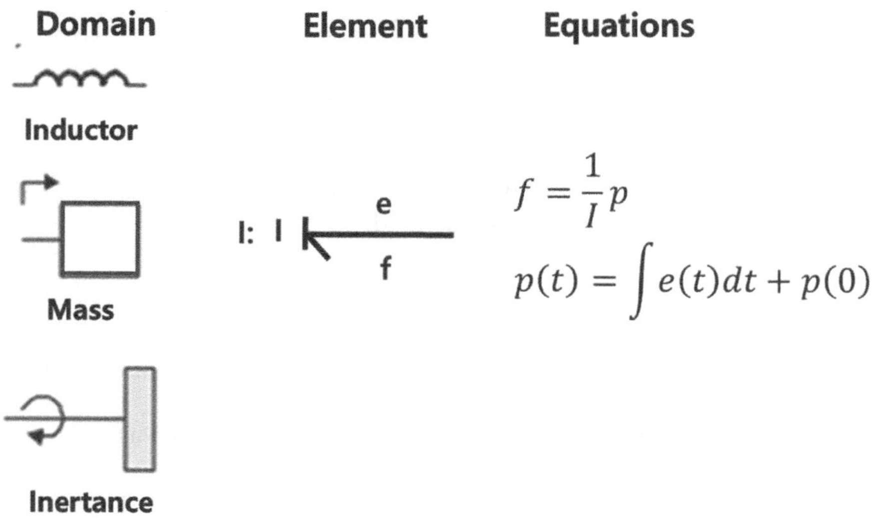

FIGURE 4.37
Instances of *I* elements.

As momentum is changed into its derivative, effort is defined as

$$p(t) = \int e(t)dt \Rightarrow \dot{p} = \frac{dp}{dt} = e(t) \tag{4.197}$$

which is regarded as the balance equation. If Equation (4.197) is interchanging variables from t to p, the linear case will be expressed:

$$H(p)-H(p_0)=\frac{1}{2I}\left(p^2-p_0^2\right) \tag{4.198}$$

Resistor elements describe situations where energy dissipates, such as mechanical dampers, Coulomb friction, and electrical resistors. In these sorts of elements, there is a connection between effort and flow. The value of R could be stable or a function of any system parameter involving a period:

$$e(t)=R\cdot f(t) \tag{4.199}$$

Compliance elements describe the situations where energy is kept—mechanical springs, electrical capacitors, etc. In these sorts of elements, there is a connection between effort and displacement variables. The value of K could be well-balanced or a function of any system parameter involving period:

$$e(t)=K\cdot q(t) \tag{4.200}$$

Inertia elements describe the relation between the "flow" and momentum as Equation (4.201) shows. The value of I tends to be well-balanced:

$$p(t)=I\cdot f(t) \tag{4.201}$$

A transformer builds no power but alters it, such as a lever or an electrical transformer. Transformers manifest those engineering phenomena that are dissimilarities in the gains of output flow and effort on the gains of input flow and effort. If the transformation ratio is expressed by the TF value, the connection between input and output may be defined as in Equations (4.202) and (4.203):

$$e_{\text{output}}(t)=TF\cdot e_{\text{input}}(t) \tag{4.202}$$

$$f_{\text{output}}(t)=\frac{1}{TF}\cdot f_{\text{input}}(t) \tag{4.203}$$

The process of deciding the computational orientation of the bond variables is expressed as causal analysis. One is the "half-arrow" sign agreement. It expresses the assumed orientation of positive energy flow. As with electrical circuit diagrams and free body diagrams, choosing positive orientation is random, with the warning that the analyst might be compatible all over with the selected definition. The other attribute is the causal stroke, which is a vertical bar put on only one end of the bond. It is not random (Figure 4.38).

On each bond, one of the two variables might be the cause and the other one the effect. This shall be deduced by the relation designated by the arrow orientation.

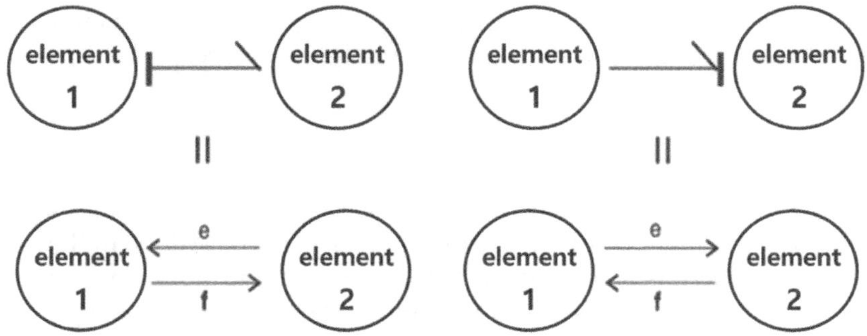

FIGURE 4.38
"Half-arrow" sign convention and meaning of the causal stroke.

Effort and flow causalities always act face-to-face with orientations in a bond. The causality allotment process chooses the sets for each bond. Causality allotment is necessary to transform the bond graph into a computable code.

Any port (single, double, or multiple) added to the bond may express either "effort" or "flow" by its causal stroke but not both. The port added to the ending of the bond with the causal stroke states the flow of the bond. The bond indicates effort upon that port. Likewise, the port on the end with no causal stroke indicates effort on the bond, while the bond expresses flow on that port.

If the system is depicted in the configuration of a bond graph, the state equations governing its behavior shall be obtained straightforwardly as first-order differential formulations with regard to the generalized variables just expressed, using uncomplicated and standardized processes, irrespective of the engineering field to which it links, even when attached to one another across fields.

Example 4.22

The mass/damper/spring system for a mechanical system is described in Figure 4.39. In mechanical diagrams, the port variables of the bond graph elements are the force on the element port and velocity of the element port. These two variables are linked with each other. The power being altered by a port with the rest of the system is the product of force and velocity: $P = Fv$. The equations of mass, damper, and spring are (we use damping coefficient a, spring coefficient K_s, mass m, and applied force F_a):

$$F_d = \alpha v,\ F_s = K_s \int v dt = \frac{1}{C_s} \int v dt,\ F_m = m\frac{dv}{dt} \tag{4.204}$$

As the loose ends of the example all have the same velocity, the common v is modified to a 1, a so-called 1-junction. This junction element also requires that the forces sum to zero, thinking about the sign (related to the power direction). The force is described as an effort, and the velocity is described as a flow.

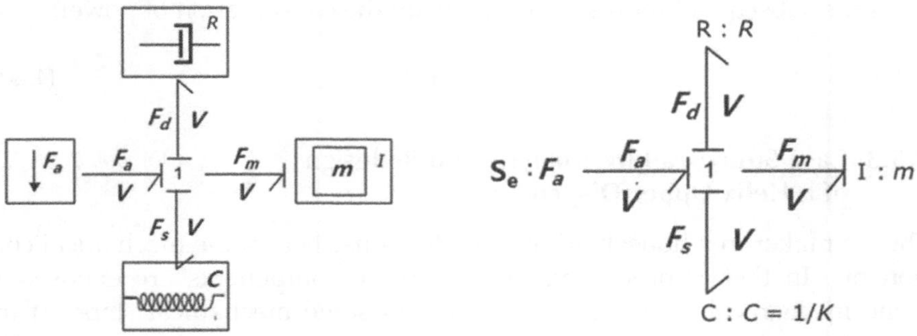

FIGURE 4.39
Damped mass/spring system for a mechanical system (instance).

Example 4.23

To perceive the intended meaning of a transformer, a mass-less ideal lever is contemplated. As shown in Figure 4.40, the lever is rigid in order to set up a linear relationship between power variables at both ends.

From the geometry, we have

$$V_2 = (b/a)V_1 \tag{4.205}$$

The power transmission specifies

$$F_2 = (a/b)F_1 \text{ so that } V_2F_2 = V_1F_1 \tag{4.206}$$

In bond graphs, such a circumstance may be defined as TF (transformer). The *r* above the transformer designates the modulus of the transformer, which can be a constant (*b*/*a*):

$$f_j = rf_i \text{ and } e_j = (1/r)e_i \tag{4.207}$$

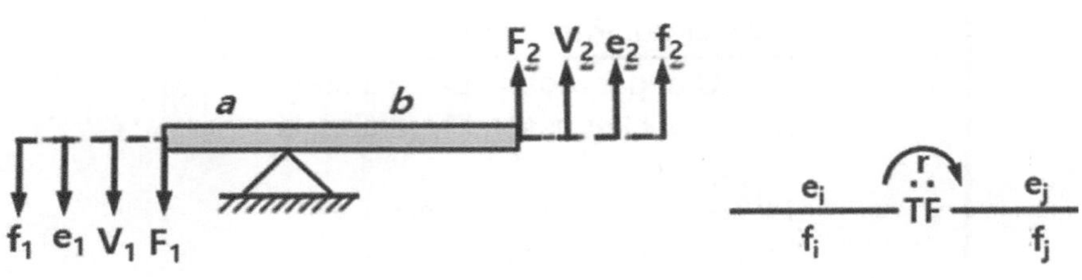

FIGURE 4.40
Perfect lever system (instance).

Thus the subsequent representation sets up the conservation of power:

$$e_j f_j = e_i f_i \tag{4.208}$$

4.5.3 Case Study: Failure Analysis and Redesign of a Helix Upper Dispenser

The ice-maker in a household refrigerator consists of some mechanical components. In the ice-producing process, these components are exposed to some mechanical loads. Ice-making needs some mechanical steps: 1) the water through tap line is raised to supply the tray; 2) the air cooled in the heat exchanger provides the water tray; 3) ice cubes are created in the bucket until it is full; and 4) when the customer presses the lever, crushed or cubed ice is distributed. In the United States, consumers need an SBS refrigerator to gather as a daily harvest of 10–200 cubes. The ice harvest could be influenced by uncontrollable customer usage situations—water pressure, ice consumption, notch settings, and the number of door opening. When the refrigerator is plugged in, the cubed ice mode is selected by itself with no control. On the other hand, a crusher smashes the cubed ice in the crushed mode. The load of the ice-maker is small because it functions without webbed or fused ice.

However, for Asian end users, webbed or fused ice will form crystals in the tray because, in cubed mode, the refrigerator does not dispense ice. If ice is dispensed under the circumstances, a notable mechanical overload occurs in the ice crusher. However, in the United States or Europe, the ice-maker system works non-stop as it is repetitively used in both crushed and cubed ice modes. This could produce a mechanical overload.

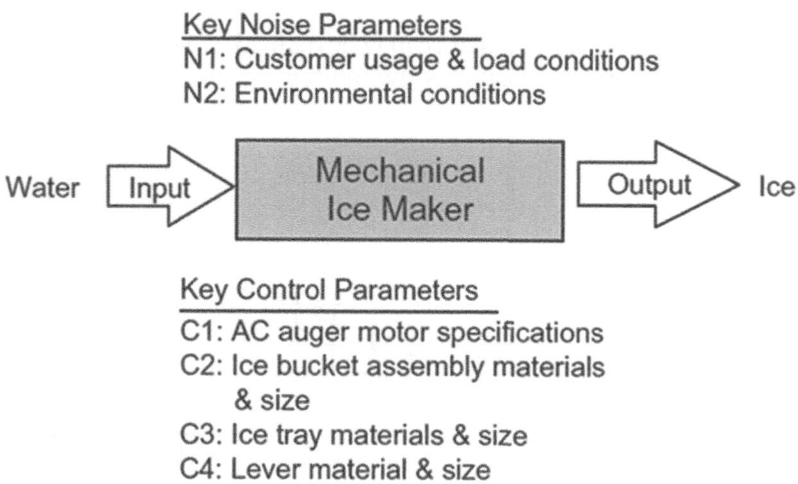

FIGURE 4.41
Simplified robust design illustration of ice-maker.

Figure 4.41 gives a clarified sketch of a robustly design ice-maker. Figure 4.42 presents a clarified diagram of the mechanical load transfer in the ice bucket and its bond graphs. An AC auger motor generates power. To attain adequate torque to squash the ice, motor power is moved through the gear system to the ice bucket fabrication—namely, to the helix upper dispenser, the blade dispenser, and the ice crusher (Figure 4.43).

The bond graph in Figure 4.43 can be defined:

$$dfE_2/dt = 1/L_a \times eE_2 \tag{4.209}$$

$$dfM_2/dt = 1/J \times eM_2 \tag{4.210}$$

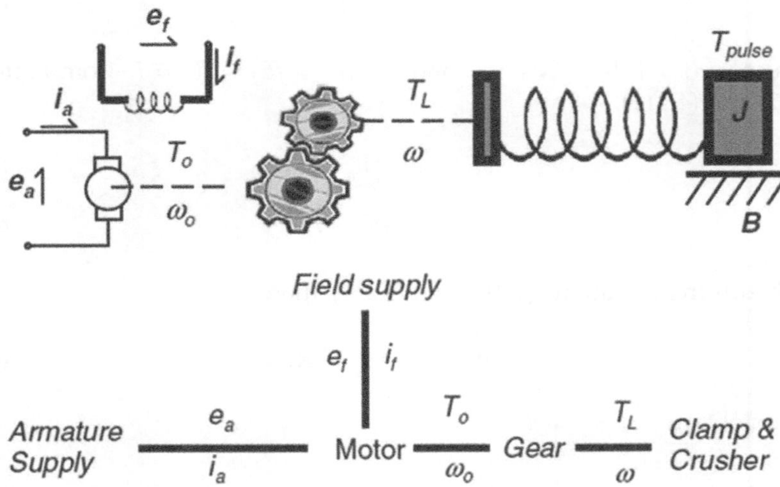

Figure 4.42 Clarified sketch for mechanical ice bucket assembly.

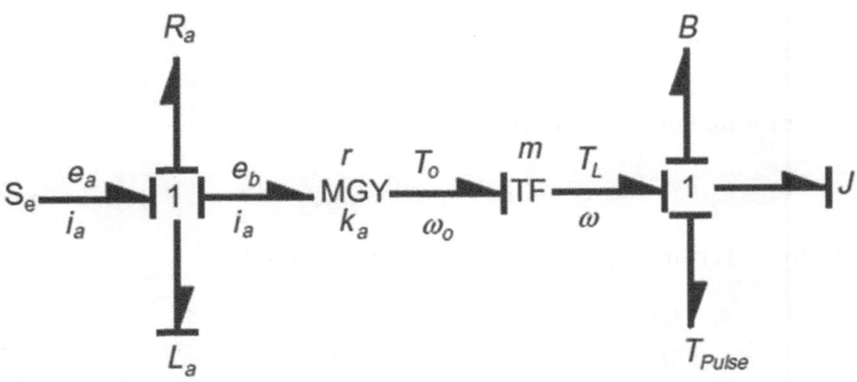

Figure 4.43 Bonding graph of ice bucket assembly.

The junction from Equation (4.209):

$$eE_2 = e_a - eE_3 \tag{4.211}$$

$$eE_3 = R_a \times fE_3 \tag{4.212}$$

The junction from Equation (4.210):

$$eM_2 = eM_1 - eM_3 \tag{4.213}$$

$$eM_1 = (K_a \times i) - T_{\text{Pulse}} \tag{4.214}$$

$$eM_3 = B \times fM_3 \tag{4.215}$$

Because $fM_1 = fM_2 = fM_3 = \omega$ and $i = fE_1 = fE_2 = fE_3 = i_a$ from Equations (4.211) and (4.212),

$$eE_2 = e_a - R_a \times fE_3 \tag{4.216}$$

$$fE_2 = fE_3 = i_a \tag{4.217}$$

If substituting Equations (4.216) and (4.217), then

$$di_a/dt = 1/L_a \times (e_a - R_a \times i_a) \tag{4.218}$$

We can obtain

$$eM_2 = [(K_a \times i) - T_{\text{Pulse}}] - B \times fM_3 \tag{4.219}$$

$$i = i_a \tag{4.220}$$

$$fM_3 = fM_2 = \omega \tag{4.221}$$

If substituting Equations (4.219)–(4.221), then

$$d\omega/dt = 1/J \times [(K_a \times i) - T_{\text{Pulse}}] - B \times \omega \tag{4.222}$$

Thus the governing equation (or state equation) could be:

$$\begin{bmatrix} di_a/dt \\ d\omega/dt \end{bmatrix} = \begin{bmatrix} -R_a/L_a & 0 \\ mk_a & -B/J \end{bmatrix} \begin{bmatrix} i_a \\ \omega \end{bmatrix} + \begin{bmatrix} 1/L_a \\ 0 \end{bmatrix} e_a + \begin{bmatrix} 1 \\ -1/J \end{bmatrix} T_{Pulse} \tag{4.223}$$

When Equation (4.223) is integrated, the angular velocity of the ice bucket assembly is obtained:

$$y_p = \begin{bmatrix} 0 & 1 \end{bmatrix} \begin{bmatrix} i_a \\ \omega \end{bmatrix} \tag{4.224}$$

References

1. Dugas, R. *A History of Classical Mechanics*. Dover Publications Inc.: New York, NY, 1988.
2. Tsai, L.-W. *Mechanism Design: Enumeration of Kinematic Structures According to Function*. CRC Press: Boca Raton, FL, 2001.
3. Lanczos, C. *Variational Principles of Mechanics*. 4th ed. University of Toronto Press: Toronto, Ontario, 1964, p. 92.
4. Fraser, C.J.L. Lagrange's early contributions to the principles and methods of mechanics. *Arch. Rational Mech.* 28(3), 197–241 (1983).
5. Karnopp, D.C.; Margolis, D.L.; Rosenberg, R.C. *System Dynamics: Modeling, Simulation, and Control of Mechatronic Systems*. 5th ed. John Wiley & Sons, Inc.: York, PA, 2000.

5

Mechanical System Design (Strength and Stiffness)

5.1 Introduction

The strength of materials, called the mechanics of deformable bodies or mechanics of materials, is a branch of engineering that studies how well a material can withstand mechanical forces [1]. It usually deals with the question in numerous ways by calculating the stresses and strains in the structural elements, such as beams, columns, and shafts. Thus strength of materials is important for mechanical and civil engineers because it deals with how objects deform under loading. Some topics that can be covered in strength of materials are stress, strain, deformation, torsion, flexure, shear diagram, and moment diagram. These structured procedures can be used to predict the response of a construction under loading and its vulnerability to many failure modes and to think about the properties of the materials, such as its yield strength, ultimate strength, Young's modulus, Poisson's ratio, etc., in the stress–stain curve.

Leonardo da Vinci (1452–1519) and Galileo Galilei (1564–1642) made the first endeavor to discover a beam hypothesis. Although Leonardo da Vinci made the pivotal discoveries, they still are evolving to complete their theories with the help of calculus and Hooke's law. Galileo was also held back by an inadequate assumption. He performed tests to determine the strength of beams, bars, and wires, although he did not provide a correct hypothesis to express the innovative results.

The widely known mathematician Leonhard Euler (1707–1783) provided the concepts of columns and calculated the critical load of a column in 1744. With unsuitable tests to complete his hypothesis, the Euler–Bernoulli beam hypothesis had been undeveloped for over a hundred years. Buildings and bridges were frequently designed as good examples until the latter part of the 19th century, while the Ferris wheel and Eiffel Tower described the success of the hypothesis on big scales. At this point, they are the premise for the analysis and design of most columns, such as tip-loaded cantilever beams.

In mechanical engineering, elasticity is a material's capacity to come back to its original shape after being deformed by an outer force. Elasticity is a

DOI: 10.1201/9781003451471-5

critical property in engineering design, especially for structures that might be regularly stressed but need to recover. The stress–strain curve is a popular method to quantify a material's elasticity, which shows how it responds to stress. The elasticity hypothesis originated from the idea that one- and two-dimensional components of structures will function durably, whose stress might be approximated as two-dimensional and thus was broadly made as three dimensions to develop a more whole hypothesis of the behavior of materials. A chief researcher in the mechanics of materials was Stephen Timoshenko [2].

A load on a mechanical part will bring about internal forces in the member, which can be named stresses when they are accompanied by units of measurement. Through a number of different ways, the stresses cause deformation of the material, which can develop to crack and completely fracture. A material has a rest shape, which goes away from it on the execution of an outer force. The amount of leaving from the rest shape is named as deformation. Deformation of the materials can be defined as strain when those deformations are expressed as a unit basis. That is, the ratio of deformation to original size is called strain. In elasticity, force and deformation are jointly connected by means of the stiffness relationship. Deformation and strain are linked jointly by compatibility equations. Stress and strain are linked jointly by constitutive equations (Figures 5.1 and 5.2).

The design of a mechanical product exposed to numerous (random) loads needs to possess good quality, such as sufficient strength or stiffness. Strength—the capacity to withstand an applied stress or load with no structural failure—is frequently required to be high throughout the product design of the material or its shape. If there is a design flaw in the construction, a permanent deformation of material due to loads may start to crack, fracture, and ultimately lose the ability to perform their intended functions. On the

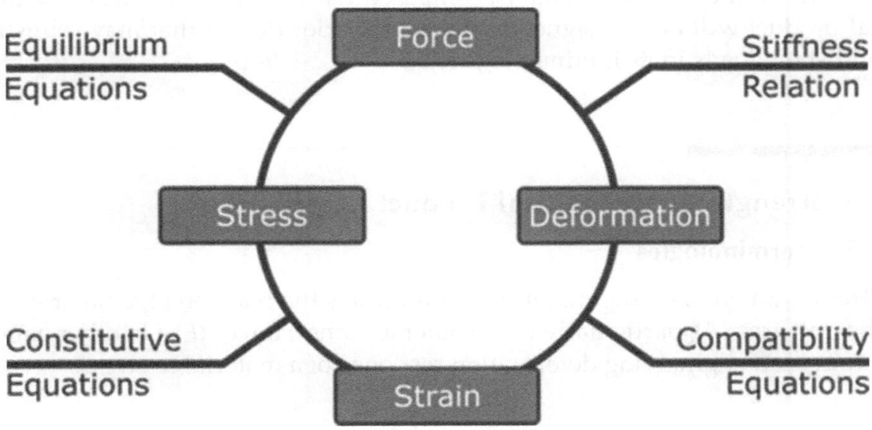

FIGURE 5.1
Connection between force, stress, strain, and deformation.

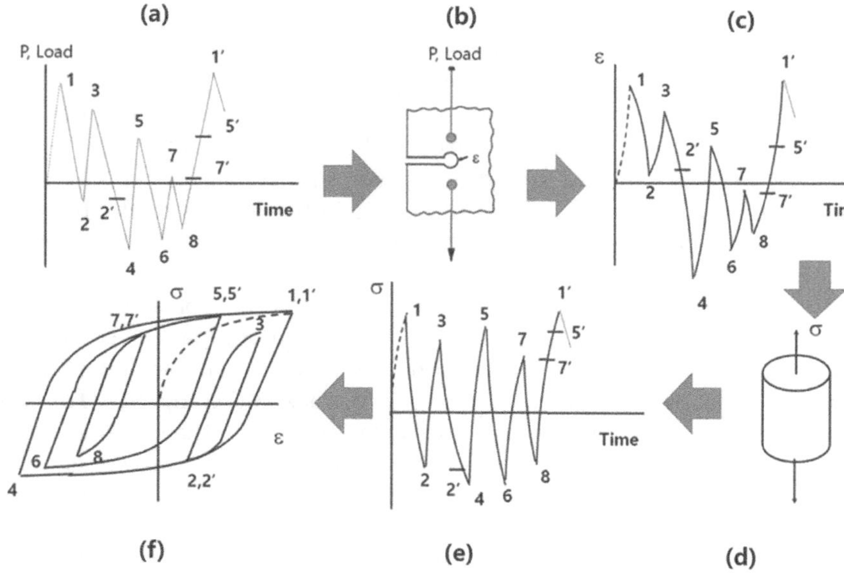

FIGURE 5.2
Load history exposed to tensile and compression in the field.

other hand, the requirement for stiffness—the resistance against reversible deformation—may alter over a large range of the product's implementation.

The failure mechanics—mainly fatigue or fracture—of components might no longer function and be differentiated by the two following factors: 1) the loads on the structure and 2) the type of product materials and their forms employed in the structure, involving the mechanism. To prevent failures from the field, the engineer might replicate them. Product can optimally alter designs by choosing a correct material and form, which has strength and stiffness in the structure. That is, compared with competitors, the mechanical product will be redesigned to have a superior design that may robustly withstand loads in its lifetime (Figure 5.3).

5.2 Strength of Mechanical Product

5.2.1 Terminologies

Stress is an engineering quantity that indicates the reaction of a material on the unit area (A), particularly in the outer (or inner) forces (F). On other hand, strain is an engineering deformation response of a material to stress.

$$\sigma = \frac{F}{A} \tag{5.1}$$

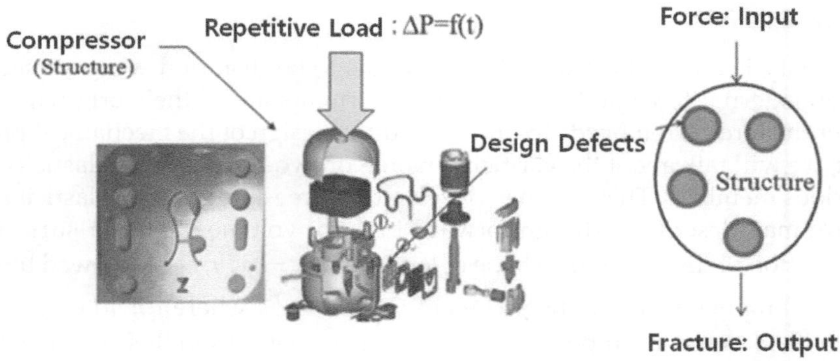

FIGURE 5.3
Failure mechanics created by a load on a component made from a particular material.

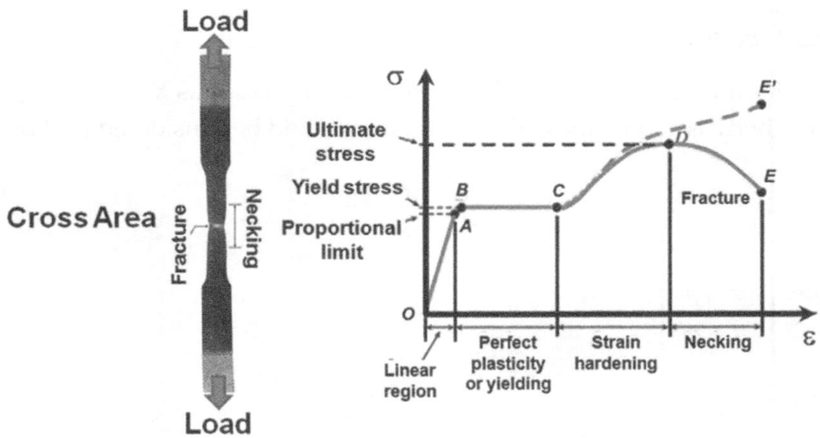

FIGURE 5.4
Terms in the strength of materials.

The linear part in the stress–strain curve is the elastic district, and the slope is Young's modulus. The elastic range stops when the material stretches out its yield strength. After the yield point, the curve slope lessens a little, and deformation does not stop. Strain hardening and plastic deformation continues until it stretches out the ultimate tensile stress. Deformation is a reaction to any changes in the shape of an object due to an exerted force. Elastic deformation is that, once the forces are no longer exerted, the object comes back to its original form. This sort of deformation requires the stretching of the atom bonds. Linear elastic deformation is ruled by Hooke's law, which states (Figure 5.4):

$$\sigma = E\varepsilon \tag{5.2}$$

where σ is the exerted stress, E is Young's modulus, and ε is the strain.

5.2.2 Elasticity

Elasticity is an attribute of (solid) materials to be distorted when subjected to an external force and to return to a normal state of their original form after the force is removed. To understand the design of the mechanical product, we will talk about the elasticity in one- or two-dimensional elasticity for various members. Then we may develop the three-dimensional elasticity [3].

We may describe a straightforward body of volume V and of surface S. The action of distorting the shape at local point $x=[x,y,z]^T$ is allowed by the three components of its displacement $u=[u,v,w]^T$, where $u=u(x,y,z)$, i.e., each displacement component is a function of position. On this occasion, two fundamental types of external forces act on a body: 1) body force (force per unit volume), e.g., weight, and 2) surface traction (force per unit surface area), e.g., friction (Figure 5.5).

5.2.2.1 Body Force

For volume element dV, the body force may be expressed as $\underline{X} = [X_a, X_b, X_c]^T$. As the body is accelerating, the inertia force could be considered part of $\underline{X}$:

$$\underline{X} = \underline{\tilde{X}} - \rho\underline{\ddot{u}} \tag{5.3}$$

where $\rho\underline{\ddot{u}} = \begin{Bmatrix} \rho\ddot{u} \\ \rho\ddot{v} \\ \rho\ddot{w} \end{Bmatrix}$.

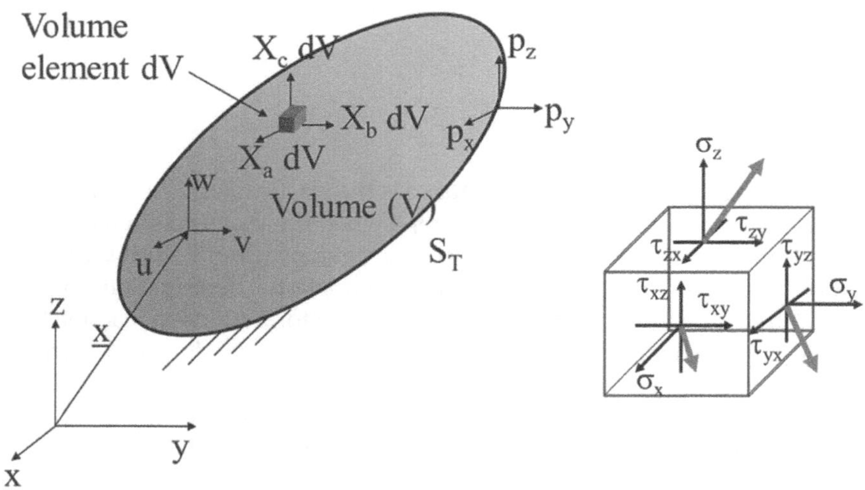

FIGURE 5.5
Uncomplicated body exposed to body force and surface traction.

5.2.2.2 Surface Traction

As the distributed force per unit of surface area, surface traction may be expressed:

$$\underline{T}_S = \begin{Bmatrix} p_x \\ p_y \\ p_z \end{Bmatrix} \tag{5.4}$$

There is an inner force behaving in a particular way in response to a body. As defined as the internal reaction forces per unit area, we may define it as stress.

5.2.2.3 Internal Forces

As a solid slice of material from the body is eliminated, some reaction forces are due to the external forces applied on it. For the hexahedron, on each surface, the inner reaction force per unit area is recognized as stress and can be broken down into three orthogonal components. Presume a body lined up in the Cartesian coordinate system with several forces exerted on it, such that the vector whole of all the forces is zero. Take a piece orthogonal to the x orientation and describe a little area on this slice as ΔA_x. Let the whole force exerted on this small area be

$$\underline{\Delta F} = \Delta F_x \cdot \hat{i} + \Delta F_y \cdot \hat{j} + \Delta F_z \cdot \hat{k} \tag{5.5}$$

We may describe the subsequent scalar amounts:

$$\sigma_{xx} = \lim_{\Delta A_x \to 0} \frac{\Delta F_x}{\Delta A_x}, \tau_{xy} = \lim_{\Delta A_x \to 0} \frac{\Delta F_y}{\Delta A_x}, \tau_{xz} = \lim_{\Delta A_x \to 0} \frac{\Delta F_z}{\Delta A_x} \tag{5.6}$$

Identically, thinking about slices at right angles to the y and z orientations, we define

$$\tau_{yx} = \lim_{\Delta A_y \to 0} \frac{\Delta F_x}{\Delta A_y}, \sigma_{yy} = \lim_{\Delta A_y \to 0} \frac{\Delta F_y}{\Delta A_y}, \tau_{yz} = \lim_{\Delta A_y \to 0} \frac{\Delta F_z}{\Delta A_y} \tag{5.7}$$

and

$$\tau_{zx} = \lim_{\Delta A_z \to 0} \frac{\Delta F_x}{\Delta A_z}, \tau_{zy} = \lim_{\Delta A_z \to 0} \frac{\Delta F_y}{\Delta A_z}, \sigma_{zz} = \lim_{\Delta A_z \to 0} \frac{\Delta F_z}{\Delta A_z} \tag{5.8}$$

We can say that σ_x, σ_y, and σ_z are normal stresses. On the other hand, the existing six are the shear stresses. As $\tau_{xy} = \tau_{yx}$, $\tau_{yz} = \tau_{zy}$ and $\tau_{zx} = \tau_{xz}$, we can suppose that only six stress components are independent. The stress vector is expressed:

$$\underline{\sigma} = \left[\sigma_x \, \sigma_y \, \sigma_z \, \tau_{xy} \, \tau_{yz} \, \tau_{zx}\right]^T \tag{5.9}$$

On the other hand, the amount of deformation is expressed as the strain and can be broken up into three orthogonal components. Strains have six independent strain components:

$$\underline{\varepsilon} = \left[\varepsilon_x \varepsilon_y \varepsilon_z \gamma_{xy} \gamma_{yz} \gamma_{zx}\right]^T \tag{5.10}$$

Presume the balance of a differential volume element to attain the three equilibrium equations of elasticity:

$$\begin{aligned} &\frac{\partial \sigma_x}{\partial x} + \frac{\partial \tau_{xy}}{\partial y} + \frac{\partial \tau_{xz}}{\partial z} + X_a = 0 \\ &\frac{\partial \tau_{xy}}{\partial x} + \frac{\partial \sigma_y}{\partial y} + \frac{\partial \tau_{yz}}{\partial z} + X_b = 0 \\ &\frac{\partial \tau_{xz}}{\partial x} + \frac{\partial \tau_{yz}}{\partial y} + \frac{\partial \sigma_z}{\partial z} + X_c = 0 \end{aligned} \tag{5.11}$$

In a way that is brief but comprehensive, we can express the equilibrium equations:

$$\underline{\partial}^T \underline{\sigma} + \underline{X} = \underline{0} \tag{5.12}$$

where $\underline{\partial} = \begin{bmatrix} \frac{\partial}{\partial x} & 0 & 0 \\ 0 & \frac{\partial}{\partial y} & 0 \\ 0 & 0 & \frac{\partial}{\partial z} \\ \frac{\partial}{\partial y} & \frac{\partial}{\partial x} & 0 \\ 0 & \frac{\partial}{\partial z} & \frac{\partial}{\partial y} \\ \frac{\partial}{\partial z} & 0 & \frac{\partial}{\partial x} \end{bmatrix}$.

5.2.2.4 Strong Formulation for the 3D Elasticity Problem

As the outer applied loads (on S_T and in V) and the expressed displacements (on S_u) are given, we may solve for the occurring as a result of displacements, strains, and stresses necessitated to continue equilibrium of the body.

Equilibrium equations in a fixed volume are expressed:

$$\underline{\partial}^T \underline{\sigma} + \underline{X} = \underline{0} \quad \text{in } V \tag{5.13}$$

As the displacement boundary expresses, displacements are defined on the boundary of S_u:

$$\underline{u} = \underline{u}^{\text{specified}} \quad \text{on } S_u \tag{5.14}$$

If we set $\sin\theta = \dfrac{dx}{ds} = n_y$ and $\cos\theta = \dfrac{dy}{ds} = n_x$ in Figure 5.6, elasticity traction in 2D may be expressed:

$$p_x ds = \sigma_x dy + \tau_{xy} dx \text{ or } p_x = \sigma_x \frac{dy}{ds} + \tau_{xy} \frac{dx}{ds} = \sigma_x n_x + \tau_{xy} n_y \tag{5.15}$$

As traction (force) boundary circumstances, tractions are expressed on the boundary of S_T. Traction may be expressed as a distributed force per unit area:

$$\underline{T_S} = \begin{Bmatrix} p_x \\ p_y \\ p_z \end{Bmatrix} \tag{5.16}$$

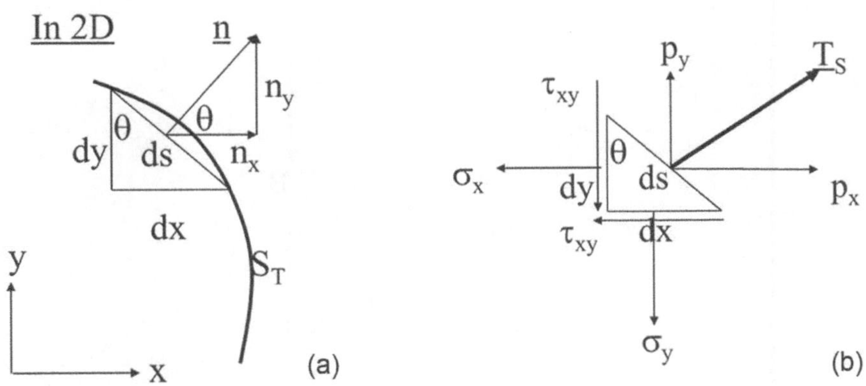

FIGURE 5.6
Uncomplicated wedge body exposed to 2D surface traction.

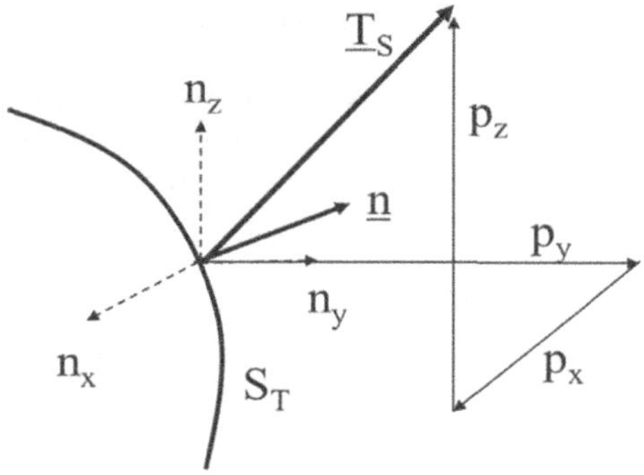

FIGURE 5.7
Uncomplicated wedge body exposed to 3D surface traction.

As n is the unit external normal to S_T in Figure 5.7, traction may be expressed:

$$\begin{aligned} p_x &= \sigma_x n_x + \tau_{xy} n_y + \tau_{xz} n_z \\ p_y &= \tau_{xy} n_x + \sigma_y n_y + \tau_{yz} n_z \\ p_z &= \tau_{xz} n_x + \tau_{zy} n_y + \sigma_z n_z \end{aligned} \tag{5.17}$$

5.2.2.5 *Strain–Displacement Relationships*

As shown in Figure 5.8, in 2D elasticity, the strain–displacement relationships from little fragments may be attained:

$$\varepsilon_x = \frac{A'B' - AB}{AB} = \frac{\left(dx + \left(u + \dfrac{\partial u}{\partial x} dx\right) - u\right) - dx}{dx} = \frac{\partial u}{\partial x} \tag{5.18a}$$

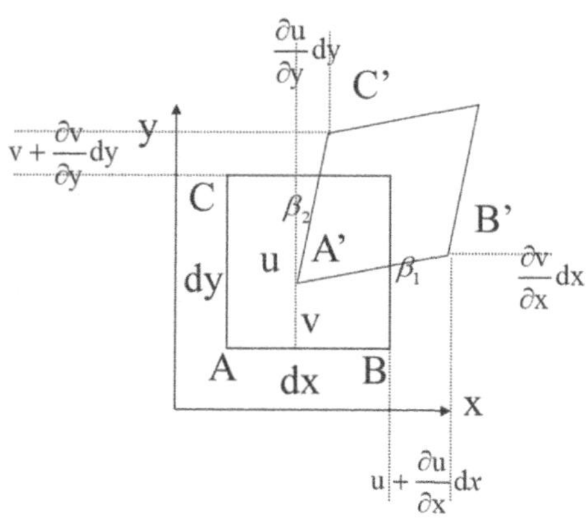

FIGURE 5.8
Strain–displacement relationships of small fragments in 2D elasticity.

$$\varepsilon_y = \frac{A'C' - AC}{AC} = \frac{\left(dy + \left(v + \frac{\partial v}{\partial y}dy\right) - v\right) - dy}{dy} = \frac{\partial v}{\partial y} \tag{5.18b}$$

$$\gamma_{xy} = \frac{\pi}{2} - \text{angle}\left(C'A'B'\right) = \beta_1 + \beta_2 \approx \tan\beta_1 + \tan\beta_2 \approx \frac{\partial v}{\partial x} + \frac{\partial u}{\partial x} \tag{5.18c}$$

For 3D elasticity, strain–displacement relationships may be enlarged:

$$\begin{aligned} \varepsilon_x &= \frac{\partial u}{\partial x} \\ \varepsilon_y &= \frac{\partial v}{\partial y} \\ \varepsilon_z &= \frac{\partial w}{\partial z} \\ \gamma_{xy} &= \frac{\partial u}{\partial y} + \frac{\partial v}{\partial x} \\ \gamma_{yz} &= \frac{\partial v}{\partial z} + \frac{\partial w}{\partial y} \\ \gamma_{zx} &= \frac{\partial u}{\partial z} + \frac{\partial w}{\partial x} \end{aligned} \tag{5.19}$$

Expressed compactly:

$$\varepsilon = \partial u \tag{5.20}$$

where $\underline{\varepsilon} = \begin{Bmatrix} \varepsilon_x \\ \varepsilon_y \\ \varepsilon_z \\ \gamma_{xy} \\ \gamma_{yz} \\ \gamma_{zx} \end{Bmatrix}$ $\underline{\partial} = \begin{bmatrix} \frac{\partial}{\partial x} & 0 & 0 \\ 0 & \frac{\partial}{\partial y} & 0 \\ 0 & 0 & \frac{\partial}{\partial z} \\ \frac{\partial}{\partial y} & \frac{\partial}{\partial x} & 0 \\ 0 & \frac{\partial}{\partial z} & \frac{\partial}{\partial y} \\ \frac{\partial}{\partial z} & 0 & \frac{\partial}{\partial x} \end{bmatrix}$ $\underline{u} = \begin{Bmatrix} u \\ v \\ w \end{Bmatrix}$.

5.2.2.6 Stress–Strain Relationship

As General Hooke's law is exerted, linear elastic material may be expressed:

$$\underline{\sigma} = \underline{D}\,\underline{\varepsilon} \tag{5.21}$$

where

$$\underline{D} = \frac{E}{(1+\nu)(1-2\nu)} \begin{bmatrix} 1-\nu & \nu & \nu & 0 & 0 & 0 \\ \nu & 1-\nu & \nu & 0 & 0 & 0 \\ \nu & \nu & 1-\nu & 0 & 0 & 0 \\ 0 & 0 & 0 & \frac{1-2\nu}{2} & 0 & 0 \\ 0 & 0 & 0 & 0 & \frac{1-2\nu}{2} & 0 \\ 0 & 0 & 0 & 0 & 0 & \frac{1-2\nu}{2} \end{bmatrix}$$

5.2.2.7 Principle of Minimum Potential Energy

For a linear elastic frame exposed to surface tractions $\underline{T_S} = \left[p_x, p_y, p_z\right]^T$ and body forces $\underline{X} = \left[X_a, X_b, X_c\right]^T$, leading to displacements $u = \left[u, v, w\right]^T$, strains $\underline{e}$, and stresses $\underline{s}$, the feasible energy P is defined as the strain energy minus the potential energy of the loads needing $\underline{X}$ and $\underline{T_S}$:

$$\Pi = U - W \tag{5.22}$$

where $U = \frac{1}{2}\int_V \underline{\sigma}^T \underline{\varepsilon}\, dV$ and $W = \int_V \underline{u}^T \underline{X}\, dV + \int_{S_T} \underline{u}^T \underline{T_S}\, dS$.

The one that fulfills the equilibrium equations in the middle of all permissible displacement fields also holds the potential energy P to a minimum. The so-called allowable displacement field has the subsequent properties:

1. The first derivative of the displacement components prevails.
2. It accomplishes the boundary states on S_u.

5.2.3 Beam

A beam in mechanical engineering is a constructional component whose length is longer, compared to its cross-sectional area. Explicitly, it is a bar-like constructional component whose principal function is to support cross-wise loading and transport it to the supports. The Euler–Bernoulli beam hypothesis plays a crucial part in constructional examination because it may help mechanical engineers to calculate the load-carrying and deflection attributes of beams in an uncomplicated manner.

As a beam is bent toward a lower level, the materials in the lowest portion of the beam expand, but those in the top portion shrink. These changes in the lengths of the materials create stresses in the materials. Those that are extended have tensile stresses exerted on the materials in the direction of the longitudinal axis of the beam, while those that are compressed are exposed to compressive stresses. There always exists a neutral surface of the beam holding materials that do not undergo any compression or extension and thus are not exposed to any compressive or tensile stress. It thus can be concluded that, from the neutral axis, the value of the bending stress will alter linearly with distance. At the present moment, we begin the analysis of the beam from the viewpoint of elasticity.

5.2.3.1 Euler–Bernoulli Beam Theory

As shown in Figure 5.9, before bending:

$$AB = CD = EF = \Delta x \tag{5.23}$$

As the bending line part *AB* is squeezed, line segment *CD* is elongated, and line segment *EF* does not differ in size. As a consequence, line segment *EF* is mentioned as the neutral surface. The junction of the neutral surface with

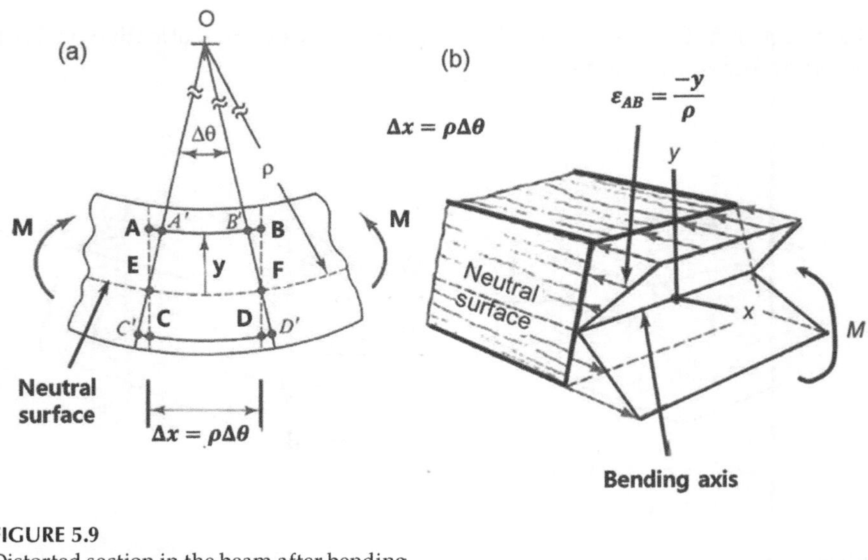

FIGURE 5.9
Distorted section in the beam after bending.

the longitudinal plane of symmetry is expressed as the neutral axis. From the calculus and analytic geometry in Figure 5.9, we may express the curvature:

$$\kappa = \frac{1}{\rho} = \frac{\Delta\theta}{\Delta x} \tag{5.24}$$

where κ is the curvature, and ρ is the radius of the curvature.

5.2.3.2 Strain–Displacement Relationships

As $\rho = \dfrac{\Delta x}{\Delta\theta}$ and $\Delta x = \rho\Delta\theta$, along line piece AB, the normal strain in the longitudinal orientation may be defined:

$$\varepsilon_{AB} = \frac{\delta}{L} = \frac{L_f - L_i}{L_i} = \frac{(\rho - y)\Delta\theta - \rho\Delta\theta}{\rho\Delta\theta} = \frac{-y}{\rho} \tag{5.25}$$

5.2.3.3 Stress–Strain Relationship

As Hooke's law states, the stresses are proportional to the interval y from the neutral axis:

$$\sigma = E\varepsilon = \frac{-Ey}{\rho} \tag{5.26}$$

5.2.3.4 Equilibrium Equations

From Figure 5.10, we discover that internal stresses are statically similar to the outer forces and moment:

$$\sum F_{x(I)} = \sum F_{x(\mathrm{II})} \Rightarrow \int \sigma\, dA = 0 \tag{5.27}$$

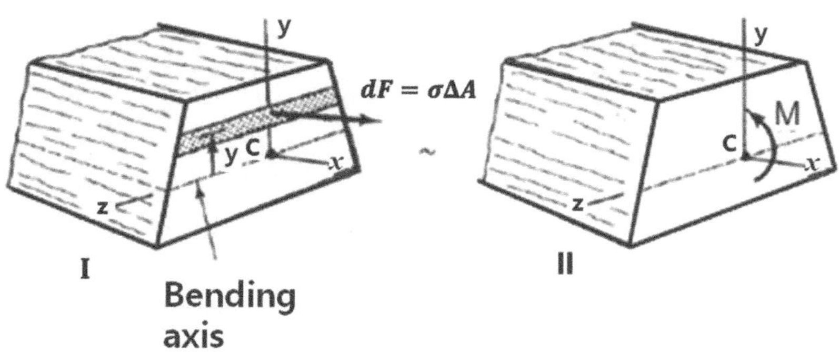

FIGURE 5.10
Bending stresses on a beam cross section.

$$\sum M_{z(I)} = \sum M_{z(\text{II})} \Rightarrow -\int y\sigma \, dA = M \tag{5.28}$$

If Equation (5.26) is substituted, we can obtain:

$$-\int y \frac{-Ey}{\rho} dA = M \tag{5.29}$$

$$M = \frac{E}{\rho} \int y^2 dA = \frac{EI}{\rho} \tag{5.30}$$

where the moment of inertia is $I = \int y^2 dA$.

As curvature κ is equal to $1/\rho$, the bending momentum may be restated:

$$M = \frac{EI}{\rho} = EI\kappa = EIy'' \tag{5.31}$$

If merging to Equations (5.28) and (5.29), we may define the bending equation:

$$\sigma = \frac{My}{I} \tag{5.32}$$

5.2.4 Flat Plate

Plates are flat constructional members having thicknesses much smaller than that of other sizes. To find the distribution of displacement and stress for a plate exposed to a stated set of forces, fundamental presumptions such as the Kirchhoff theory are necessitated:

1. The deflecting action of the middle surface is minute contrasted with the thickness of the plate. The midplane stays unstrained by the subsequent bending.
2. First plane segments are perpendicular to the midsurface and perpendicular to that surface after bending.
3. The stress component perpendicular to the midplane is small-scale contrasted with the other stress components.

Based on these presumptions, a three-dimensional plate problem may be lessened to a two-dimensional one. Think about a plate without load, in which the xy plane corresponds with the midplane (Figure 5.11). The components of displacement at a point are indicated by u, v, and w.

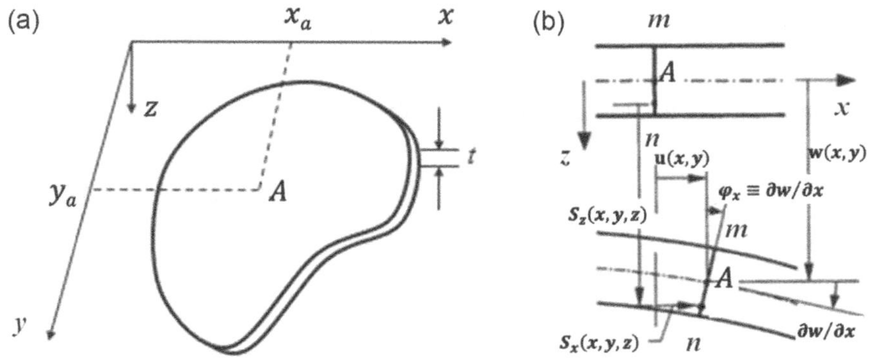

FIGURE 5.11
Plate of constant thickness: (a) plate; (b) plate part before and after deflection.

As stated by the previous assumptions, the displacement field can be defined:

$$u(x,y,z) = -z\frac{\partial w}{\partial x} \tag{5.33a}$$

$$v(x,y,z) = -z\frac{\partial w}{\partial y} \tag{5.33b}$$

$$w(x,y,z) = w(x,y) \tag{5.33c}$$

5.2.4.1 Strain–Displacement Relationships

The non-zero linear strains attached with the displacement field are expressed:

$$\varepsilon_x = \frac{\partial u}{\partial x} = -z\frac{\partial^2 w}{\partial x^2} \tag{5.34a}$$

$$\varepsilon_y = \frac{\partial v}{\partial y} = -z\frac{\partial^2 w}{\partial y^2} \tag{5.34b}$$

$$\gamma_{xy} = \frac{\partial u}{\partial y} + \frac{\partial v}{\partial x} = -2z\frac{\partial^2 w}{\partial x \partial y} \tag{5.34c}$$

where γ_{xy} is the shear strain, and $(\varepsilon_x, \varepsilon_y)$ comprise the normal strain.

5.2.4.2 Stress–Strain Relationship

From Hooke's law, we may attain stress because $\varepsilon_z = \gamma_{yz} = \gamma_{xz} = 0$:

$$\sigma_x = \frac{E}{1-\nu^2}\left[\varepsilon_x + \nu\varepsilon_y\right] \tag{5.35a}$$

$$\sigma_y = \frac{E}{1-\nu^2}\left[\varepsilon_y + \nu\varepsilon_x\right] \tag{5.35b}$$

$$\tau_{xy} = G\gamma_{xy} \tag{5.35c}$$

From Equation (5.34), we can clarify:

$$\sigma_x = -\frac{E}{1-\nu^2} z\left[\frac{\partial^2 w}{\partial x^2} + \nu\frac{\partial^2 w}{\partial y^2}\right] \tag{5.36a}$$

$$\sigma_x = -\frac{E}{1-\nu^2} z\left[\frac{\partial^2 w}{\partial y^2} + \nu\frac{\partial^2 w}{\partial x^2}\right] \tag{5.36b}$$

$$\tau_{xy} = -\frac{E}{1+\nu} z\frac{\partial^2 w}{\partial x \partial y} \tag{5.36c}$$

5.2.4.3 Equilibrium Equations

The stresses distributed over the width of the plate yield bending moments, twisting moments, and vertical shear forces. These moments and forces per unit length are also expressed as stress resultants:

$$\begin{Bmatrix} M_x \\ M_y \\ M_{xy} \end{Bmatrix} = \int_{\frac{-t}{2}}^{\frac{t}{2}} \begin{Bmatrix} \sigma_x \\ \sigma_y \\ \tau_{xy} \end{Bmatrix} zdz \tag{5.37}$$

Exchanging Equation (5.36) into Equation (5.37), we may obtain the subsequent formula for the bending and twisting moments:

$$M_x = -D\left(\frac{\partial^2 w}{\partial x^2} + \nu\frac{\partial^2 w}{\partial y^2}\right)$$

$$M_y = -D\left(\frac{\partial^2 w}{\partial y^2} + \nu\frac{\partial^2 w}{\partial x^2}\right) \tag{5.38}$$

$$M_{xy} = -D(1-v)\frac{\partial^2 w}{\partial x \partial y}$$

where $D = \frac{Et^3}{12(1-v^2)}$.

Thus, for the bending of a thin plate, the differential equation of equilibrium is expressed:

$$\frac{\partial^2 M_x}{\partial x^2} + 2\frac{\partial^2 M_{xy}}{\partial x \partial y} + \frac{\partial^2 M_y}{\partial y^2} = -p \tag{5.39}$$

Exchanging Equation (5.38) into Equation (5.39), the governing equation for the plate may be defined:

$$\frac{\partial^4 w}{\partial x^4} + 2\frac{\partial^4 w}{\partial x^2 \partial y^2} + \frac{\partial^4 w}{\partial y^4} = \frac{p}{D} \tag{5.40}$$

5.2.4.4 Boundary Conditions

The boundary circumstances on the outside limit $x = a$ of the rectangular plate with edges parallel to the x- and y-axes are (Figure 5.12):

1. *Clamped, fixed, or built-in edge*: $w = 0, \frac{\partial w}{\partial x} = 0$ $(x = a)$
2. *Straightforwardly supported edge*: $w = 0, M_x = -D\left(\frac{\partial^2 w}{\partial x^2} + \nu\frac{\partial^2 w}{\partial y^2}\right) = 0$ $(x = a)$
3. *Free edge*: $\frac{\partial^2 w}{\partial x^2} + \nu\frac{\partial^2 w}{\partial y^2} = 0, \frac{\partial^3 w}{\partial x^3} + (2-\nu)\frac{\partial^3 w}{\partial x \partial y^2} = 0$ $(x = a)$

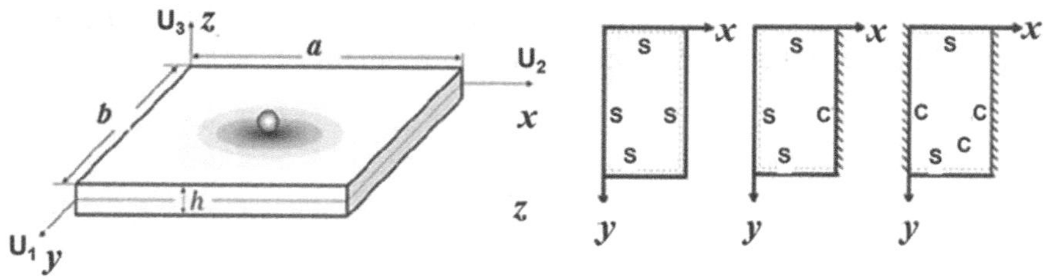

FIGURE 5.12
Various boundary conditions of a plate: S—simply supported edge, F—free edge, C—clamped or built-in edge.

5.2.5 Torsion Member

Torsion is a distorted shape of an object due to an exerted torque. As the shear stresses for noncircular sections no longer denote the perimeter of a curved geometry, the torsional theory of circular sections may not be extended to the torsion of noncircular sections. Besides, plane cross sections do not have a plane and do not distort the execution of torque, and warping of the cross section occurs.

As shown in Figure 5.13, presume a prismatic bar of isotropic homogeneous noncircular segments exposed to twisting action. Established on the mechanics convention; that is, the axis of the bar is consistent with the *z*-axis, the cross section is set in the *x*-*y* plane. The bar is fastened at $z = 0$ base, and the opposite base $z = 1$ is twisted by angle 1 α.

We may attain the subsequent presumptions:

- The cross sections in the *x*-*y* plane rotate like a rigid body. Regarding a noncircular shape, the cross section is not planar but is distracted in the *z* orientation.
- The deflection and twist rate α are constant along the entire length of the bar. Thus it is reduced to a two-dimensional problem.

Assume the subsequent geometrical behavior: Based on the Saint-Venant theory, think about any point P in the section, which may revolve and warp owing to the implementation of *T*, as seen in Figure 5.13b. The displacements *u*, *v*, and *w* in orientations *x*, *y*, and *z* under these assumptions may be defined:

$$u = -r\beta \sin\theta = -\beta y = -\alpha yz$$

$$v = -r\beta \cos\theta = \beta x = \alpha xz \tag{5.41}$$

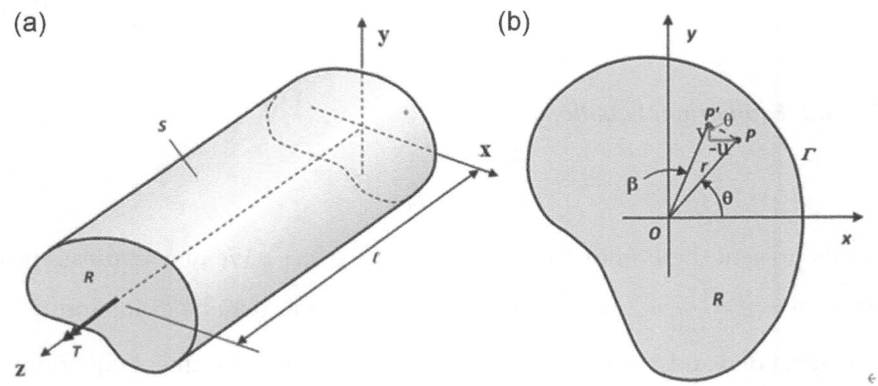

FIGURE 5.13
(a) Noncircular section under twist; (b) cross section.

$$w = \alpha\psi(x, y)$$

where $\beta = \alpha z$, and $\psi(x, y)$ is an unknown function that defines the deflection.

5.2.5.1 Strain–Displacement Relationships

This is assumed to be differentiable. A straightforward calculation builds the identical strain (small-scale deformation). So we may presume:

$$\varepsilon_x = \varepsilon_y = \varepsilon_z = \gamma_{xy} = 0 \tag{5.42}$$

Thus the solely shearing strains that exist are γ_{xz} and γ_{yz}, which are defined:

$$\gamma_{xz} = \frac{\partial w}{\partial x} + \frac{\partial u}{\partial z} = \alpha\left(\frac{\partial \psi}{\partial x} - y\right) \tag{5.43}$$

$$\gamma_{yz} = \frac{\partial w}{\partial y} + \frac{\partial v}{\partial z} = \alpha\left(\frac{\partial \psi}{\partial y} + x\right) \tag{5.44}$$

5.2.5.2 Stress–Strain Relationship

From Hooke's law, we may attain stress from Equations (5.42)–(5.44):

$$\sigma_x = \sigma_y = \sigma_z = \tau_{xy} = 0 \tag{5.45}$$

$$\tau_{xz} = G\alpha\left(\frac{\partial \psi}{\partial x} - y\right) \tag{5.46}$$

$$\tau_{yz} = G\alpha\left(\frac{\partial \psi}{\partial y} + x\right) \tag{5.47}$$

5.2.5.3 Equilibrium Equations

$$\frac{\partial \tau_{xz}}{\partial x} + \frac{\partial \tau_{yz}}{\partial y} = 0 \tag{5.48}$$

Let us present the Prandt stress function $\phi = \phi(x, y)$. We may express shear stress as $\tau_{xz} = \frac{\partial \phi}{\partial y}$, $\tau_{yz} = -\frac{\partial \phi}{\partial x}$. By differentiating Equations (5.46) and (5.47) in respect of y and x, respectively, we may attain the subsequent equation:

$$\frac{\partial \tau_{xz}}{\partial y} - \frac{\partial \tau_{yz}}{\partial x} = \frac{\partial^2 \varphi}{\partial y^2} - \left(-\frac{\partial^2 \varphi}{\partial x^2}\right) = G\alpha\left(\frac{\partial^2 \psi}{\partial x \partial y} - 1\right) - G\alpha\left(\frac{\partial^2 \psi}{\partial x \partial y} + 1\right) = -2G\alpha \tag{5.49}$$

The preceding equality will be stated differently into an inhomogeneous second-order partial differential equation, which is called the Poisson equation:

$$\nabla^2\phi = \frac{\partial^2\phi}{\partial y^2} + \frac{\partial^2\phi}{\partial x^2} = -2G\alpha \tag{5.50}$$

5.2.5.4 Boundary Conditions

Because zero surface forces are inspected, the traction vector $T = \left(T_x, T_y, T_z\right)$ on the boundary has zero components. Putting τ_{xy} and τ_{xz} from Equations (5.46) and (5.47) into equality $T_z = 0$:

$$T_z = \tau_{xz}n_x + \tau_{yz}n_y = \frac{\partial\phi}{\partial y}n_x - \frac{\partial\phi}{\partial x}n_y = \frac{\partial\phi}{\partial y}\frac{dy}{ds} + \frac{\partial\phi}{\partial x}\frac{dx}{ds} = \frac{\partial\phi}{\partial s} = 0 \tag{5.51}$$

Thus Equation (5.46) requires the tangent derivative of φ, which is equal to zero, and thus φ is constant along each component of the boundary.

5.2.5.5 Torque, Section Moment, and Shear Stress

The torsional moment or torque M is computed as the double integral over the cross section:

$$M = \iint_{R^2}\left(-\tau_{xz}y + \tau_{yz}x\right)dxdy \tag{5.52}$$

where R^2 is the cross section of the bar.

$$M = -\iint_{R^2}\left(\frac{\partial\phi}{\partial y}y + \frac{\partial\phi}{\partial x}x\right)dxdy \tag{5.53}$$

Using the way of integration by components, we can attain the subsequent outcomes:

$$\iint_{R^2}\frac{\partial\phi}{\partial y}ydxdy = \int_{\Gamma}\phi y n_y ds - \iint_{R^2}\phi(x,y)dxdy = -\iint_{R^2}\phi(x,y)dxdy \tag{5.54}$$

$$\iint_{R^2}\frac{\partial\phi}{\partial x}xdxdy = -\iint_{R^2}\phi(x,y)dxdy \tag{5.55}$$

Thus we may attain the torsional moment:

$$M = 2\iint_{R^2}\phi(x,y)\,dxdy \tag{5.56}$$

5.3 Stiffness of Mechanical Product—Vibration

5.3.1 Introduction

In this section, the vibrations of a mechanical system will be discussed [4]. We realize that all human beings require vibrations in one form or another to feel something. For example, we perceive through eardrum vibration, perceive by light wave vibration, breath by lung vibration, and talk through tongue vibration. Vibration may be expressed as a time-dependent displacement of a particle system or a particle with respect to an equilibrium position. It is a restoring force in a mechanical system that frequently pulls the system to its equilibrium location by the springing or elastic property of the system. Namely, as the spring is extended, energy is kept in reserve as a potential energy, and a spring force evolves. As the block is allowed to leave, the potential energy will be altered to kinetic energy so that the block is pulled to the equilibrium location. This motion will continually swing around the equilibrium position.

A system relating to or causing vibration is a dynamic system for which variables such as the input excitations and output response are time dependent. In addition to the initial conditions, it relies on the outer excitation. As most observed vibrating systems are very complex, inspecting all the characteristics for mathematical examination is not feasible. Only the crucial characteristics are inspected in the analysis to roughly calculate the system attributes under specified input states. The analysis of a vibrating system involves 1) mathematical modeling, 2) deduction of the governing equation, 3) solution of the equation, and 4) explanation of the outcome. Thus a dynamic system may be examined 1) to give some account for the motion, 2) to make a formal application for physics $\sum_X F_{ext} = M\ddot{x}$, or 3) to draw the forces in an orientation where they act (free body diagram), presuming position values. After that, we can set up the linearized N-DOF system: 1) $M\ddot{x} + C\dot{x} + Kx = F(t)$, 2) identify undamped frequenciesω_i and mode shapes $[u]$, and 3) invoke the modal expansion theorem: $x = [u]\, q(t) = \sum_{i=1}^{N} \{u\}^{(i)} q_i(t)$ and $M_i\ddot{q} + C_i\dot{q} + C_i q = Q_i(t)$.

To perceive vibration, some fundamental terms must be understood: (1) periodic motion is a movement that repeats itself after equal intervals of a period, (2) time period is a time needed to complete one cycle, (3) frequency is the number of cycles per unit of period, (4) amplitude is the maximum displacement of a vibrating body from its equilibrium location, (5) natural frequency is the frequency of free vibration with no outer excitation, and (6) degree of freedom is the minimum numeric of independent coordinates required to describe the movement of a system at any given moment.

5.3.2 Analysis of the Free Vibratory Motion in Mechanical Systems

We will discuss the response of only one DOF system in which movement is ruled by a second-order differential equation. As vibration happens, the system has one natural frequency. Only two variables, position and velocity, are required to state the movement of the system. Thus numerous structures may be modeled as single DOF systems. Perceiving how only one degree-of-freedom system responds supplies the engineer with a critical understanding of the basic behavior of general systems. Firstly, the response of a single DOF system will be defined with free vibration.

The motion of equation for the vibration of the mechanical system (Figure 5.14) is

$$\Sigma F_x = M\ddot{x} = K_1\left(\frac{L}{2} - x\right) - K_2\left(\frac{L}{2} + x\right) \text{ or } M\ddot{x} + (K_1 + K_2)x = \frac{L}{2}(K_1 - K_2) \quad (5.57)$$

Displacement in x can be static x_s or dynamic x_d:

Let $x = x_s + x_d$ where $x_s = \left[\frac{\frac{L}{2}(K_1 - K_2)}{K_1 + K_2}\right]$. We know that $\dot{x} = \dot{x}_d$ or

$\ddot{x} = \ddot{x}_d$

So Equation (5.57) is restated:

$$M\ddot{x}_d + (K_1 + K_2)x_d + (K_1 + K_2)x_s = \frac{L}{2}(K_1 - K_2) \quad (5.58)$$

As $(K_1 + K_2)x_s = \frac{L}{2}(K_1 - K_2)$, Equation (5.58) is

$$M\ddot{x}_d + \left(K_{eq}\right)x_d = 0 \text{ where } K_{eq} = (K_1 + K_2) \quad (5.59)$$

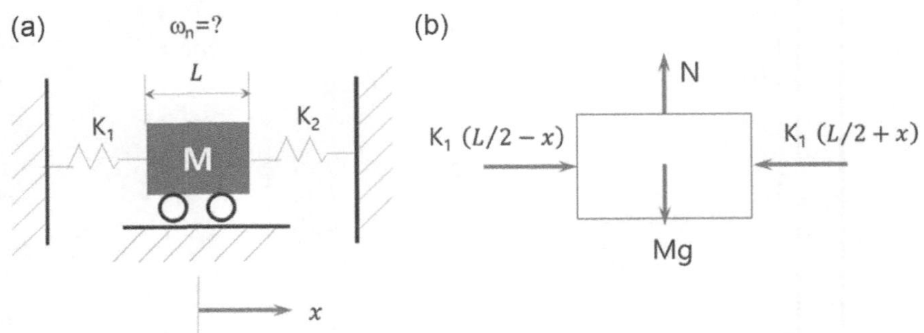

FIGURE 5.14
Vibration of the mechanical system without damping: (a) mechanical system; (b) free body diagram.

If $x_d(t) = A_1 cos(\omega t) + B_1 sin(\omega t)$, substitution into Equation (5.59) gives

$$\underbrace{(-M\omega^2 + K_{eq})}_{=0}\underbrace{(A_1 cos(\omega t) + B_1 sin(\omega t)}_{\neq 0} = 0 \tag{5.60}$$

The undamped natural frequency is

$$\omega_n = \sqrt{\frac{K_{eq}}{M}} \tag{5.61}$$

If $x_d(t) = x(t=0)$ in Figure 5.15, $A_1 = x_0$

If $\dot{x} = -A_1 \omega sin(\omega t) + B_1 \omega sin(\omega t)$ and $v_0 = \dot{x}(t=0) = B_1 \omega_n$, $B_1 = \frac{v_0}{\omega_n}$

$$x(t) = x_0 cos(\omega t) + \frac{v_0}{\omega_n} sin(\omega t) = A cos(\omega_n t - \phi) \tag{5.62}$$

where $A = \sqrt{x_0^2 + \left(\frac{v_0}{\omega_n}\right)^2}$, $\phi = tan^{-1}\left(\frac{B}{A}\right) = tan^{-1}\left(\frac{v_0}{x_0 \omega_n}\right)$

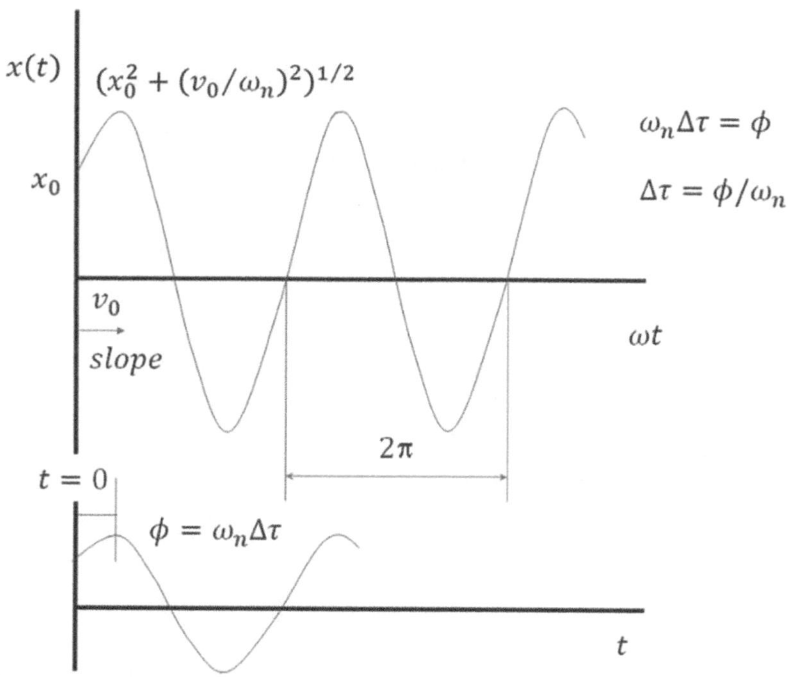

FIGURE 5.15
Steady-state solution $x(t)$.

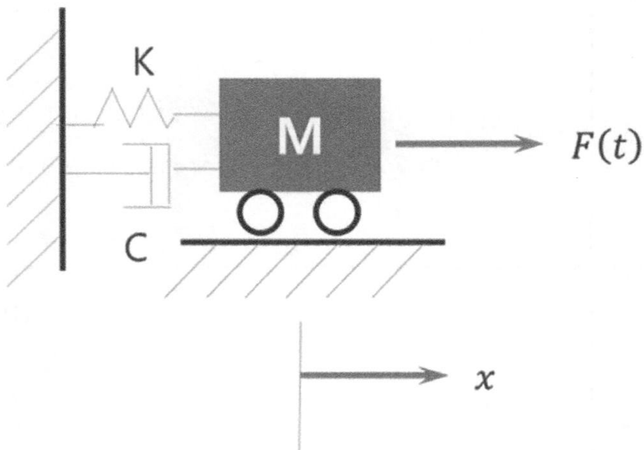

FIGURE 5.16
Vibration of the mechanical system with damping and outer force.

The motion of equation in Figure 5.16 is

$$M\ddot{x} + C\dot{x} + Kx = 0 \tag{5.63}$$

If $C\dot{x} - F(t) = 0$ and Equation (5.63) is solved, then $\omega_n = \sqrt{\frac{K}{M}}$.

Because $A\cos(\omega_n t + \phi) = \textit{real part of } Ae^{i(\omega_n t - \phi)}$ in Equation (5.63), $x(t) = e^{-st}$ may be presumed. Equation (5.63) can be altered:

$$\left(Ms^2 + Cs + K\right)e^{-st} = 0 \text{ or } s^2 + \frac{C}{M}s + \frac{K}{M} = 0 \tag{5.64}$$

The roots in Equation (5.64) are

$$s_{1,2} = -\frac{C}{2M} \pm \sqrt{\left(\frac{C}{2M}\right)^2 - \frac{K}{M}} = -\left(\frac{C}{2M\omega_n}\right)\omega_n \pm \omega_n\sqrt{\left(\frac{C}{2M\omega_n}\right)^2 - 1} \tag{5.65}$$
$$= -\zeta\omega_n \pm \omega_n\sqrt{\zeta^2 - 1}$$

where $\zeta = \frac{C}{2M\omega_n}$.

There are several cases in Equation (5.65):

$$st = \begin{cases} -\zeta\omega_n t \pm \omega_n\sqrt{\zeta^2 - 1}\,t \textit{ for } \zeta > 1 \\ -\omega_n t \textit{ for } \zeta = 1 \\ -\zeta\omega_n t \pm i\omega_n\sqrt{1 - \zeta^2}\,t \textit{ for } \zeta < 1 (\textit{oscillation}) \end{cases} \tag{5.66}$$

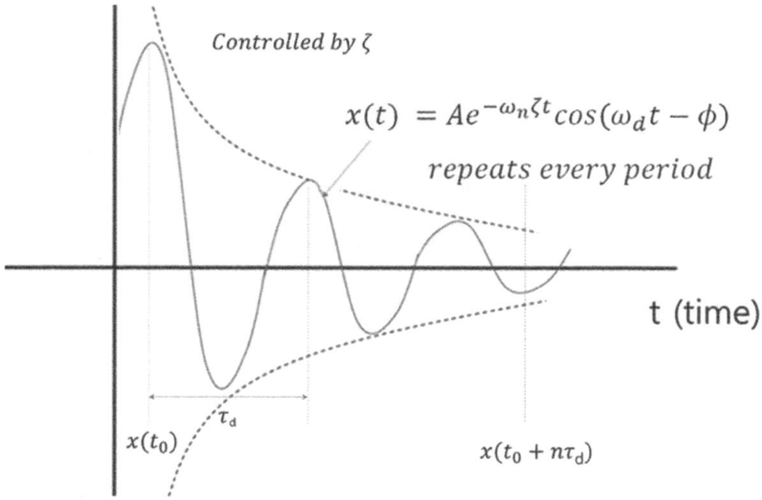

FIGURE 5.17
Homogeneous solution $x(t)$ for $\zeta<1$.

The response$x(t)$ for$\zeta < 1$ (Figure 5.17) is

$$x(t) = Ae^{-\zeta\omega_n t}\cos(\omega_d t - \phi) \tag{5.67}$$

where $\omega_d \equiv$ damped natural frequency $= \omega_n\sqrt{1-\zeta^2}$, $\phi \equiv$ phased angle

Equation (5.67) is redefined:

$$x(t) = \left[\frac{x_0}{\sqrt{1-\zeta^2}}\cos(\omega_d t - \Psi) + \frac{v_0}{\omega_d}\sin(\omega_d t)\right]e^{-\zeta\omega_n t} \tag{5.68}$$

$$x(t) = \left[x_0\cos(\omega_d t) + \frac{v_0}{\omega_d}\sin(\omega_d t)\right]e^{-\zeta\omega_n t} \textit{ for small damping } (\zeta < 0.1) \tag{5.69}$$

The ratio $x(t)/x(t+n\tau_d)$ from Equation (5.67) is

$$\frac{x(t)}{x(t+n\tau_d)} = \frac{e^{-\omega_n\zeta t_0}}{e^{-\omega_n\zeta(t_0+n\tau_d)}} = e^{\omega_n\zeta n\tau_d} \tag{5.70}$$

If Equation (5.70) takes the logarithm,

$$\ln\frac{x(t)}{x(t+n\tau_d)} = \omega_n\zeta n\tau_d = \omega_n\zeta n\frac{2\pi}{\omega_d} = \omega_n\zeta n\frac{2\pi}{\omega_n\sqrt{1-\zeta^2}} \tag{5.71}$$

If ζ is small, $\sqrt{1-\zeta^2} \approx 1$. Equation (5.71) is

$$\ln \frac{x(t)}{x(t+n\tau_d)} = n2\pi\zeta \tag{5.72}$$

If Equation (5.72) is rearranged, damping ζ is

$$\zeta = \frac{1}{2\pi n} ln \frac{x(t)}{x(t+n\tau_d)} \tag{5.73}$$

If n is such that $x(t+n\tau) = 50\%\, of\, x(t)$, Equation (5.73) is

$$\zeta = \frac{1}{2\pi n_{50\%}} ln(2) = \frac{0.11}{n_{50\%}} \tag{5.74}$$

Example 5.1: Damping Analysis for a Straightforward Bar

For an uncomplicated bar in Figure 5.18, when 4 or 8 cycles begin to be halved in amplitude, determine the damping proportion.

Solution

$$\theta(t) \cong \left(\theta_0 \cos(\omega_d t) + \frac{\dot{\theta}_0}{\omega_d} \sin\omega_d t\right) e^{-\zeta\omega_n t}$$

$$\theta(t) = \frac{\dot{\theta}_0}{\omega_d} \sin\omega_d t$$

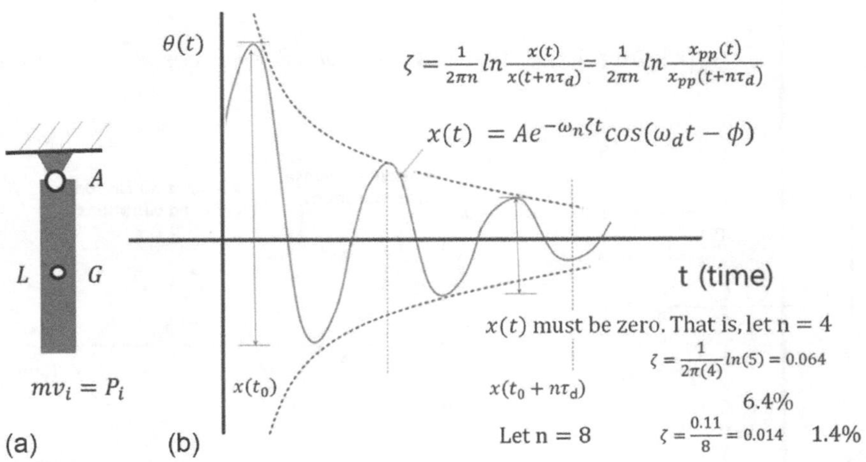

FIGURE 5.18
Straightforward bar: (a) model; (b) response.

$$I_{zz} = \frac{ML^2}{3}$$

$$\zeta = \frac{c}{2m\omega_n} \Rightarrow \zeta = \frac{c_t}{2\left(I_{zz} + mL^2\right)\omega_n}$$

$$^A H_2 = {}^A H_1 = {}^A\left(I_{zz} + mL^2\right)\dot{\theta}_0 = mv_i L$$

$$\dot{\theta}_0 = \left(\frac{mv_i L}{I_{zz} + mL^2}\right)$$

$$^A H_i = r \times P_i = mv_i L\hat{I}_z$$

If $n = 4, \zeta = \frac{1}{2\pi n} ln(5) = 0.064$ or 6.4%.

If $n = 8, \zeta = \frac{0.11}{8} = 0.014$ or 1.4%.

5.3.3 Analysis of the Forced Vibration Motion in Mechanical Systems

For forced vibration, pay particular attention to harmonic excitation of the form $\cos(\omega t)$ (outer excitation) and steady-state response. The transfer function $H_{X/F}(\omega)$ is shown in Figure 5.19,

The motion of equation in Figure 5.19 is

$$M\ddot{x} + C\dot{x} + Kx = F(t) = F_o \cos(\omega t) \tag{5.75}$$

Presume that the steady-state solution $x_{s.s.} = x_0 \cos(\omega t - \phi)$.
Plug in Equation (5.75)

$$x_0 \left[\left(k - M\omega^2\right)\cos(\omega t - \phi) - C\omega sin(\omega t - \phi)\right] = F_0 \cos \omega t \tag{5.76}$$

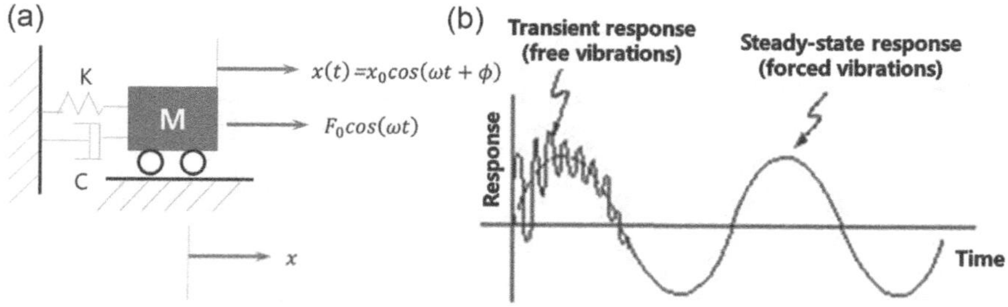

Figure 5.19
Vibration of the mechanical system with outer force $F_o\cos(\omega\tau)$: (a) model with forced vibration; (b) response.

By dividing K in Equation (5.75), we obtain

$$x_0\left[\left(1-\frac{\omega^2}{\omega_n^2}\right)\cos(\omega t-\phi)-2\zeta\frac{\omega}{\omega_n}\sin(\omega t-\phi)\right]=\frac{F_0}{K}\cos\omega t \tag{5.77}$$

where $\omega_n=\sqrt{\frac{K}{M}}$, $\zeta=\frac{C}{2M\omega_n}$.

Because $\cos(\omega t-\phi)=\cos\omega t\cos\phi+\sin\omega t\sin\phi$ and $\sin(\omega t-\phi)=\sin\omega t\cos\phi-\cos\omega t\sin\phi$, we insert them into Equations (5.76) and (5.77):

$$x_0\left[\left(1-\frac{\omega^2}{\omega_n^2}\right)\cos\phi-2\zeta\frac{\omega}{\omega_n}\sin\phi\right]\cos\omega t=\frac{F_0}{K}\cos\omega t \tag{5.78}$$

$$x_0\left[\left(1-\frac{\omega^2}{\omega_n^2}\right)\sin\phi-2\zeta\frac{\omega}{\omega_n}\cos\phi\right]\sin\omega t=0 \tag{5.79}$$

The solutions are

$$x_0=\frac{\frac{F_0}{k}}{\left[\left(1-\frac{\omega^2}{\omega_n^2}\right)^2+\left(2\zeta\frac{\omega}{\omega_n}\right)^2\right]^{\frac{1}{2}}}=\left(\frac{F_0}{k}\right)\left|H_{X/F}(\omega)\right|e^{-i\phi}\text{ and }\phi=\tan^{-1}\left(\frac{2\zeta\frac{\omega}{\omega_n}}{1-\frac{\omega^2}{\omega_n^2}}\right) \tag{5.80}$$

If $x_{static}=\frac{F_0}{k}$, $\left|H_{X/F}(\omega)\right|(=M)=\frac{x_0}{x_{static}}=\frac{x_0}{\frac{F_0}{k}}=\frac{\frac{F_0}{k}}{\left[\left(1-\frac{\omega^2}{\omega_n^2}\right)^2+\left(2\zeta\frac{\omega}{\omega_n}\right)^2\right]^{\frac{1}{2}}}$ from Equation (5.80)

The variation of M and ϕ with the frequency proportion $r=\frac{\omega}{\omega_n}$ is plotted as shown in Figure 5.20.

5.3.4 Vibration Isolation of Mechanical Systems for Robust Design

The vibration design of a mechanical product is systemically attached to the vibration isolation, which reduces the disagreeable results of vibration (robust design). Thus proposed strategies for a mechanical design need the introduction of a mounting rubber between the source of vibration and the mechanical product so that a reduction in the dynamic response of the system is achieved under the defined conditions of vibration excitation.

Vibration isolation will be used when 1) the foundation of the vibrating machinery is fixed against its big unstable forces and 2) the product in

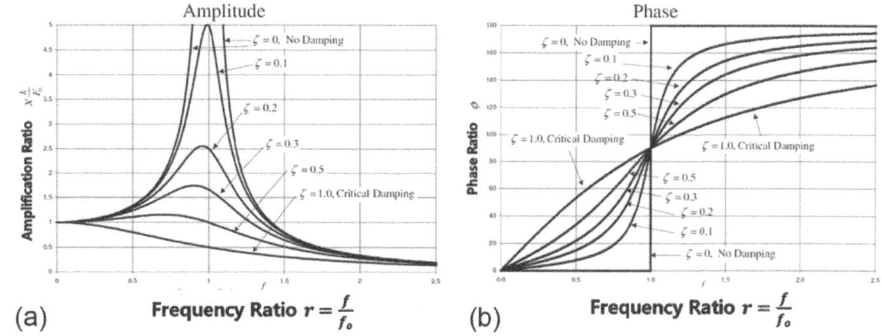

FIGURE 5.20
Variation of M and ϕ with frequency proportion r: (a) amplification ratio; (b) phase ratio.

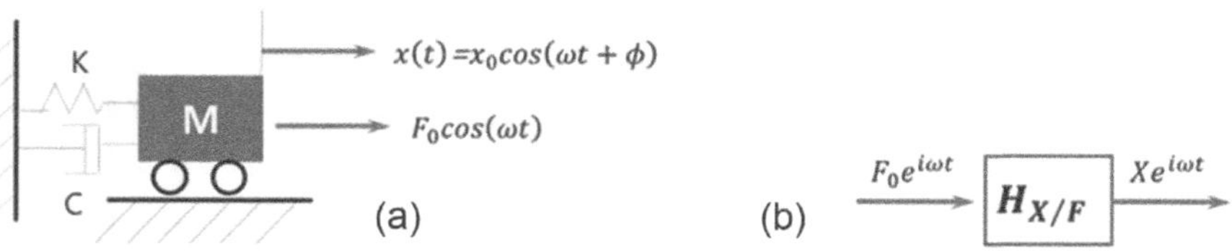

FIGURE 5.21
(a) Model and (b) steady-state response (transfer function) of mechanical system.

transportation is fixed from environmental situations. The first type of isolation is used when a mechanical system is exposed to an excitation force. On the other hand, the second pattern of isolation may be discovered by chance: When a household refrigerator is in transit, it is exposed to random vibration from the foundation. Vibration might be isolated as the refrigerator properly utilizes the mounting rubber. Vibration (or noise) may also have a feeling of discomfort to consumers.

The response of a mechanical system with mass M, damping C, and spring K in the motion of equation (Figure 5.21) is defined:

$$M\ddot{x}+C\dot{x}+Kx=F(t)=F_0\cos(\omega t)=\text{Real part}\left[F_0e^{i\omega t}\right] \tag{5.81}$$

First, the steady-state response is

$$x(t)=x_0\cos(\omega t-\phi)=\text{Re}\left[x_0e^{i(\omega t-\phi)}\right]=\text{Re}\left[\underbrace{x_0e^{-i\phi}}_{X}\cdot e^{i\omega t}\right]=\text{Re}\left[Xe^{i\omega t}\right] \tag{5.82}$$

By inserting Equation (5.82) into Equation (5.81), we reach at

$$\left[\left(K-M\omega^2\right)+ic\omega\right]Xe^{i\omega t}=F_0e^{i\omega t} \tag{5.83}$$

From Figure 5.21(b), steady state response is

$$H_{X/F}(\omega) = \frac{X}{F_0} = \frac{1}{\left[\left(K - M\omega^2\right) + ic\omega\right]} = \frac{1/K}{\left[\left(1 - \frac{\omega^2}{\omega_n^2}\right) + 2i\zeta\frac{\omega}{\omega_n}\right]} = \left|H_{X/F}(\omega)\right| e^{-i\phi} \quad (5.84)$$

where $\left|H_{X/F}(\omega)\right| = \frac{\frac{1}{K}}{\left[\left(1 - \frac{\omega^2}{\omega_n^2}\right)^2 + \left(2\zeta\frac{\omega}{\omega_n}\right)^2\right]^{\frac{1}{2}}}$, $\phi = \tan^{-1}\left(\frac{2\zeta\frac{\omega}{\omega_n}}{1 - \frac{\omega^2}{\omega_n^2}}\right)$.

Assume $y \ll x$, the transmitted force F_t from the foundation excitation in Figure 5.22 is

$$F_t = C\dot{x} + Kx = [K + i\omega C]Xe^{i\omega t} \quad (5.85)$$

In this case, the response $\frac{F_t}{F_o}$ is

$$\frac{F_t}{F_o} = \frac{[K + i\omega C]Xe^{i\omega t}}{F_o e^{i\omega t}} = [K + i\omega C]H_{X/F}(\omega) \quad (5.86)$$

Transfer function is

$$\left|H_{\frac{F_t}{F_o}}\right| = \left|H_{X/Y}(\omega)\right| = \frac{\left[1 + \left(2\zeta\frac{\omega}{\omega_n}\right)^2\right]}{\left[\left(1 - \frac{\omega^2}{\omega_n^2}\right)^2 + \left(2\zeta\frac{\omega}{\omega_n}\right)^2\right]^{\frac{1}{2}}} \quad (5.87)$$

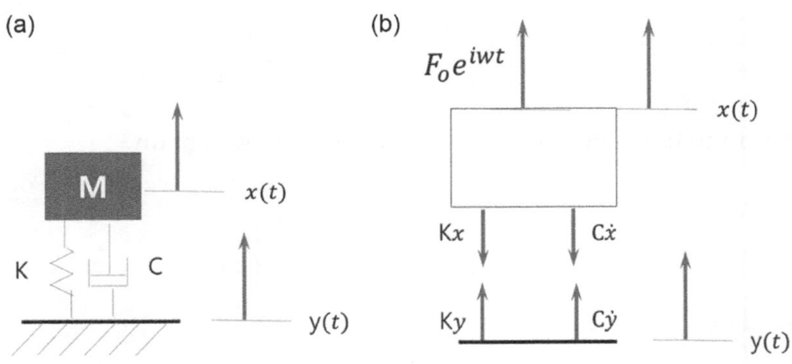

FIGURE 5.22
Vibratory mechanical system excited from the foundation: (a) model; (b) free body diagram.

Example 5.2: Vibration Insulation

As the motor (source @ 20 Hz) is fastened on the ending of the bar (whole weight = 20 kg), some vibration is designed to lessen 90%. Determine $\frac{\omega}{\omega_n}$ such that $\left|H_{X/y}\right| = 0.1$

Solution

$$\text{For } \zeta = 0,\ \left|H_{X/y}\right| = \frac{1}{\left|1-\frac{\omega^2}{\omega_n^2}\right|} = \frac{1}{\frac{\omega^2}{\omega_n^2}-1} = 0.1 \Rightarrow \frac{\omega}{\omega_n} = x = \sqrt{11} = 3.31$$

$$\text{If } f = 20 \text{ Hz},\ f_n = \frac{f}{3.31} = \frac{20Hz}{3.31} = 6.04\text{Hz},\ \omega_n = 37.96\frac{rad}{s} = \sqrt{\frac{K}{M}}.$$

If M = 20 kg, K=28,827 N/m

Example 5.3

An exhaust fan revolving at 1,000 RPM is to be borne by four springs, each possessing a stiffness of K. As only 10% of the unbalanced force of the fan is to be transferred to the foundation, what might be the value of K? Presume the mass of the exhaust fan to be 40 kg.

Solution

As the transmissibility must be 0.1, the forcing frequency is specified by

$$\omega = (1{,}000 \times 2\pi)/60 = 104.72 \text{ rad/s}$$

The natural frequency of the system is specified by

$$\omega_n = \left(\frac{k}{m}\right)^{1/2} = \left(\frac{4K}{40}\right)^{1/2} = \frac{\sqrt{K}}{3.1623}$$

By assuming the damping proportion to be $\zeta = 0$, we obtain

$$T_f = 0.1 = \frac{1}{r^2-1} = \frac{1}{\left(\frac{104.72 \times 3.1623}{\sqrt{K}}\right)^2 - 1}$$

$$\frac{331.1561}{\sqrt{K}} = 3.3166 \qquad K = 9.97 \text{ N/m}$$

5.3.5 Multiple Degree-of-Freedom (MDOF) Systems

Multiple DOF systems are expressed as those necessitating two or more coordinates to express their movement. Namely, MDOF systems need N equations of motion (N mode shapes) which is made up of ordinary DOF differential equations. Multi-DOF systems fall into two groups: 1) straightforward lumped-parameter systems, consisting of only a few DOF, and 2) structures that are now almost always established by utilizing the finite element method.

The motion of equation for multi-DOF systems with free motion (no damping) is

$$[M]\{\ddot{X}\}+[K]\{X\}=0 \tag{5.88}$$

The solutions of MDOF are presumed:

$$\{X\}=\begin{Bmatrix} x_1(t) \\ \vdots \\ x_N(t) \end{Bmatrix}=\begin{bmatrix} a_1 \\ \vdots \\ a_N \end{bmatrix}\cos(\omega_i t-\phi_i) \tag{5.89}$$

Doubly differentiating Equation (5.89):

$$\{\ddot{X}\}=\begin{Bmatrix} \ddot{x}_1(t) \\ \vdots \\ \ddot{x}_N(t) \end{Bmatrix}=-\omega^2\begin{bmatrix} a_1 \\ \vdots \\ a_N \end{bmatrix}\cos(\omega t-\phi) \tag{5.90}$$

After plugging in Equation (5.88):

$$-\omega^2[M]\begin{bmatrix} a_1 \\ \vdots \\ a_N \end{bmatrix}\cos(\ddot{\omega}t-\phi)+[K]\begin{bmatrix} a_1 \\ \vdots \\ a_N \end{bmatrix}\cos(\omega_i t-\phi_i)=0 \text{ or } \left[-\omega^2[M]+[K]\right]=0 \tag{5.91}$$

After Equation (5.91) is multiplied by $[M]^{-1}$:

$$-\omega^2\begin{bmatrix} a_1 \\ \vdots \\ a_N \end{bmatrix}\cos(\ddot{\omega}t-\phi)+\underbrace{[M]^{-1}[K]}_{[A]}\begin{bmatrix} a_1 \\ \vdots \\ a_N \end{bmatrix}\cos(\omega_i t-\phi_i)=0 \tag{5.92}$$

$$[A]\begin{bmatrix} a_1 \\ \vdots \\ a_N \end{bmatrix}\cos(\omega_i t-\phi_i)=\omega^2\begin{bmatrix} a_1 \\ \vdots \\ a_N \end{bmatrix}\cos(\ddot{\omega}t-\phi) \text{ or } [A]X=\omega^2X \Leftrightarrow \underbrace{[A-\omega^2 I]}_{det(A)=0}X=0 \tag{5.93}$$

The natural frequencies from the characteristic Equation (5.93) are

$$A-\omega^2 I=0 \Rightarrow \omega^2=? \tag{5.94}$$

The equation of motion of the 2-DOF mechanical system in Figure 5.23 is

$$M_1\ddot{x}_1+(K_1+K_2)x_1-K_2x_2=F_1(t)=0 \tag{5.95}$$

$$M_2\ddot{x}_2-K_1x_1+K_2x_2=F_1(t)=0 \tag{5.96}$$

Equations (5.95) and (5.96) may be defined in matrix form:

$$\begin{bmatrix} M_1 & 0 \\ 0 & M_2 \end{bmatrix}\begin{Bmatrix} \ddot{x}_1 \\ \ddot{x}_2 \end{Bmatrix}+\begin{bmatrix} K_1+K_2 & -K_2 \\ -K_2 & K_2 \end{bmatrix}\begin{Bmatrix} x_1 \\ x_2 \end{Bmatrix}=\begin{Bmatrix} F_1(t) \\ F_2(t) \end{Bmatrix}=0 \text{ or } M\ddot{X}+KX=0 \tag{5.97}$$

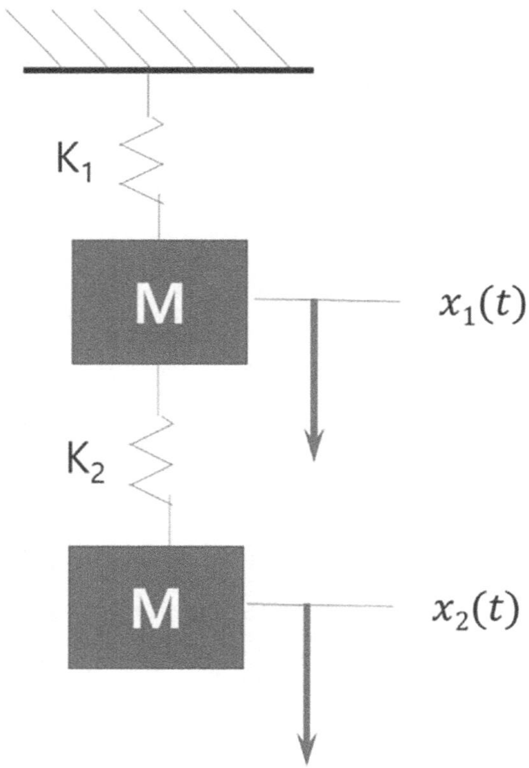

FIGURE 5.23
Two-degree-of-freedom (2-DOF) mechanical system (instance).

To solve for ω_1 & ω_2 (mode shape), think about a special case: 1) $M_1 = M_2 = M$ and 2) $K_1 = K_2 = K$. Equation (5.97) is

$$\begin{bmatrix} M & 0 \\ 0 & M \end{bmatrix} \begin{Bmatrix} \ddot{x}_1 \\ \ddot{x}_2 \end{Bmatrix} + \begin{bmatrix} 2K & -K \\ -K & K \end{bmatrix} \begin{Bmatrix} x_1 \\ x_2 \end{Bmatrix} = \begin{bmatrix} 0 \\ 0 \end{bmatrix} \tag{5.98}$$

Presume the masses oscillate with the same frequency ω and dissimilar amplitudes a_1 and a_2. The solutions are presumed:

$$\begin{Bmatrix} x_1 \\ x_2 \end{Bmatrix} = \begin{bmatrix} a_1 \\ a_2 \end{bmatrix} \cos(\omega t - \phi) \tag{5.99}$$

Doubly differentiating Equation (5.99):

$$\begin{Bmatrix} \ddot{x}_1 \\ \ddot{x}_2 \end{Bmatrix} = -\omega^2 \begin{bmatrix} a_1 \\ a_2 \end{bmatrix} \cos(\omega t - \phi) \tag{5.100}$$

Substituting into Equation (5.100):

$$\begin{bmatrix} -M\omega^2 & 0 \\ 0 & -M\omega^2 \end{bmatrix} \begin{bmatrix} a_1 \\ a_2 \end{bmatrix} \cos(\omega t - \phi) + \begin{bmatrix} 2K & -K \\ -K & K \end{bmatrix} \begin{bmatrix} a_1 \\ a_2 \end{bmatrix} \cos(\omega t - \phi) = \begin{bmatrix} 0 \\ 0 \end{bmatrix} \tag{5.101}$$

Dividing Equation (5.101) by $\cos(\omega t - \phi)$:

$$\underbrace{\begin{bmatrix} -M\omega^2 + 2K & -K \\ -K & -M\omega^2 + K \end{bmatrix}}_{det[A]=0} \begin{bmatrix} a_1 \\ a_2 \end{bmatrix} = \begin{bmatrix} 0 \\ 0 \end{bmatrix} \tag{5.102}$$

Solving det[A] for a_1, a_2:

$$\det[A] = \det\begin{pmatrix} -M\omega^2 + 2K & -K \\ -K & -M\omega^2 + K \end{pmatrix} = 0 \text{ or } M^2\omega^4 - 3KM\omega^2 + K^2 = 0 \tag{5.103}$$

From Equation (5.103), the natural frequencies are

$$\omega^2 = \left(\frac{3 \pm \sqrt{5}}{2} \right) \frac{K}{M} \tag{5.104}$$

From the first row of Equation (5.103), we can attain the amplitude ratio:

$$\left(-M\omega^2 + 2K\right) a_1 = K a_2 \text{ or } \frac{a_1}{a_2} = \frac{K}{\left(-M\omega^2 + 2K\right)} \tag{5.105}$$

For $\omega^2 = \left(\frac{3-\sqrt{5}}{2}\right)\frac{K}{M} \Rightarrow \omega_1 = 0.618\sqrt{\frac{K}{M}}$, the amplitude ratio is

$$\frac{a_1}{a_2} = 0.618 \tag{5.106}$$

For $\omega^2 = \left(\frac{3+\sqrt{5}}{2}\right)\frac{K}{M} \Rightarrow \omega_2 = 1.618\sqrt{\frac{K}{M}}$, the amplitude ratio is

$$\frac{a_1}{a_2} = -1.618 \tag{5.107}$$

These are natural modes. That is, we realize the facts: 1) ratios, not absolute magnitudes; 2) each natural mode is with a specific natural frequency ω_n; 3) these explain the circumstances in which the whole system is swinging at one frequency; and 4) natural modes may be thought as special initial circumstances that bring the system to decouple into independent subsystems.

$$\begin{Bmatrix} x_1 \\ x_2 \end{Bmatrix} = A_1 \begin{Bmatrix} 0.618 \\ 1 \end{Bmatrix} \cos(\omega_1 t - \phi_1) + A_2 \begin{Bmatrix} 0.618 \\ -1 \end{Bmatrix} \cos(\omega_2 t - \phi_2) \tag{5.108}$$

where A_1, A_2, ϕ_1, ϕ_2 are determined by initial conditions.

5.3.6 Modal Analysis of Vibration Systems (Mode Superposition)

Modal analysis can decide the vibration characteristics (natural frequencies and mode shapes) of a mechanical structure or component, manifesting the motion of dissimilar components of the structure under loading circumstances. The motion of equation is

$$M\ddot{X} + C\dot{X} + KX = F(t) \tag{5.109}$$

The solution of Equation (5.109) is presumed:

$$X = uq(t) \tag{5.110}$$

where X is the generalized coordinate, and $q(t)$ is the modal coordinate.

By the modal expansion theorem, Equation (5.110) is stated differently:

$$X = \{u\}^{(n)} q_n(t) = \left[\{u\}^{(1)} \quad \cdots \quad \{u\}^{(n)}\right] \begin{Bmatrix} q_1(t) \\ \cdots \\ q_n(t) \end{Bmatrix} \tag{5.111}$$

For the 2-DOF system:

$$\{x(t)\} = \{u\}^{(1)} q_1(t) + \{u\}^{(2)} q_2(t) \tag{5.112}$$

Differentiating Equation (5.112):

$$\dot{X} = u\dot{q}(t) \tag{5.113}$$

$$\ddot{X} = u\ddot{q}(t) \tag{5.114}$$

After plugging them into Equation (5.109) for $X, \dot{X}, \ddot{X}$ and multiplying by u^T:

$$u^T Mu\ddot{q}(t) + u^T Cu\dot{q}(t) + u^T Kuq(t) = u^T F \text{ for ideal conditions} \tag{5.115}$$

We realize that mode shapes are at right angles to one another.

Because $u^T Mu = \begin{bmatrix} \ddots & & \\ & M_i & \\ & & \ddots \end{bmatrix}, u^T Ku = \begin{bmatrix} \ddots & & \\ & K_i & \\ & & \ddots \end{bmatrix}, u^T Cu = \begin{bmatrix} \ddots & & \\ & C_i & \\ & & \ddots \end{bmatrix}$, then

$$[M]\ddot{q} + [C]\dot{q} + [K]q = u^T F = Q(t) \text{ (N independent DOF systems)} \tag{5.116}$$

Start with the response to initial conditions:

$$X = uq(t) \text{ or } X = \sum_i \{u\}^i q_i(t) \text{ (practical for critical modes)} \tag{5.117}$$

From the solution form of 1-DOF system, $q_i(t)$ may be anticipated:

$$x(t) = \left[x_0 \cos(\omega_d t) + \frac{(v_0 + \zeta\omega_n x_0)}{\omega_d} \sin(\omega_d t)\right] e^{-\zeta\omega_n t} \tag{5.118}$$

Likewise, the forms of *N* modal solutions are

$$q_1(t) = \left[q_{10} \cos(\omega_{1d} t) + \frac{(\dot{q}_{10} + \zeta_1\omega_1 q_{10})}{\omega_d} \sin(\omega_{1d} t)\right] e^{-\zeta_1\omega_1 t} \tag{5.119}$$

$$q_2(t) = \left[q_{20} \cos(\omega_{2d} t) + \frac{(\dot{q}_{20} + \zeta_2\omega_2 q_{20})}{\omega_d} \sin(\omega_{2d} t)\right] e^{-\zeta_2\omega_2 t} \tag{5.120}$$

The (Rayleigh) damping model is defined:

$$[C] = \alpha[M] + \beta[K] \tag{5.121}$$

$$u^T Cu = \alpha u^T Mu + \beta u^T Ku = \begin{bmatrix} \ddots & & \\ & C_i & \\ & & \ddots \end{bmatrix} = \alpha \begin{bmatrix} \ddots & & \\ & M_i & \\ & & \ddots \end{bmatrix} + \beta \begin{bmatrix} \ddots & & \\ & K_i & \\ & & \ddots \end{bmatrix} \quad (5.122)$$

For 2 DOF, the model is

$$C_1 = \alpha M_1 + \beta K_1 \quad (5.123)$$

$$C_2 = \alpha M_2 + \beta K_2 \quad (5.124)$$

$$x(t) = \sum_{i=1}^{n} \{u\}^{(1)} q_1 + \{u\}^{(2)} q_2 + \cdots + \{u\}^{(n)} \quad (5.125)$$

5.3.7 Response of 2-DOF Systems by the Use of Transfer Functions

To distinguish the response of an *N*-DOF system, such as 2-DOF systems to forced vibrations, it is practical to specify a transfer function or frequency response function between the input and output of the system.

The motion of equation in Figure 5.24(a) is

$$(m_1 + m_2)\ddot{x} + b\dot{x} + kx = -\frac{1}{2} m_2 L \left(\ddot{\theta} \cos\theta - \dot{\theta}^2 \sin\theta \right) \quad (5.126)$$

If $\dot{\theta} = \omega = constant$, Equation (5.126) is

$$(m_1 + m_2)\ddot{x} + b\dot{x} + kx = F(t) = \frac{1}{2} m_2 L \omega^2 \sin(\omega t) \quad (5.127)$$

The motion of equation in Figure 5.24(b) is

$$M\ddot{x} + C\dot{x} + Kx = F(t) = \begin{Bmatrix} F_1 \\ F_2 \end{Bmatrix} e^{i\omega t} \quad (5.128)$$

If $x = Xe^{i\omega t}$, $\dot{x} = i\omega X e^{i\omega t}$ and $\ddot{x} = -\omega^2 X e^{i\omega t}$

If plugging them into Equation (5.128):

$$\underbrace{[-\omega^2 M + i\omega C + K]}_{Z(\omega)\rightarrow impedence\ matrix} Xe^{\ i\omega t} = F_e^{\ i\omega t} \quad (5.129)$$

where $Z(\omega) = -\omega^2 \begin{bmatrix} m_1 & 0 \\ 0 & m_2 \end{bmatrix} + i\omega \begin{bmatrix} c_1 + c_2 & -c_2 \\ -c_2 & c_2 \end{bmatrix} + \begin{bmatrix} k_1 + k_2 & -k_2 \\ -k_2 & k_2 \end{bmatrix} = \begin{bmatrix} Z_{11} & Z_{12} \\ Z_{21} & Z_{22} \end{bmatrix}$

$$ZX = F \text{ or } X = Z^{-1} F \equiv HF \quad (5.130)$$

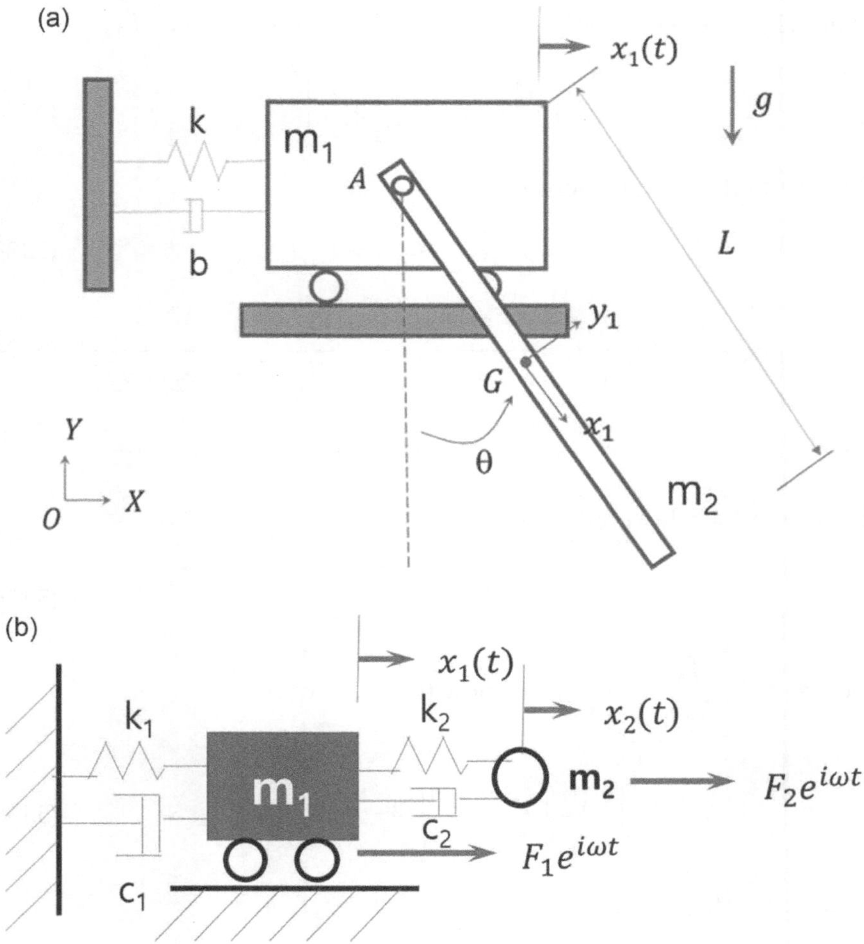

FIGURE 5.24
2-DOF systems (example): (a) with pendulum; (b) with small mass.

where H is $N \times N$ matrix of transfer functions, $H_{ij} \equiv$ response at i per unit force j

For $N = 2(2 \times 2)$, the inverse Z^{-1} is

$$Z^{-1} = \frac{\begin{bmatrix} Z_{22} & -Z_{12} \\ -Z_{21} & Z_{11} \end{bmatrix}}{det|Z|} = \frac{\begin{bmatrix} Z_{22} & -Z_{12} \\ -Z_{21} & Z_{11} \end{bmatrix}}{Z_{11}Z_{22} - Z_{12}^{\ 2}} = \begin{bmatrix} H_{11} & H_{12} \\ H_{21} & H_{22} \end{bmatrix} \tag{5.131}$$

where $Z_{11} = -\omega^2 m_{11} + i\omega C_{11} + K_{11} = -\omega^2 m_1 + i\omega(c_1 + c_2) + k_1 + k_2$.

$$\begin{Bmatrix} x_1 \\ x_2 \end{Bmatrix} = \begin{bmatrix} H_{11} & H_{12} \\ H_{21} & H_{22} \end{bmatrix} \begin{Bmatrix} F_1 \\ F_2 \end{Bmatrix} \tag{5.132}$$

If $F_2 = 0$ and $C = 0$, response x_1 is

$$x_1 = H_{11}F_1 + H_{12}F_2 = H_{11}F_1 \tag{5.133}$$

$$x_1 = \frac{Z_{22}}{Z_{11}Z_{22} - Z_{12}^2} F_1 = \frac{\left(-\omega^2 m_2 + k_2\right)}{\left[\left(-\omega^2 m_1 + k_1 + k_2\right)\left(-\omega^2 m_2 + k_2\right) - k_2^2\right]} F_1 \tag{5.134}$$

where $Z_{11} = -\omega^2 m_1 + k_1 + k_2$, $Z_{11} = -\omega^2 m_2 + k_2$, $Z_{12} = k_2$.

If $-\omega^2 m_2 + k_2 = 0$, $x_1 = 0$. That is,

$$\omega^2 = \frac{k_2}{m_2} \tag{5.135}$$

Because $\Omega_n = \sqrt{\frac{k_1}{m_1}} = \sqrt{\frac{k_2}{m_2}}$, we know that $x_1 = x_{1s}$:

$$\left|\frac{x_1(\omega)}{x_{1s}}\right| = \frac{F_1 \left|H_{11}(\omega)\right|}{F_1 / k_1} = k_1 \left|H_{11}(\omega)\right| \tag{5.136}$$

$x_2 = H_{21}F_1$ where $F_1 = 0$.

$$\begin{aligned} x_2\big|_{\omega=\Omega_N} &= \frac{-Z_{12}}{Z_{11}Z_{22} - Z_{12}^2} F_1 \\ &= \frac{-k_2 F_1}{\left[\left(-\omega^2 m_1 + k_1 + k_2\right)\left(\underbrace{-\omega^2 m_2 + k_2}_{0}\right) - k_2^2\right]} \\ &= \frac{F_1}{k_2} \end{aligned} \tag{5.137}$$

$$\frac{x_2}{x_{1s}} = \frac{F_1 / k_2}{F_1 / k_1} = \frac{k_1}{k_2} = \frac{m_1}{m_2} = \frac{1}{\alpha} \tag{5.138}$$

where μ ≡ mass ratio (Figure 5.25).

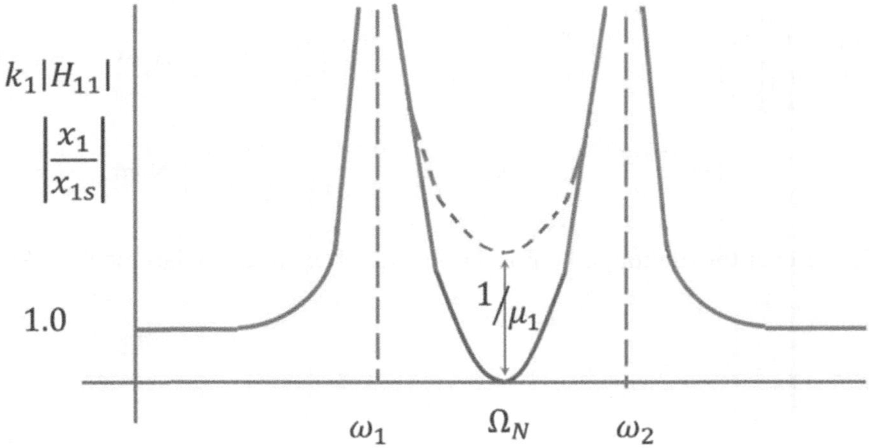

FIGURE 5.25
Response of 2-DOF systems (example).

Example 5.4 Modal Analysis 1

Solution

If $m_1 = 0.3193$ kg, $m_2 = 0.6309$ kg, $k_1 = 55.87$ N/m, and $k_2 = 36.047$ N/m in Figure 5.20, then carry out modal analysis of the system in Figure 5.26.

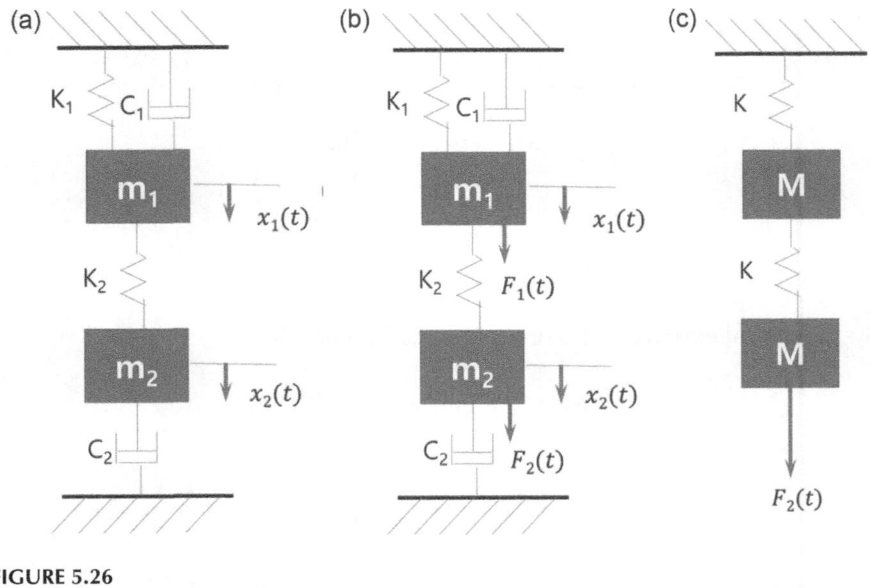

FIGURE 5.26
Two-degree-of-freedom mechanical system with damping (instance): (a) not forced; (b) forced (case 1); (c) forced (case 2).

$$\begin{bmatrix} m_1 & 0 \\ 0 & m_2 \end{bmatrix}\begin{Bmatrix} \ddot{x}_1 \\ \ddot{x}_2 \end{Bmatrix}+\begin{bmatrix} C_1 & 0 \\ 0 & C_2 \end{bmatrix}\begin{Bmatrix} \dot{x}_1 \\ \dot{x}_2 \end{Bmatrix}+\begin{bmatrix} K_1+K_2 & -K_2 \\ -K_2 & K_2 \end{bmatrix}\begin{Bmatrix} x_1 \\ x_2 \end{Bmatrix}=0 \text{ or } M\ddot{x}+C\dot{x}+Kx=0$$

$$M=\begin{bmatrix} 0.3193 & 0 \\ 0 & 0.630 \end{bmatrix}\text{kg, } K=\begin{bmatrix} 91.97 & -36.047 \\ -36.047 & 36.047 \end{bmatrix}\text{ N/m}$$

To discover the undamped natural frequencies ω_ns and displacements $\{u\}^i$:

$$M\ddot{X}+KX=0$$

$$X=\{u\}e^{i\omega t}$$

$$\underbrace{[-\omega^2 M+K]}_{\det(A)=0}\{u\}e^{i\omega t}=0$$

The natural frequencies are

$$\omega_1=5.6546\ rad/s, \omega_2=17.695\ rad/s$$

$$u=\begin{bmatrix} \begin{Bmatrix} 1.0 \\ 2.2667 \end{Bmatrix}^{(1)} & \begin{Bmatrix} 1.0 \\ -0.2236 \end{Bmatrix}^{(2)} \end{bmatrix}$$

Verify $\omega_1=\sqrt{\dfrac{K_1}{M_1}}=\sqrt{\dfrac{113.71}{3.556}}=5.6546\ rad/s,\ \omega_2=\sqrt{\dfrac{K_2}{M_2}}=17.695\ rad/s$

$$u^T Mu=\begin{bmatrix} M_1 & 0 \\ 0 & M_2 \end{bmatrix}=\begin{bmatrix} 3.5562 & 0 \\ 0 & 0.3508 \end{bmatrix}$$

$$u^T Ku=\begin{bmatrix} K_1 & 0 \\ 0 & K_2 \end{bmatrix}=\begin{bmatrix} 113.71 & 0 \\ 0 & 109.839 \end{bmatrix}$$

By the modal expansion theorem, the modal equation is

$$M_1\ddot{q}_1+C_1\dot{q}_1+K_1q_1=Q_1(t)$$

$$u^T Cu=\begin{bmatrix} C_1 & 0 \\ 0 & C_2 \end{bmatrix}$$

$$Q=u^T F$$

If $F = 0$, the initial conditions are

$$x(t=0) = \begin{Bmatrix} x_{10} \\ \vdots \\ x_{20} \end{Bmatrix}$$

$$\dot{x}(t=0) = \begin{Bmatrix} v_{10} \\ \vdots \\ v_{20} \end{Bmatrix} = 0$$

Because $x = uq$:

$$x(0) = uq(0)$$

$$q(0) = u^{-1}x(0)$$

$$\dot{q}(0) = u^{-1}\dot{x}(0)$$

$$M_1\ddot{q}_1 + C_1\dot{q}_1 + K_1 q_1 = 0$$

$$\text{Let } \begin{Bmatrix} x_{10} \\ x_{20} \end{Bmatrix} = \begin{Bmatrix} 1 \\ 2.260 \end{Bmatrix}^{(1)}, \begin{Bmatrix} \dot{x}_{10} \\ \dot{x}_{20} \end{Bmatrix} = 0$$

$$\begin{Bmatrix} q_{10} \\ q_{20} \end{Bmatrix} = u^{-1}x(0) = \begin{bmatrix} 0.898 & 0.4016 \\ 0.9102 & -0.4016 \end{bmatrix} \begin{Bmatrix} 1 \\ 2.260 \end{Bmatrix}^{(1)} = \begin{Bmatrix} 1 \\ 0 \end{Bmatrix}$$

$$[C] = \alpha[M] + \beta[K]$$

$$u^T C u = \alpha[M] + \beta[K]$$

$$[C] = \begin{bmatrix} C_1 & 0 \\ 0 & C_2 \end{bmatrix}$$

$$C_1 = \alpha M_1 + \beta K_1$$

$$C_2 = \alpha M_2 + \beta K_2$$

$$\zeta_1 = \frac{C_1}{2\omega_1 M_1} = \frac{\alpha M_1}{2\omega_1 M_1} + \frac{\beta K_1}{2\omega_1 M_1} = \frac{\alpha}{2\omega_1} + \frac{\beta\omega_1}{2}$$

$$\zeta_2 = \frac{C_2}{2\omega_2 M_2} = \frac{\alpha}{2\omega_2} + \frac{\beta\omega_2}{2}$$

For the test:

$$\zeta_1 \approx \frac{0.11}{n_{50\%}} = \frac{0.11}{2.5} = 0.044 \; or \; 4.4\%$$

$$\zeta_2 \approx \frac{0.11}{n_{50\%}} = \frac{0.11}{10} = 0.011 \; or \; 1.1\%$$

If solving them, $\alpha = 10.71$, $\beta = -0.033$.

$$[C] = 10.71[M] - 0.033[K]$$

$$\begin{bmatrix} \ddots & & \\ & C_i & \\ & & \ddots \end{bmatrix} = u^T C u = \alpha \begin{bmatrix} \ddots & & \\ & M_i & \\ & & \ddots \end{bmatrix} + \beta \begin{bmatrix} \ddots & & \\ & K_i & \\ & & \ddots \end{bmatrix}$$

$$x(t) = \sum_{i=1}^{n} \{u\}^{(1)} q_1 + \{u\}^{(2)} q_2$$

$$q_1(t) = \left[q_{10}\cos(\omega_{1d}t) + \frac{(\dot{q}_{10} + \zeta_1\omega_1 q_{10})}{\omega_d} \sin(\omega_{1d}t) \right] e^{-\zeta_1\omega_1 t}$$

$$q_2(t) = \left[q_{20}\cos(\omega_{2d}t) + \frac{(\dot{q}_{20} + \zeta_2\omega_2 q_{20})}{\omega_d} \sin(\omega_{2d}t) \right] e^{-\zeta_2\omega_2 t}$$

For two-degree-of-free system in Figure 5.26(c):

$$F = \begin{Bmatrix} 0 \\ F_2 \end{Bmatrix} e^{i\omega t}$$

$$Q = u^T F = \begin{bmatrix} 1.0 & 2.2667 \\ 1.0 & -0.2236 \end{bmatrix} \begin{Bmatrix} 0 \\ F_2 \end{Bmatrix} \text{ or } Q_1 = 2.2667 F_2 e^{i\omega t} \quad Q_2 = -0.2236 F_2 e^{i\omega t}$$

$$M_1 \ddot{q}_1 + C_1 \dot{q}_1 + K_1 q_1 = 2.2667 F_2 e^{i\omega t}$$

$$|q_1| = |Q_1| \left| H_{q_1/Q_1}(\omega) \right| = \frac{|Q_1| \left(1/K_1 \right)}{\left[\left(1 - \frac{\omega^2}{\omega_1^2} \right)^2 + \left(2\zeta_1 \frac{\omega}{\omega_1} \right)^2 \right]^{\frac{1}{2}}}$$

$$\text{If } \omega = \omega_1, |q_1| = \frac{2.2667F_2}{K_1} \cdot \frac{1}{2\zeta_1} = \frac{2.2667F_2}{113.71} \cdot 11.36 = 0.227F_2.$$

$$\text{If } \frac{\omega}{\omega_2} = \frac{\omega_1}{\omega_2}, |q_2| = |Q_2| \left| H_{q_2/Q_2}(\omega) \right|$$

$$= \frac{-0.2236F_2}{\underbrace{109.84}_{K_2}} \cdot \frac{1}{\underbrace{\left[\underbrace{\left(1 - \frac{\omega^2}{\omega_1^2}\right)^2}_{0.806} + \underbrace{\left(2\xi_1 - \frac{\omega}{\omega_1}\right)^2}_{5\times 10^{-5}} \right]^{\frac{1}{2}}}_{0.898}} = 0.0023F_2$$

$$x = uq$$

$$q_1 = q_1 \cos\left(\omega_1 t - \frac{\pi}{2}\right) = 0.227F_2 \cos\left(\omega_1 t - \frac{\pi}{2}\right)$$

$$q_2 = q_2 \cos(\omega_1 t - \phi_2) = 0.0023F_2 \cos(\omega_1 t - \phi_2)$$

Example 5.5: Modal Analysis 2

Perform modal analysis on the double pendulum system in Figure 5.27.

Solution

The motion of equation for the double pendulum system in Figure 5.27 (a) is

$$\ddot{\theta}_1 + \left(\frac{g}{L} + \frac{k}{m_1}\right)\theta_1 - \frac{k}{m_1}\theta_2 = 0$$

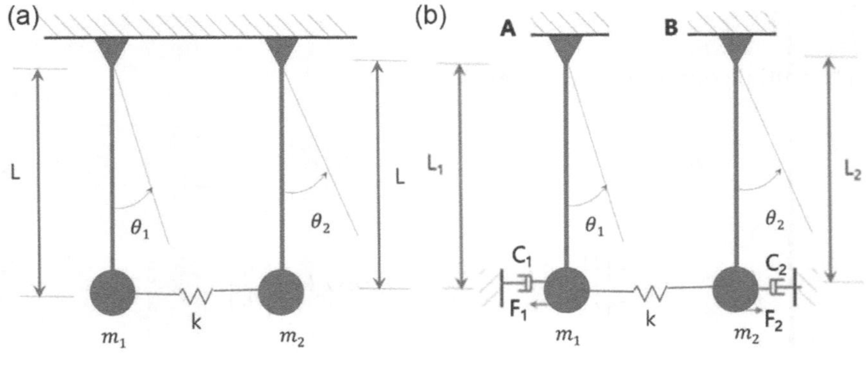

FIGURE 5.27
Double pendulum system (example): (a) no damping and force; (b) with damping and force.

$$\ddot{\theta}_2 - \frac{k}{m_2}\theta_1 + \left(\frac{g}{L} + \frac{k}{m_2}\right)\theta_2 = 0$$

The matrix form is

$$\begin{bmatrix} 1 & 0 \\ 0 & 1 \end{bmatrix}\begin{Bmatrix} \ddot{\theta}_1 \\ \ddot{\theta}_2 \end{Bmatrix} + \begin{bmatrix} \left(\frac{g}{L} + \frac{k}{m_1}\right) & -\frac{k}{m_1} \\ -\frac{k}{m_2} & \left(\frac{g}{L} + \frac{k}{m_2}\right) \end{bmatrix}\begin{Bmatrix} \theta_1 \\ \theta_2 \end{Bmatrix} = 0 \text{ or } M\ddot{\theta} + K\theta = 0$$

Let $\theta = \{u\}e^{i\omega t}$. Plug in the motion of equation.

$$\underbrace{\left[-\omega^2 M + K\right]}_{\det(A)=0}\{u\}e^{i\omega t} = 0$$

Let $m_1 = m_2 = m$ and $h = \left(\frac{g}{L} + \frac{k}{m}\right)$. The characteristic equation is

$$\omega^2 - 2h\omega^2 + h^2 - \left(\frac{k}{m}\right)^2 = 0$$

$$\omega_1 = \sqrt{g/L},\ \omega_2 = \sqrt{g/L + 2k/m}$$

$$\left[\begin{bmatrix} -\omega^2 & 0 \\ 0 & -\omega^2 \end{bmatrix} + \begin{bmatrix} h & -k/m \\ -k/m & h \end{bmatrix}\right] = \begin{bmatrix} -\omega^2 + h & -k/m \\ -k/m & -\omega^2 + h \end{bmatrix}\begin{Bmatrix} u_1 \\ u_2 \end{Bmatrix} = 0$$

For $\omega = \omega_1$, $u_1 = u_2$
For $\omega = \omega_2$, $u_1 = -u_2$
The initial conditions θ_{10}, θ_{20} are

$$\theta_{i0} = \begin{Bmatrix} 1 \\ 0 \end{Bmatrix} = C_1\begin{Bmatrix} 1 \\ 1 \end{Bmatrix} + C_2\begin{Bmatrix} 1 \\ -1 \end{Bmatrix} = 1/2\begin{Bmatrix} 1 \\ 1 \end{Bmatrix} + 1/2\begin{Bmatrix} 1 \\ -1 \end{Bmatrix}$$

$$1/2\begin{Bmatrix} 1 \\ 1 \end{Bmatrix}\cos\omega_1 t + 1/2\begin{Bmatrix} 1 \\ -1 \end{Bmatrix}\cos\omega_2 t = \begin{Bmatrix} \theta_1 \\ \theta_2 \end{Bmatrix}$$

$$1/2\cos\omega_1 t + 1/2\cos\omega_2 t = \theta_1(t) = \cos\left(\frac{\omega_1 - \omega_2}{2}\right)t\cos\left(\frac{\omega_1 + \omega_2}{2}\right)t$$

$$\tfrac{1}{2}\cos\omega_1 t-\tfrac{1}{2}\cos\omega_2 t=\theta_1(t)=\sin\left(\frac{\omega_2-\omega_1}{2}\right)t\sin\left(\frac{\omega_1+\omega_2}{2}\right)t$$

The motion of equation of the double pendulum system in Figure 5.27(b) is

$$\begin{bmatrix} m_1L_1^2 & 0 \\ 0 & m_2L_2^2 \end{bmatrix}\begin{Bmatrix} \ddot{\theta}_1 \\ \ddot{\theta}_2 \end{Bmatrix}+\begin{bmatrix} C_1L_1^2 & 0 \\ 0 & C_2L_2^2 \end{bmatrix}\begin{Bmatrix} \dot{\theta}_1 \\ \dot{\theta}_2 \end{Bmatrix}$$

$$+\begin{bmatrix} m_1gL_1+kL_1^2 & -kL_1L_2 \\ -kL_1L_2 & m_2gL_2+kL_2^2 \end{bmatrix}\begin{Bmatrix} \theta_1 \\ \theta_2 \end{Bmatrix}=\begin{Bmatrix} F_1L_1 \\ F_2L_2 \end{Bmatrix}$$

If $m_1=m_2=0.5\ \text{kg},\ L_1=1.1\ \text{m},\ L_2=1.0\ \text{m}, k=0.5\ \text{N}/\text{m}$, it is

$$M=\begin{bmatrix} 0.605 & 0 \\ 0 & 0.50 \end{bmatrix},\ K=\begin{bmatrix} 6.005 & -0.55 \\ -0.55 & 5.405 \end{bmatrix}$$

The natural frequencies and displacements are

$$\begin{Bmatrix} \omega_1 \\ \omega_2 \end{Bmatrix}=\begin{Bmatrix} 3.0445 \\ 3.3851 \end{Bmatrix},\ u=\begin{bmatrix} \begin{Bmatrix} 1.0 \\ 0.7139 \end{Bmatrix}^{(1)} & \begin{Bmatrix} 1.0 \\ -1.6949 \end{Bmatrix}^{(2)} \end{bmatrix}$$

By the modal expansion theorem, $x=uq$:

$$\begin{Bmatrix} \theta_1 \\ \theta_2 \end{Bmatrix}=uq=\sum_{i=1}^{N}\{u\}^{(i)}q_i(t)=\begin{Bmatrix} 1.0 \\ 0.7139 \end{Bmatrix}^{(1)}q_1(t)+\begin{Bmatrix} 1.0 \\ -1.6949 \end{Bmatrix}^{(2)}q_2(t)$$

The modal mass, stiffness matrices, and modal forces are

$$[M]=u^TMu=\begin{bmatrix} 0.8599 & 0 \\ 0 & 2.0414 \end{bmatrix}\left(=\{1\ \ -1.6949\}\begin{bmatrix} 0.605 & 0 \\ 0 & 0.50 \end{bmatrix}\begin{Bmatrix} 1 \\ -1.6949 \end{Bmatrix}=m_2\right)$$

$$[K]=u^TKu=\begin{bmatrix} 9.9699 & 0 \\ 0 & 23.3918 \end{bmatrix}$$

Let $F_1=F_1\cos\omega t,\ F_2=0$:

$$Q=u^TF=\begin{bmatrix} 1 & 0.7139 \\ 1 & -1.6949 \end{bmatrix}\begin{Bmatrix} F_1L_1 \\ 0 \end{Bmatrix}\cos\omega t$$

Let $q_1=q_1\cos(\omega t-\phi)$ and $\omega=\omega_1$. The modal equation is

$$M_1\ddot{q}_i + C_1\dot{q}_1 + K_1 q_1 = \left[\underbrace{-M_1\omega_1^2 q_1 + K_1 q_1}_{o}\right] - C_1\omega_1 q_1 \sin(\omega t - \phi) = Q_1(t) = F_1 L_1 \cos\omega t$$

$$C_1\omega_1 q_1 = F_1 L_1$$

Let $\omega = \omega_2$. The second mode contribution is

$$\begin{Bmatrix}\theta_1\\ \theta_2\end{Bmatrix}^{(2)} = \begin{Bmatrix} \\ \end{Bmatrix}^{(2)} q_2(t)$$

$$q_2(t) = |Q_2|\left|H_{q_2/Q_2}\right|\cos(\omega_2 t - \phi_2) = \frac{F_1 L_1\left(\frac{1}{K_2}\right)}{\left[\underbrace{\left(1-\frac{\omega_2^2}{\omega_2^2}\right)^2}_{=0} + \left(2\zeta\frac{\omega_2}{\omega_2}\right)^2\right]^{\frac{1}{2}}} = \frac{F_1 L_1\left(\frac{1}{K_2}\right)}{2\zeta}$$

$$u^T C u = \alpha[M] + \beta[K] \approx \alpha[M] \text{ or } C_2 = \alpha M_2 = \alpha(2.04)$$

$$\zeta_2 = \frac{C_2}{2\omega_2 M_2} = \frac{\alpha(2.04)}{2\omega_2 M_2} = 0.02$$

References

1. Hibbeler, R.C. *Mechanics of Materials,* 11th ed., Pearson: London, 2022.
2. Weingart, R.G.; Stephen, P. Timoshenko: Father of Engineering Mechanics in the U.S., *Struct. Mag.,* **2007**, *7*, 2–3.
3. Landau, L.D.; Lipshitz, E.M. *Theory of Elasticity: Volume 7 (Theoretical Physics),* 3rd ed., Butterworth-Heinemann: Oxford, 1970.
4. Tongue, B.H. *Principles of Vibration,* Oxford University Press: Oxford, 2001.

6

Mechanical System Failures and Quantum Mechanics

6.1 Introduction

A material is (a combination of) substances that forms a non-living or living object. Materials may be categorized on the foundation of their biological, chemical, and physical function attributes. They can generally be split into classifications: 1) metals, 2) polymers/plastics, 3) ceramics, etc. Metals are ductile, opaque, conductive (electrical and thermal), etc. Polymers/plastics are transparent, semitransparent, insulators, etc. Ceramics are compounds of nonmetallic and metallic materials (carbides, oxides, sulfides, nitrides).

Numerous materials structurally arranged at scales of dissimilar lengths are formed from a large number of grains, each of which has grain boundaries and second-phase particles. Therefore, after deleting the outcomes of grain boundaries and second-phase particles, it is uncomplicated to look into plastic deformation in a single crystal. When only one crystal in a metal is stressed in tension beyond its elastic limit, it stretches out a little, which is expressed as plastic deformation. It necessitates the breakup and remaking of various chemical bonds. Plastic deformation that may not be returned to its initial state will result from a slip, twinning, or a combination of both. Under tensile stress, plastic deformation is discerned by a strain-hardened region, a necking region, and a fracture. The mechanical properties of a material—fatigue, strength, ductility, hardness, impact resistance, fracture toughness, etc.—include its reaction to an exerted load (Figure 6.1).

A typical failure mechanism of a mechanical product is fatigue and fracture. Based on unconventional research performed by Battelle in 1982, roughly 80–90% of all structural failures occur via a fatigue mechanism [1]. The quality cost that has to be paid due to these failures is estimated to be $1.5 billion yearly. Battelle also proposed that these may be reduced by 29% by the proper application of fatigue technology. The necessitated prices are critical, and injury and hazards to human life due to numerous failures are dominant to a possibly infinite extent. Fatigue in a material is the starting and spreading of cracks due to regularly repeated loading. As

DOI: 10.1201/9781003451471-6

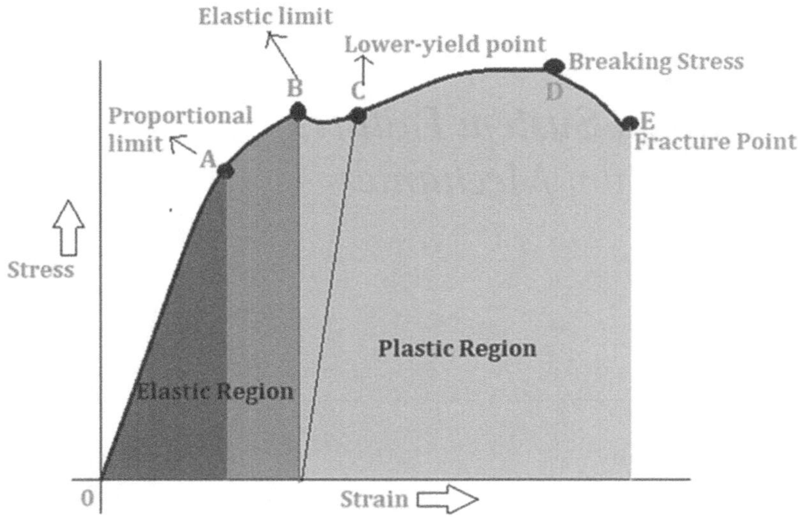

FIGURE 6.1
Mechanical properties of engineering materials.

coined by the French engineer Jean-Victor Poncelet in the middle of the 19th century, the word "fatigue" signifies that the material begins to be fragile, which finally leads to the process of losing cohesion or fracturing. It develops in a flawed component of a product as a result of regularly repeated plastic deformation in a restricted region on the external surface of the component where there is a small-scale crack or an existing flaw from an earlier time.

Fracture is a kind of developing fault of a single member into breaking the whole due to an active stress. Investigations in the area of fracture mechanics includes the response in materials for (dynamic) loading and, as a consequence, the occurrence and spreading of its cracks. The emergence of cracks in a material covers the following fracture mechanisms: 1) fatigue, 2) brittle (cleavage) fracture, 3) ductile (shear) fracture, 4) de-adhesion, and 5) crazing. The classification of fracture mechanics clearly relies on the material's characteristics. That is, a brittle fracture is the presiding failure mechanism of low-strength materials. Coming between strength levels, there is a movement between a fracture under linear elastic circumstances and a brittle ductile overload. That is, fracture failure at very high-strength values is decided by the flow properties of the material. Because understanding these complex phenomena—fatigue and fracture—they may be described in terms of quantum mechanics and from the perspective of micro- or nano-size.

Quantum mechanics is a basic theory that explains the behavior of physics on very small scales, i.e., at and below the scale of atoms. Particles such as electrons have wavelike properties that can be governed by the Schrodinger equation. There are numerous subtle things about quantum mechanics—the measurement problem, hidden variables along with Bell's theorem, and wave-particle duality. However, it is comparatively uncomplicated to apply quantum mechanics to a physical system and to make predictions that are consistent with experimentation. Thus quantum mechanics is also effective in figuring out why the product breaks down the (metal or ionic/covalent) bonds of a material in the structure might be exposed to various loads.

6.2 Materials Science and Engineering for Mechanical Products

6.2.1 Introduction

Materials science originally derived from the production of ceramics and its metallurgy (the use of fire). In the late 19th century, the Josiah Willard Gibbs (an American scientist) indicated that the thermodynamic properties of atomic structure in several phases are connected to the physical properties of a material. Materials science has become the oldest interdisciplinary area of researching and discovering materials that can be used in other areas and industries. The materials scientist emphasizes comprehension of how material processing affects its structure and consequently its properties and performance. It can be utilized to move forward comprehension in many research arenas, including biomaterials, nanotechnology, and metallurgy. Forensic engineering and failure analysis in materials science are also main avenues for inquiring into materials, products, structures, or components that might bring about distinctive injury or damage to property. Such examinations are key to the comprehension, for instance, of the sources of diverse aviation accidents and incidents (Figure 6.2).

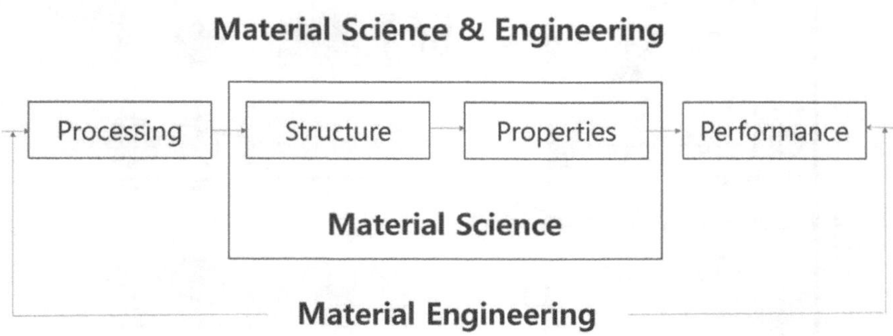

FIGURE 6.2
Materials science and engineering for mechanical products.

Materials in our society can be defined according to their eras:

- Stone age (2.5 million BCE)
- Bronze Age (3500 BCE): the world's first alloy (Cu+Sn)
- Iron (or Steel) Age (1000 BCE)
- Polymer Age? Silicon Age? (The present)

Metals have metallic bonds. They are:

- Strong, ductile, high density;
- Highly thermal and conductive of electricity;
- Opaque, reflective.

Polymers/plastics have covalent bonding that shares electronics. They are:

- Softly ductile, low strength, low density;
- Thermal and electrical insulators;
- Optically translucent or transparent.

Ceramics have ionic bonding (refractory) and are compounds of metallic and nonmetallic elements (oxides, carbides, nitrides, sulfides). They are:

- Brittle, glassy, elastic;
- Nonconductive (insulators).

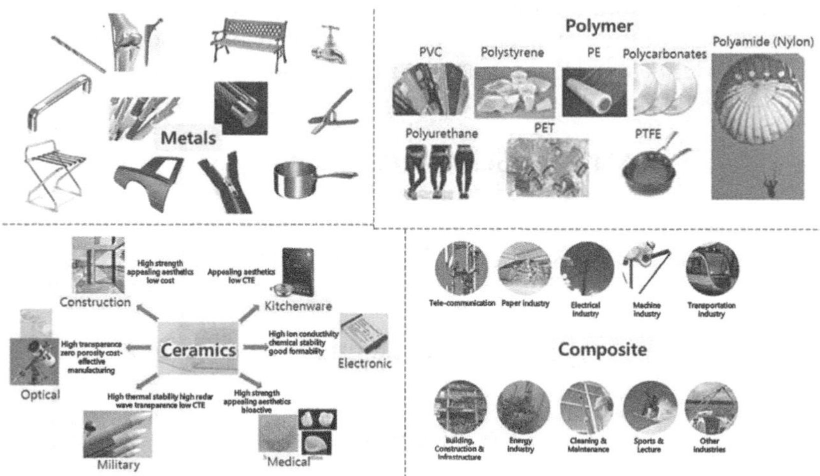

FIGURE 6.3
Application of materials—metals, polymer, ceramics, and composite.

Composites are combinations of two or more materials (Figure 6.3).

The process of carefully choosing appropriate materials is as follows:

1. *Pick implementation*→ decide the required properties (mechanical, electrical, thermal, magnetic, optical, deteriorative).
2. *Properties*→ recognize the candidate materials (structure, composition).
3. *Material*→ recognize the needed processing (changes in structure and overall sharing: casting, sintering, vapor deposition, doping, forming, joining, annealing).

6.2.2 Atomic Structure and Interatomic Bonding

To comprehend the bond formation in materials, it is required to realize the arrangement of electrons around the central nucleus. The numeric of protons in the nucleus is the atomic number of the element. A proton is charged in a positive way, and an electron carries an equal but opposite negative charge. Atomic structure has the following properties:

- *Atom*: Electrons: 9.11×10^{-28} g; protons and neutrons: 1.67×10^{-24} g
- *Atomic number*: numeric of protons in the nucleus of the atom (=number of electrons of neutral species)
- *Atomic mass unit (amu)*: 1/12 mass of ^{12}C
- *Atomic weight*: weight of 6.022×10^{23} atoms (1 amu/atom = 1 g/mol)

Valence electrons decide all of the 1) electrical, 2) chemical, 3) thermal, and 4) optical attributes. Electrons have wavelike and particle properties in orbitals specified by a probability. Each orbital is determined by quantum numbers as follows (Figure 6.4):

H 2.1																	He N/A
Li 1.0	Be 1.5											B 1.5	C 2.5	N 3.0	O 3.5	F 4.0	Ne N/A
Na 0.9	Mg 1.2											Al 1.5	Si 1.8	P 2.1	S 2.5	Cl 3.0	Ar N/A
K 0.8	Ca 1.0	Sc 1.3	Ti 1.5	V 1.6	Cr 1.6	Mn 1.5	Fe 1.8	Co 1.9	Ni 1.8	Cu 1.9	Zn 1.6	Ga 1.6	Ge 1.8	As 2.0	Se 2.4	Br 2.8	Kr N/A*
Rb 0.8	Sr 1.0	Y 1.2	Zr 1.4	Nb 1.6	Mo 1.8	Tc 1.9	Ru 2.2	Rh 2.2	Pd 2.2	Ag 1.9	Cd 1.7	In 1.7	Sn 1.8	Sb 1.9	Te 2.1	I 2.5	Xe N/A*
Cs 0.7	Ba 0.9		Hf 1.3	Ta 1.5	W 1.7	Re 1.9	Os 2.2	Ir 2.2	Pt 2.2	Au 2.4	Hg 1.9	Tl 1.8	Pb 1.9	Bi 1.9	Po 2.0	At 2.2	Rn N/A
Fr 0.7	Ra 0.9																

FIGURE 6.4
Electronegativity periodic table.

- *Quantum number*: n = principal (energy level shell: K, L, M, N, O), l = subsidiary (orbitals: s, p, d, f), m_l = magnetic (1, 3, 5, 7), m_s = spin (1/2, −1/2)
- Distinct energy levels
- Tendency to occupy the lowest available energy state
- *Examples*: C (atomic number = 6) $1s^2 2s^2 2p^2$ and Fe (atomic number = 26) $1s^2 2s^2 2p^6 3s^2 3p^6 3d^6 4s^2$

As seen in Figure 6.5, primary bonding of materials consists of 1) ionic bonding, 2) covalent bonding, 3) metallic bonding. Ionic bonding occurs between metal (+ions) and nonmetal (-ions) which has dissimilar electronegativities and requires electronic transfer. Ionic bonding is the total move of valence electron(s) between atoms. It is a kind of chemical bond that produces the pair of oppositely charged ions. In ionic bonds, the metal loses electrons to be a positively charged cation. On the other hand, the nonmetal receives those electrons to become a negatively charged anion. For instance, magnesium oxide (MgO) is Mg (metal) $1s^2 2s^2 2p^6 3s^2$ and O (nonmetal) $1s^2 2s^2 2p^4$. If magnesium donates two electrons and oxide receives them, Mg^{2+} (metal) $1s^2 2s^2 2p^6$ and O^{2-} (nonmetal) $1s^2 2s^2 2p^6$. Minimum energy for stable bonding is required when attractive and repulsive terms balance.

A covalent bond establishes when the two nonmetal atoms share two electrons. The electrons required are in the external shells of the atoms. An atom that splits one or more of its electrons may finish its external shell. Covalent bonds are powerful—a large quantity of energy is needed to break them. That is, because covalent bonding requires alike electronegativity and share

Bonding Type	Substance	Bonding Energy (kcal/mol)	Melting Temperature (°C)
Ionic	NaCl	153	801
	MgO	239	1000
Covalent	Si	108	1410
	C	170	> 3550
Metallic	Hg	16	-39
	Al	77	660
	Fe	97	1538
	W	203	3410
vander Waals	Ar	1.8	-189
	Cl_2	7.4	-101

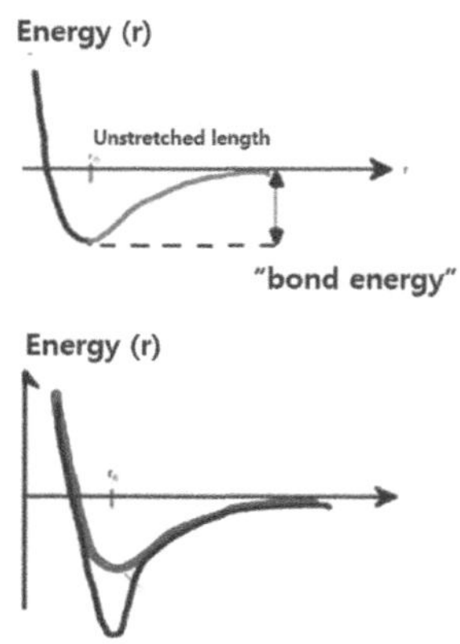

FIGURE 6.5
Primary bonding of materials.

electrons, it is determined by valence at *s* and *p* orbitals. Metallic bond is delocalized as electron cloud (or sea of electrons). Metals and alloys have one, two, or three valence electrons that are free to drift throughout the entire metal. So metals are powerful, and they can support a uniform structure and usually have high melting and boiling points and are fine conductors of electricity and heat.

6.3 Materials in Quantum Mechanics

6.3.1 Brief History

Quantum mechanics deals with the characteristics of material and light on the atomic and subatomic scale. It tries to explain and consider the properties of molecules and atoms and their component parts—protons, electrons, neutrons, and other more obscure particles such as quarks and gluons. With the discovery of subatomic particles, the necessity for a new kind of mechanics began to emerge. Quantum mechanics supplies a single and complete theory that describes the behavior of material on an atomic (nanometer or matter) scale. These properties cover the particle interactions with one another and with electromagnetic radiation (i.e., light, X-rays, and gamma rays). First of all, let's take a brief glance at the historical timeline of how quantum mechanics came about.

Max Planck (1900) proposed that light with frequency v is released in quantized lumps of energy, which arise from essential multiples of the quantity,

$$\mathrm{E} = h\nu = \hbar\omega \tag{6.1}$$

where $h \approx 6.63 \times 10^{-34}\,\mathrm{J{\cdot}s}$, ν is the number of cycles per unit time, and $\hbar \equiv h/2\pi$.

A problem designed to test knowledge in late 19th century physics was the blackbody radiation issue that was the challenge of predicting the radiation intensity of a heated cavity at a given wavelength and temperature. In a nutshell, the problem was that the traditional (continuous) theory of light forecasted which of the determined objects would radiate an energy quantity that is without limits. Planck's theory of quantized radiation not only got rid of the infinity issue but also forecasted the form of the power curve as a function of temperature. Planck's theory straightforwardly appends the information of how many lumps of energy a wave holds.

Light (electromagnetic radiation) is a periodic variation of an electric field and can be divided by wavelength λ, which is the interval between consecutive maxima and determines the color of visible light waves. Albert Einstein

(1905) defined that quantization was inherent to the light and that the lumps might be described as particles (photons). This suggestion was an outcome of his tasks on the photoelectric effect, which deals with the absorption of light and the emission of elections from a material. A beam of ultraviolet (UV) light striking a metal surface may emit electrons. For electron ejection, the frequency v ought to be identical or greater than a threshold frequency v_o. It can plot the kinetic energy (KE) of ejected electrons as a function of the frequency of incoming light:

$$\mathrm{KE} = h\nu - h\nu_o \tag{6.2}$$

Light consists of separate packages named photons (it is quantized). A photon energy is corresponding to its frequency ($E = h\nu$). The intensity of light is expressed in terms of photons per second. The wave-particle duality of light is as follows: 1) light as a wave with the properties of waves versus light as a particle, with the photoelectric effect. Light as a particle is explained by means of the photoelectric effect (continued) and photon momentum. Matter is also a kind of wave.

Niels Bohr (1913) demonstrated that electrons in atoms have wavelike characteristics. He correctly expressed a few characteristics about hydrogen, especially the quantized energy levels, that were familiar. Louis de Broglie (1924) indicated that all particles are connected to waves, where the frequency and number of the (matter) wave are defined by the identical relationship we found for photons:

$$E = \hbar\omega \text{ and } p = h\nu / c = h / \lambda (\text{since } c = \nu\lambda) \tag{6.3}$$

The bigger E and p are, the larger ω and k are. Even for little E and p, which are distinctive of a photon, ω and k are very large because $\hbar$ is so little. Thus any everyday-sized particle with large (in comparison) energy and momentum values will have very big ω and k values.

Electromagnetic radiation has a constant speed: the "speed of light" ($c = \lambda\nu = 2.9979 \times 10^8\,\mathrm{m/s}$). Werner Heisenberg (1925) and Erwin Schrodinger (1926) explained a particular version of quantum mechanics. Particularly, Schrodinger wrote down a wave equation that decides how the waves evolve in space and time. Max Born (1926) defined Schrodinger's wave as a probability amplitude. By "amplitude," we mean that the wave will be squared to obtain the required probability. This provides the probability of finding a particle at a defined location. Paul Dirac (1926) described that Heisenberg's and Schrodinger's sorts of quantum mechanics were similar in that they both could be obtained from a more general sort of quantum mechanics. Photons, also named light (electromagnetic radiation), can be explained by energy, λ, or v. Electrons, named photoelectrons, may be explained by KE, velocity, or λ. eV is a unit of energy ($1.6022 \times 10^{-19}\,\mathrm{J}$)

Example 6.1

Think about our two light sources as follows: 1) a UV lamp ($\lambda = 254$ nm) and 2) a red laser pointer ($\lambda = 700$ nm). What is the energy per photon released by the UV lamp or red laser pointer?

Solution

$$\text{Energy for UV: } E = h\nu = \frac{hc}{\lambda} = \frac{\left(6.626\text{x}10^{-34}\,\text{Js}\right)\left(2.998\text{x}10^{8}\,\text{m/s}\right)}{254\text{x}10^{-9}\,\text{m}} = 7.82\text{x}10^{-19}\,\text{J}$$

$$\text{Energy for red laser pointer: } E = \frac{\left(6.626\text{x}10^{-34}\,\text{Js}\right)\left(2.998\text{x}10^{8}\,\text{m/s}\right)}{700\text{x}10^{-9}\,\text{m}} = 2.84\text{x}10^{-19}\,\text{J}$$

What is the whole numeric of photons released by the laser pointer in 60 seconds if the intensity (I) = 1.00 mW? (1.00 mW = $\frac{1.00\text{x}10^{-3}\,\text{J}}{\text{s}}$)

Solution

$$\textit{Intensity for red laser pointer: } I = \frac{1.00\text{x}10^{-3}\,\text{J}}{\text{s}} \times \frac{1\,\text{photon}}{2.84\text{x}10^{-19}\,\text{J}} \times 60s = 2.1\text{x}10^{17}\text{ photons}$$

6.3.2 Schrodinger Equation

The Schrodinger mathematical problem is an equation of movement for particles (like electrons) that explains their wavelike properties. We obtain the Schrodinger formulation. A particle of energy is the whole of the kinetic and potential energies. We assume that the potential is a function of just x.

$$E = K + V = \frac{1}{2}mv^2 + V(x) = \frac{p^2}{2\text{m}} + V(\text{x}) \tag{6.4}$$

De Broglie declared that all minute portion of matters will be defined as waves with frequency ω and wavenumber k and that $E = \hbar \acute{E}$ and $p = \hbar k$:

$$\hbar\omega = \frac{\hbar^2 k^2}{2m} + V(\text{x}) \tag{6.5}$$

A wave with frequency ω and wavenumber k will be defined as $\psi(x,t) = Ae^{i(kx-\omega t)}$. In 3D, we have $\psi(r,t) = Ae^{i(kr-\omega t)}$.

Let us just consider 1D:

$$\frac{\partial\psi}{\partial t} = -i\omega\psi \Rightarrow \omega\psi = i\frac{\partial\psi}{\partial t} \text{ and } \frac{\partial^2\psi}{\partial x^2} = -k^2\psi \Rightarrow k^2\psi = -\frac{\partial^2\psi}{\partial x^2} \tag{6.6}$$

As we grow the energy Equation (6.5) by ψ and consequently attach these relationships, we obtain

$$\hbar(\omega\psi) = \frac{\hbar^2}{2m}\left(k^2\psi\right) + V(x)\psi \Rightarrow i\hbar\frac{\partial\psi}{\partial t} = \frac{-\hbar^2}{2m}\frac{\partial^2\psi}{\partial x^2} + V\psi \tag{6.7}$$

This is the Schrodinger equation. As we put the x and t arguments back in, the equation has the formation

$$i\hbar\frac{\partial\psi(x,t)}{\partial t} = \frac{-\hbar^2}{2m}\frac{\partial^2\psi(x,t)}{\partial x^2} + V(x)\psi(x,t) \tag{6.8}$$

In 3D, the x dependence depends on all three coordinates (x, y, z), and the word $\frac{\partial^2\psi}{\partial x^2}$ is $\nabla^2\psi$ (the whole of the second derivative). Max Born elucidates that $|\psi|^2$ takes the probability of finding the particle at particular position x.

6.3.3 Resolving the Schrodinger Equation: Infinite Square Well (Example)

By various types of (covalent or ionic) bonds, the atomic electronics are isolated and bound in the materials. Failure of a part may begin to become visible due continual fatiguing of preexisting imperfections or tiny cracks in the microscale constructional exterior of a part. If there are large stresses in the structure, a crack can increase in size from microscopic-sized voids and grow until it reaches some critical magnitude of the component. Modeling this behavior requires a quantum/transported life-stress prototype.

Think about a particle that is overly managed to move only in the x way from $x = 0$ to $x = a$. The potential energy, $V(x)$, may be expressed:

$$V(x) = \begin{cases} 0 (0 \le x \le a) \\ \infty(x < 0 \ or \ x > a) \end{cases} \tag{6.9}$$

The Schrodinger wave equation is defined:

$$\hat{H}\text{È} = E\psi \tag{6.10}$$

where $\hat{H}$ is the Hamiltonian operator in the x orientation, ψ is the wave function, and E is the (electron) energy.

As $\hat{H} = -\frac{h^2}{8\pi^2 m}\frac{d^2}{dx^2} + V$, we may put this in Equation (6.10):

$$\frac{d^2\psi}{dx^2}+\frac{8\pi^2 m}{h^2}(E-0)\psi=0 \text{ or } \frac{d^2\psi}{dx^2}+\frac{8\pi^2 m}{h^2}(E-V)\psi=0 \tag{6.11}$$

where m is the electron mass, h is the Planck constant, and V is the potential energy.

$V=\infty$ as the exterior barriers, this is feasible only when $\psi=0$; that is, particles are not exterior barriers. As $V=0$ in the walls, Equation (6.11) may be defined:

$$\frac{d^2\psi}{dx^2}+\frac{8\pi^2 m}{h^2}(E-0)\psi=0 \text{ or } \frac{d^2\psi}{dx^2}+K^2\psi=0 \tag{6.12}$$

where $K^2=\frac{8\pi^2 mE}{h^2}$.

We may presume a means of solving Equation (6.12):

$$\psi(x)=A\sin Kx+B\cos Kx \tag{6.13}$$

where A, B = constants.

As x = 0 or x = a at the walls, $\psi(0)=\psi(a)=0$, B = 0, $K=\frac{n\pi}{a}$, and $E=\frac{n^2h^2}{8ma^2}$, n = 1, 2, 3, 4.

Thus we may define Equation (6.13):

$$\psi(x)=A\sin\left(\frac{n\pi}{a}\right)x \tag{6.14}$$

The probability of finding the particle in a small-scale area between x and x + dx is given:

$$\int_0^a \psi^2(x)dx=1 \text{ or } \int_0^a\left(A\sin\left(\frac{n\pi}{a}\right)x\right)^2 dx=1 \tag{6.15}$$

Thus, we can attain the solution of Equation (6.15):

$$\psi(x)=\sqrt{\frac{2}{a}}\sin\left(\frac{n\pi}{a}\right)x \tag{6.16}$$

where $\psi(x+a)=\psi(x)$, a is the (periodic) interval, and n is the principal quantum number.

6.3.4 Hybrid Orbitals in Chemical Bonds

Chemical bonds detain molecules together and produce short-term links that are crucial to life. The sorts of chemical bonds are covalent, ionic, and hydrogen bonds. Hybrid orbitals are mathematical combinations of atomic orbitals that create new orbitals with different shapes and energies than the original atomic orbitals. They are used to explain how electrons pair to form chemical bonds and how molecules are structured. Hybrid orbitals vary only in their direction in space. For example, in CH_4 (methane), because C (carbon) has four valence e^- and needs four more, it shares electrons from hydrogen atoms. On the other hand, because H (hydrogen) has one valence e^- and requires one more, it shares electrons from the carbon atom. Methane (CH_4) has sp^3 hybrid orbitals, which are formed when a carbon atom's *2s* orbital and three *2p* orbitals combine (Figure 6.6).

Ammonia is an inorganic chemical compound of nitrogen and hydrogen with the formula NH_3 using sp^3 hybrid orbitals. It means that the nitrogen atom in ammonia mixes its 2*s* orbital with its three 2*p* orbitals to create four equivalent hybrid orbitals (Figure 6.7).

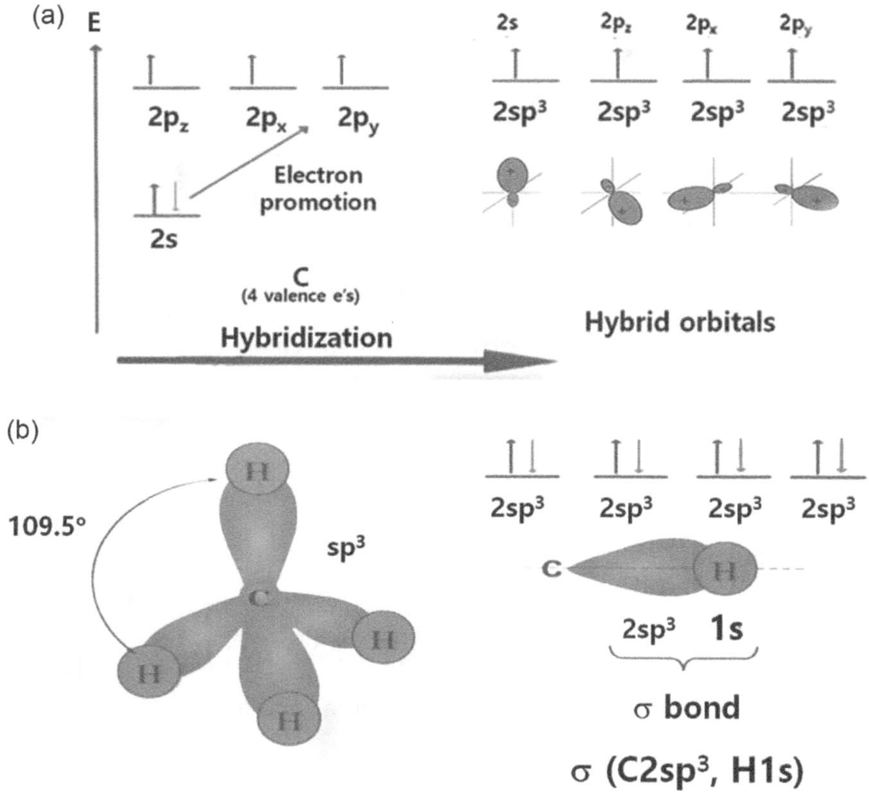

FIGURE 6.6
Covalent bonding of methane (example): (a) hybridization of methane (CH_4); (b) structure of methane (CH_4).

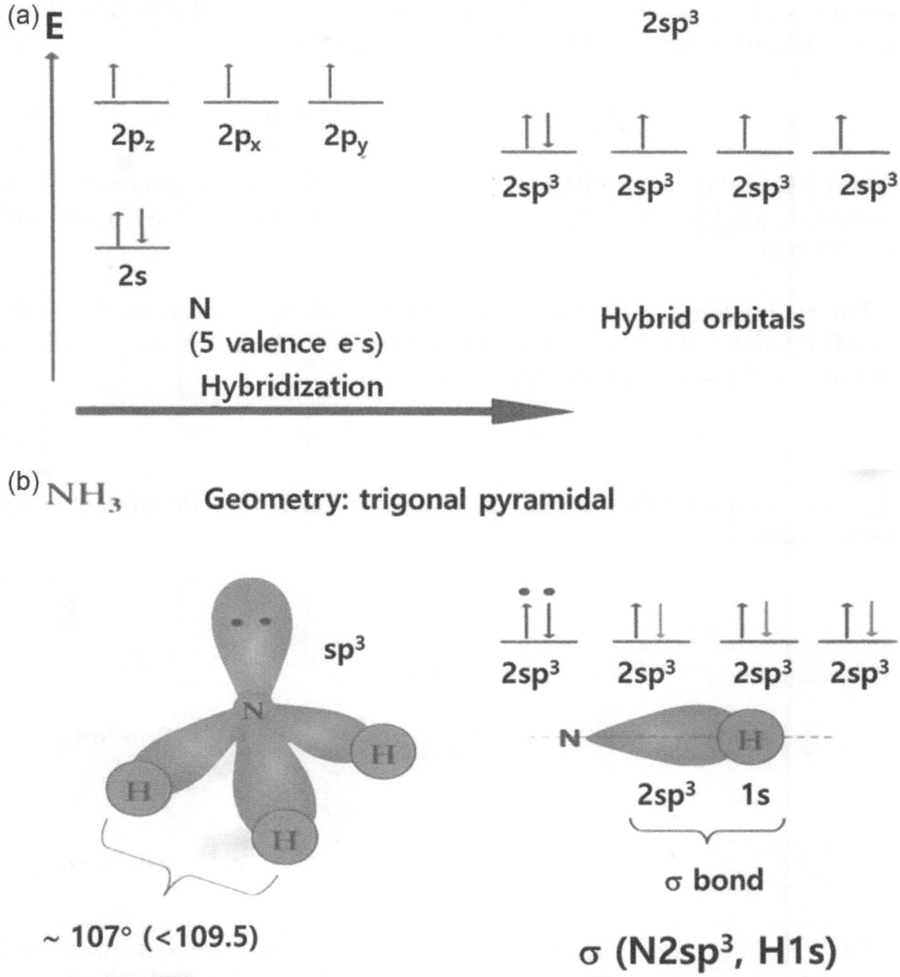

FIGURE 6.7
Covalent bonding of ammonia (example): (a) hybridization of ammonia (NH_3); (b) structure of ammonia (NH_3).

6.3.5 Simple Molecular Orbital Theory

Molecular orbital (MO) theory is a procedure for explaining the electronic structure of molecules utilizing quantum mechanics. In MO hypothesis, valence electrons are detached over the whole molecule, not enclosed in single atoms or bonds. Molecular orbitals (wave functions) originate from attaching (superimposing) atomic orbitals or wave functions. Linear combination of atomic orbitals (LCAO) creates molecular orbitals (bonding and antibonding orbitals). *N* molecular orbitals may be built by *N* atomic orbitals. It assumes that 1) valence electrons of the atoms in a molecule are the valence electrons of the whole molecule and that 2) molecular orbitals take linear combinations of atomic orbitals (AO) in the molecule.

$$\psi_n = \sum_i c_{ni}\phi_i = c_{n1}\phi_1 + c_{n2}\phi_2 + c_{n3}\phi_3 + \cdots \quad (6.17)$$

where ϕ_s are the valence AOs of the atoms making up the molecule; c_S are weighting coefficients, which tell how much of each AO is in the MO (Figure 6.8).

For H_2 (diatomic molecules), when its two electrons both reside in the bonding orbital, the molecule is more stable. Since the probability density is defined by the square of the wave function:

$$\int \psi^2 d\tau = c_1^2 \int \phi_1^2 d\tau + c_2^2 \int \phi_2^2 d\tau + 2c_1c_2 \int \phi_1\phi_2 d\tau \quad (6.18)$$

For two identical AOs on identical atoms, the electrons are shared in the same manner:

$$\psi = c_1\phi_1 + c_2\phi_2 \quad (6.19)$$

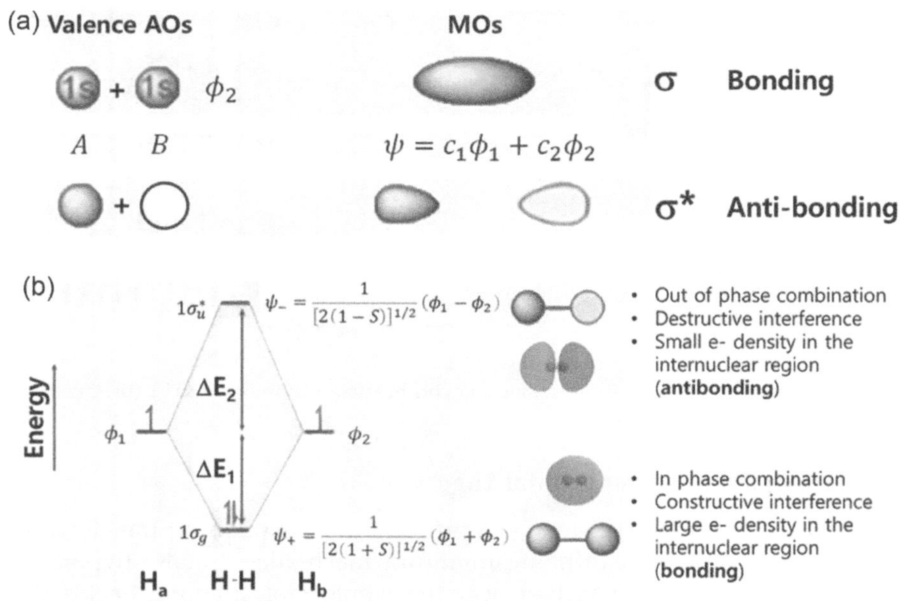

FIGURE 6.8
Simple molecular orbital theory: (a) atomic orbital and molecular orbital; (b) LCAO MO energy diagram for H_2.

For $c_1 = \pm c_2$, we have two MOs from the two AOs:

$$\psi_+ = c_{+,1}(\phi_1 + \phi_2) \qquad \psi_- = c_{-,1}(\phi_1 - \phi_2) \tag{6.20}$$

After normalization, setting $\int \psi_+^2 \, d\tau = 1$ and $\int \psi_-^2 \, d\tau = 1$

$$\psi_+ = \frac{1}{[2(1+S)]^{1/2}}(\phi_1 + \phi_2) \qquad \psi_- = \frac{1}{[2(1-S)]^{1/2}}(\phi_1 - \phi_2) \tag{6.21}$$

where S is the overlap integral $S = \int \phi_1 \phi_2 d\tau \; 0 \leq S \leq 1$.

As waves interfere in a way that is intended to have a useful effect, the amplitude grows where they overlap. Enlarged amplitude in the internuclear region translates to an increased probability density (ψ^2) between nuclei. An electron in a bonding MO may be attracted to both nuclei and will be lower in energy contrasted to an atomic orbital for an individual nucleus.

Antibonding orbitals arise from a LCAO (destructive interference). As wave functions interfere in a way that causes destruction, the amplitude lessens when they overlap. Decreased amplitude translates to a reduced probability density (ψ^2) between the nuclei and a node between the two nuclei. An electron in this antibonding orbital could be fundamentally kept out from the internuclear region and therefore have a higher energy than if in an atomic orbital.

Three circumstances should be satisfied for AOs to interrelate and establish MOs: 1) The AOs should have the identical symmetry; 2) the AOs should have alike energy; and 3) the AOs must have spatial overlap. Symmetry will permit us to treat more complex molecules by assisting us to decide which AOs merge to make MOs.

6.3.5.1 Basic Rules for MO Theory

σ bond is symmetrical about the bond axis without a nodal plane along that axis. On the other hand, π bond is electron density in the two lobes with a single nodal plane along the bond axis. Basic rules for MO theory are as follows:

1. *Rule #1*: The interaction of n AOs guides to the formation of n MOs. For example, if $n = 2$, the MO consists of a bonding and an antibonding. The bonding orbital is more balanced than the lower-energy AO. The bonding orbital is stabilized less than the antibonding orbital is destabilized (Figure 6.9).
2. *Rule #2*: As the AOs are degenerate, their interaction is proportionate to their overlap integral, S. That is, the greater the degree of overlap, the stronger the bonding/antibonding will be (Figure 6.10).

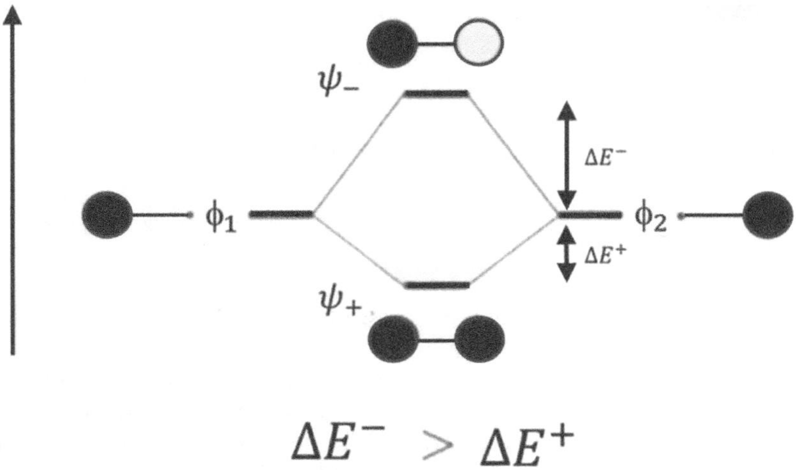

FIGURE 6.9
Rule #1 for MO theory.

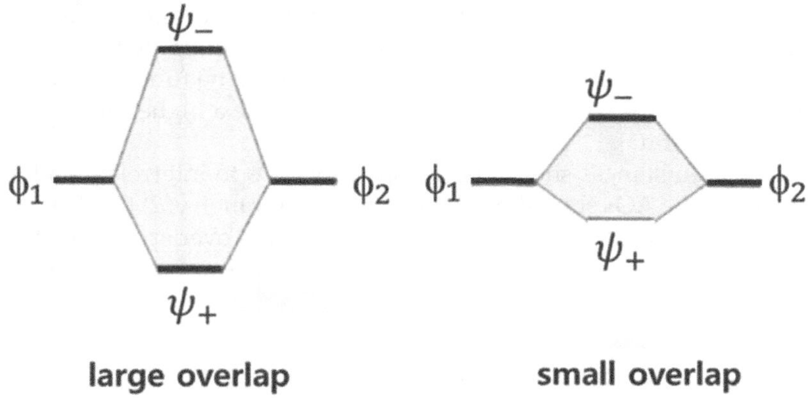

FIGURE 6.10
Rule #2 for MO theory.

3. *Rule* #3: If $S = \int \phi_1\phi_2 d\tau$, orbitals should have the same symmetry to have non-zero overlap. That is, $S = 0$ if orbitals have dissimilar irreducible representations. On other hand, If $S \neq 0$, bonding and antibonding MOs result.
4. *Rule* #4: As the AOs are non-degenerate, their interaction is proportionate to S^2 / E, where E is the energy separation between the AOs. If that happens, the bonding orbital is mainly localized on the atom with the deeper lying AO, generally the more electronegative atom. The antibonding orbital is mainly localized on the atom with the higher AO (Figure 6.11).

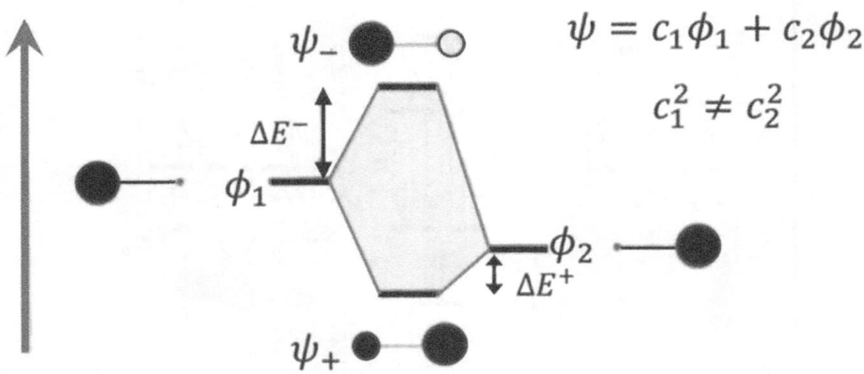

FIGURE 6.11
Rule #4 for MO theory.

6.3.5.2 Homonuclear Diatomic Molecules

Diatomic molecules are molecules that hold only the two atoms connected to each other. The atoms may be identical or dissimilar to evolve into homo- or heteronuclear diatomic molecules. The bonding of diatomic molecules is based on whether the jointly bonded atoms are metals or nonmetals.

1. Consider O_2 (one σ bond, one π bond):
   - An oxygen molecule (O_2) is made up of two oxygen atoms that are clasped jointly by a covalent bond. The Lewis dot structure excellently forecasts the incorrect electronic structure for O_2. We can utilize LCAO-MO theory to acquire a better picture (Figure 6.12).

   Just as with atoms, we can write down a molecular electron configuration for O_2

   We can also compute the O-O bond order (BO):

   $$BO = \frac{1}{2}\left[(\#\,\text{bonding}\ e) - (\#\,\text{anti}-\text{bonding}\ e)\right] = \frac{1}{2}[8-4] = 2$$

   LCAO MO theory also forecasts that O_2 has two unpaired electrons.
2. Consider N_2 (one σ bond, two π bond): A nitrogen molecule (N_2) is a colorless, odorless, and inert gas made up of two nitrogen atoms (N) that are tightly bound together. We can utilize LCAO-MO theory to take a better picture (Figure 6.13).

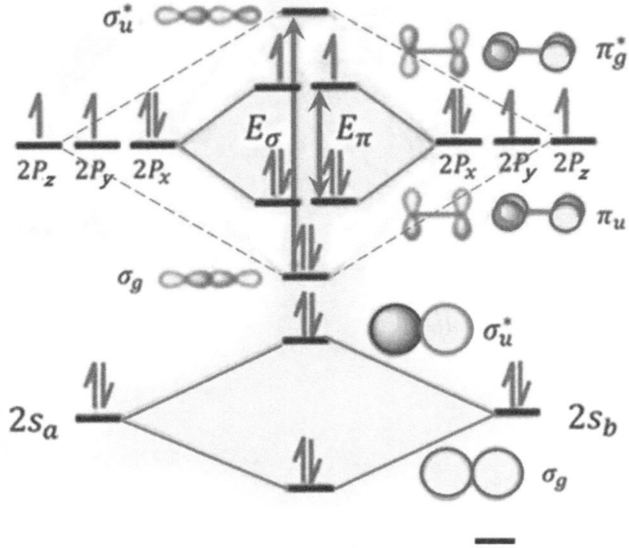

FIGURE 6.12
Oxygen molecule in LCAO-MO theory.

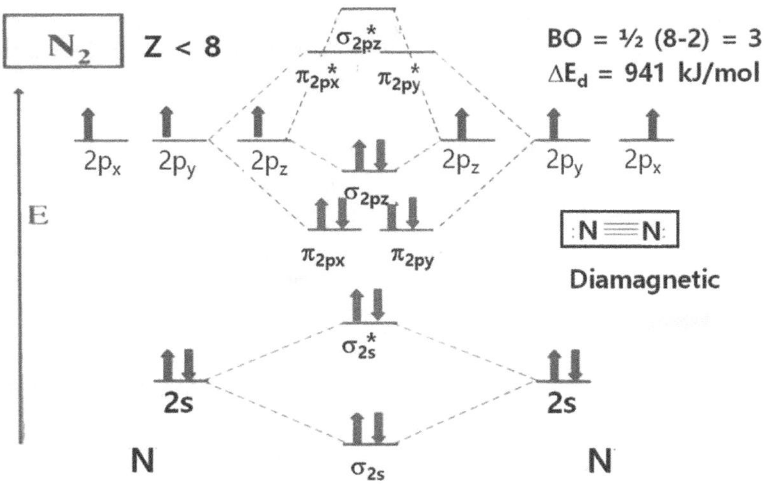

FIGURE 6.13
Oxygen molecule in LCAO-MO theory.

3. Heteronuclear diatomic molecules:
   - In molecules with more than one sort of atom, MOs are made from AOs, which have dissimilar energies. Think about carbon monoxide (CO), which is made up of one carbon atom and one oxygen atom attached by a triple bond. It is the simplest form of carbon oxide. We can utilize LCAO-MO hypothesis to take a better picture (Figure 6.14).

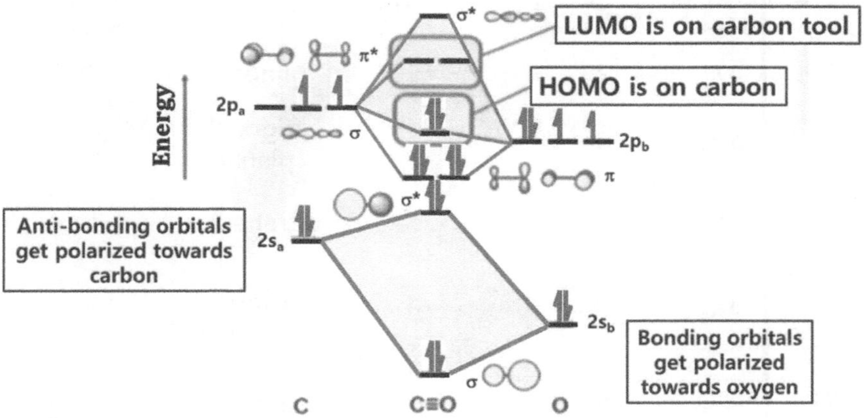

FIGURE 6.14
Oxygen molecule in LCAO-MO theory.

6.3.6 Valence Shell Electron Pair Repulsion (VSEPR) Theory

The valence shell electron pair repulsion (VSEPR) hypothesis is a prototype utilized to forecast 3D molecular geometry established on the numeric of valence shell electron bond pairs among the atoms in a molecule or ion. This model presumes that electron pairs may organize themselves to reduce repulsion effects from one another. That is, the electron pairs are as far apart as possible. Based on Lewis structures and the principles that 1) valence electrons pair each other and 2) the geometry around the central atom may be such as to reduce the repulsion.

The VSEPR prototype is practical for forecasting and imagining molecular structures: linear, trigonal planar, angled, trigonal pyramidal, tetrahedral, trigonal bipyramidal, T-shaped, disphenoidal (seesaw), square pyramidal, octahedral, square planar, and pentagonal bipyramidal (Figure 6.15).

VSEPR guidelines are as follows:

1. Steric number (SN) is utilized to forecast geometries (Figure 6.16).
   - SN = (number of atoms bonded to central atom) + (number of lone pairs)
   - With regard to electron-pair repulsion, double and triple bonds may behave like single-bonds.
   - The numeric of lone pairs and atoms bonded to the central atom is significant, not the BONDS to central atom

2. As a molecule has two or more resonance structures, and the VSEPR prototype may be applicable to any one of them (Figure 6.17).

	Formula type	SN	Lewis structure	geometry	bond angle
CO_2	AX_2	2	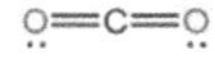	linear	180° nonpolar
BH_3	AX_3	3	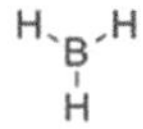	trigonal planar	120°
CH_4	AX_4	4	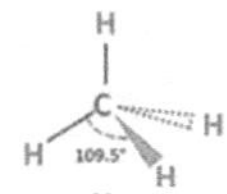	tetrahedral	109.5°
PCl_5	AX_5	5	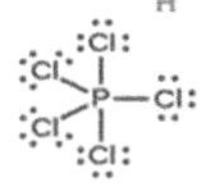	trigonal bipyramidal	120° (eq) 90° (axial)
SF_6	AX_6	6		octahedral	90°

FIGURE 6.15
Basic VSEPR models.

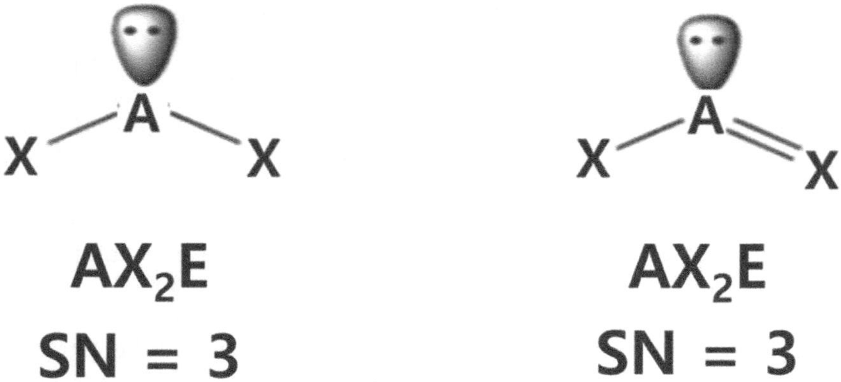

FIGURE 6.16
VSEPR Guideline #1: A = central atom; X = bonding atom; E = lone pair.

3. As there is more than one central atom in a molecule, think about the bonding about each atom independently (Figure 6.18).
4. With respect to molecules with lone pairs, electrons in bonds have less "spatial distribution" than lone pairs, meaning (Figure 6.19):
   - Electrons in bonds occupy less space.
   - On the other hand, lone-pair electrons occupy more space and thus experience more repulsion.
   - Lone-pair/lone-pair repulsion (greatest repulsion) > lone-pair/bonding-pair repulsion > bonding-pair/bonding-pair repulsion (least repulsion).

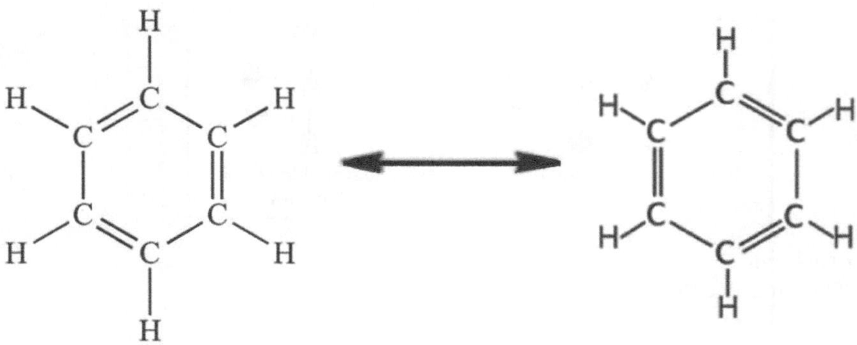

FIGURE 6.17
VSEPR Guideline #2.

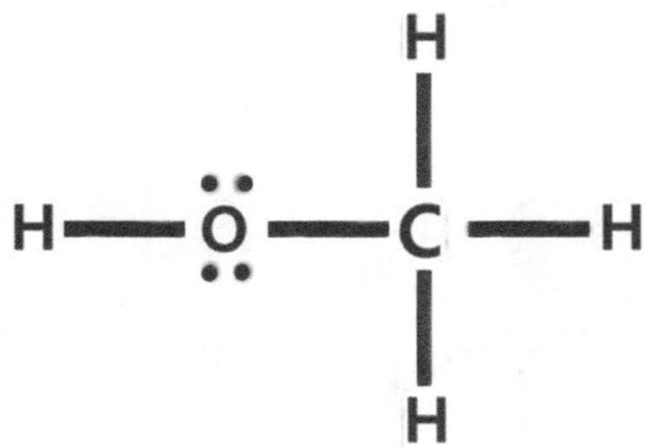

FIGURE 6.18
VSEPR Guideline #3.

FIGURE 6.19
VSEPR Guideline #4.

5. AX_4E molecules have a seesaw shape (Figure 6.20).
   - An axial lone pair might repel three bonding electron pairs with great power.
   - An equatorial lone pair repels only two with great power.

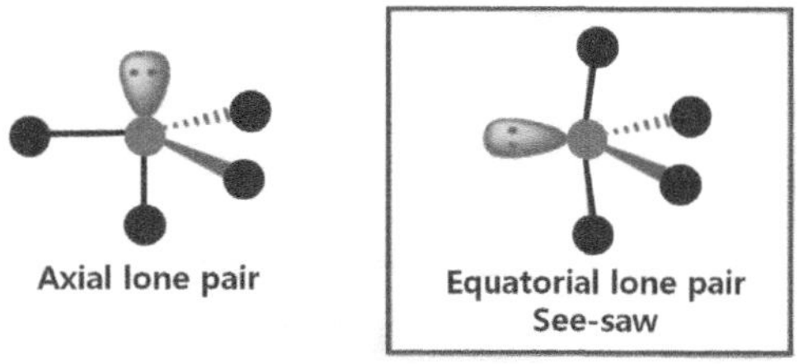

FIGURE 6.20
VSEPR Guideline #5.

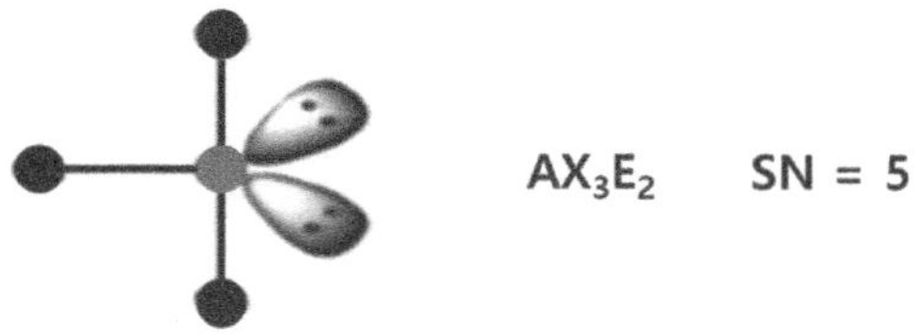

FIGURE 6.21
VSEPR Guideline #6.

FIGURE 6.22
VSEPR Guideline #7.

6. AX_4E_2 molecules have a T-shape (Figure 6.21).
7. AX_4E_2 molecules have a T-shape (Figure 6.22).
    - The two lone pairs are farthest apart when they are on facing sides of the central atom.
8. Angle between bonded atoms (Figures 6.23 and 6.24).
    - In molecules with lone-pair electrons (such as NH_3), angles between bonded atoms tend to be smaller.
    - As atomic size grows, lone pairs take up larger spatial volumes
    - Angles between bonded atoms tend to be even smaller.

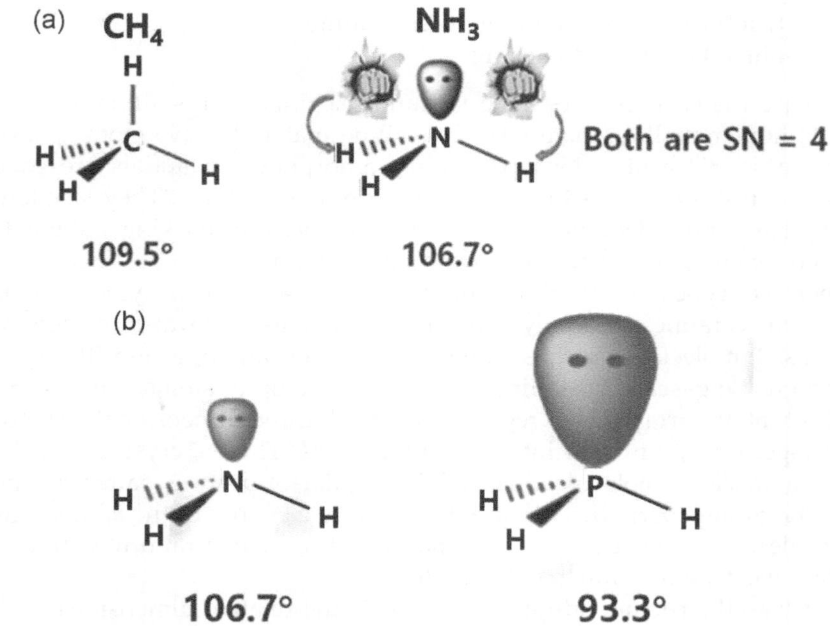

FIGURE 6.23
VSEPR Guideline #8: (a) CH_4 and NH_3; (b) NH_3 and PH_3.

	Formula type	SN	Molecular structure	Geometry
H_2O	AX_2E_2	4		Bent Polar
SF_1	AX_4E	5		See-saw
BrF_1	AX_3E_2	5		T-shaped
XeF_2	AX_2E_3	5	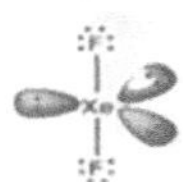	Linear
XeF_4	AX_4E_2	6		Square planner

FIGURE 6.24
VSEPR shapes.

6.3.7 Structure of Crystalline/Non-Crystalline Solids: Energy and Packing

There are the two kinds of materials and packing: 1) crystalline materials and 2) non-crystalline materials. Crystalline materials have atoms packed in periodic, 3D arrays. Non-crystalline (amorphous) materials are solids lacking a uniform three-dimensional positioning of atoms. They lack long-range structural order and have atoms without periodic packing and what is needed for complicated structures and rapid cooling.

There are types of crystalline solids that can be found in typical metals, numerous ceramics, and polymers: 1) ionic crystals, 2) covalent or network crystals, 3) molecular crystals, 4) metallic crystals, and 5) group III crystals (frozen noble gases). Crystalline solids are made up of atoms, ions, or molecules that are arrayed in a regularly ordered, three-dimensional structure held together by uniform intermolecular forces. That is, crystalline solids have atoms/ions/molecules arrayed in regular, repeating patterns. They have long-range periodicity that reduces the free energy of the atoms/ions/molecules. The unit cell is the minimal repeating structural unit with comprehensive crystal symmetry (Figure 6.25).

The crystal structure adopted relies on 1) the relative dimensions of the atoms/ions, 2) the relative numeric of the two types of atoms/ions, and 3) the electronic structure of the atoms/ions. The radius ratio is typically defined as r^* / r^-, where r^* is the radius of the cation and r^- is the radius of the anion. To understand the structure of crystalline solids, it is required to know 1) how atoms are constructed into solid structures, 2) how the density of a material depends on its structure, and 3) when material properties differ in size with the sample (i.e., part) orientation. In generally, dense and ordered packed structures are inclined to have lower energies (Figure 6.26).

For example, metals tend to be densely packed because 1) all atomic radii are identical, 2) metallic bonding is not directional, 3) nearest neighbor distances

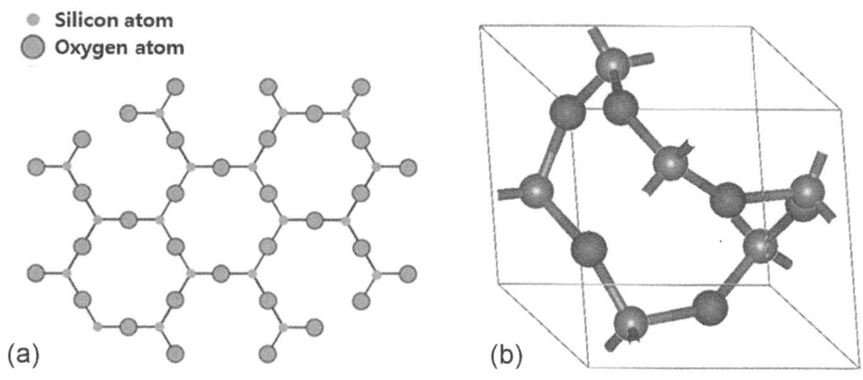

FIGURE 6.25
SiO_2 glass/crystal in crystalline solids: (a) SiO_2 glass; (b) SiO_2 crystal (α-quartz).

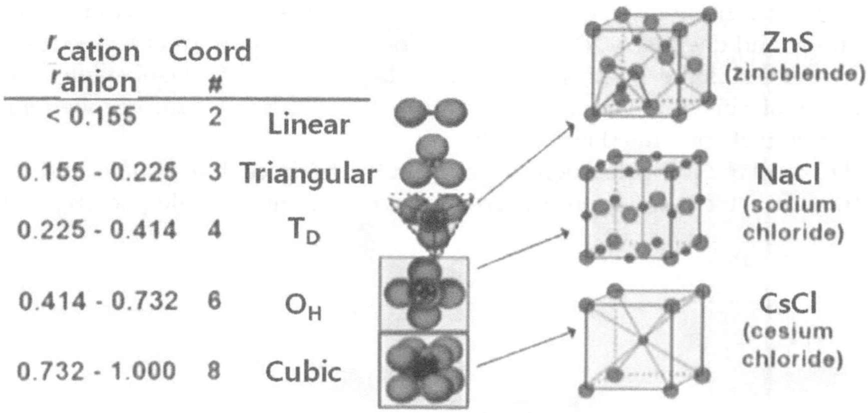

FIGURE 6.26
How many anions (bigger) can be arranged around a cation (smaller)?

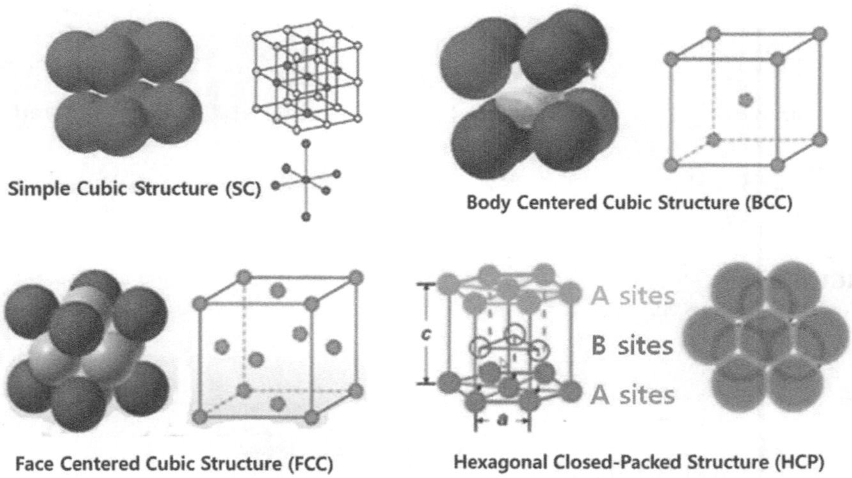

FIGURE 6.27
Simplest crystal structures.

is inclined to be small in order to lower bond energy, and 4) the electron cloud shields cores from one another. We thus will examine the simplest crystal structures such as straightforward cubic, face-centered cubic, body-centered cubic, and hexagonal close packed. Simple cubic structure (SC) has the unit cell or smallest repeating unit. Body centered cubic structure (BCC) has atoms that are in contact with each other's cubic diagonals. Face-centered cubic structure (FCC) has atoms that are in contact with each other along face diagonals. Hexagonal closed-packed structure (HCP) has the ABAB stacking sequence (Figure 6.27).

Crystallography deals with perceiving the positioning and bonding of atoms in crystalline solids and with the geometric construction of crystal lattices. Traditionally, the optical properties of crystals were of value in mineralogy and chemistry for the recognition of substances. In crystallography, the unit cell is the smallest repetitive volume that holds the absolute lattice pattern of a crystal. An important feature of the unit cell is the number of lattice points it contains (Figures 6.28 and 6.29).

Dissimilar cubic cells such as simple cubic, body-centered cubic, and face-centered cubic result from different packing atoms. Atomic packing factor

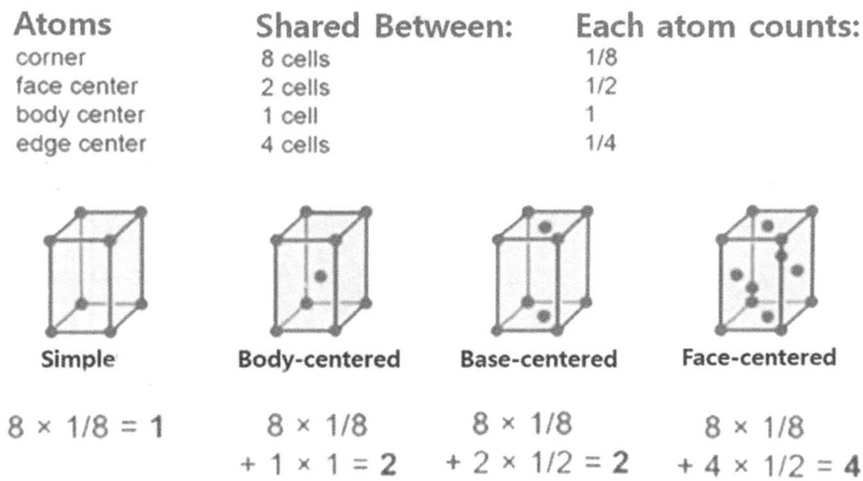

FIGURE 6.28
Contents of a unit cell.

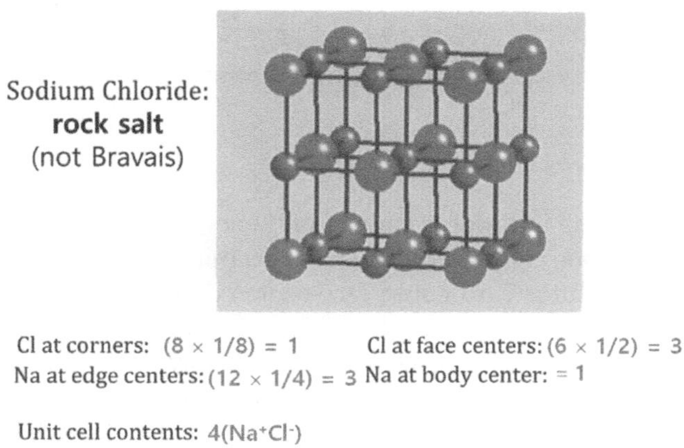

FIGURE 6.29
Unit cell of a sodium chloride.

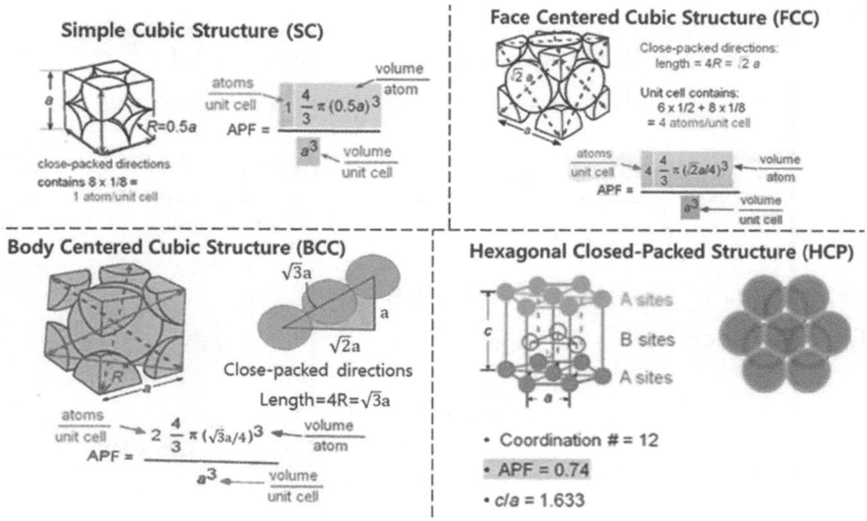

FIGURE 6.30
APF of crystal structures.

(APF), packing efficiency, or packing fraction is the fraction of volume in a crystal structure that is being used by constituent particles.

$$APF = \frac{N_a V_a}{V_{uc}} \tag{6.22}$$

where N_a is the numeric of atoms in the unit cell, V_a is the volume of one atom, V_{uc} is the volume of the unit cell.

The APF is a dimensionless quantity and always less than unity. In atomic systems, by convention, the APF is decided by presuming that atoms are rigid spheres. The radius of the spheres is taken to be the maximum value such that the atoms do not overlap. Usual sphere packings taken on by atomic systems are as follows: 1) simple cubic: 0.52; 2) face-centered cubic (FCC): 0.74 (also called cubic close-packed, CCP); 3) body-centered cubic (BCC): 0.68; and 4) hexagonal close-packed (HCP): 0.74. Atomic packing factor (APF) in the simplest crystal structures can be calculated (Figure 6.30).

Example 6.2

Iodine has an orthorhombic unit cell for which the *a*, *b*, *c* lattice parameters are 0.479, 0.725, and 0.978 nm, separately. As the APF and R are 0.547 and 0.177 nm, decide the numeric of atoms in each unit cell. If the atomic weight of iodine is 126.91 g/mol, calculate its theoretical density.

Solution

$$V_{\text{cell}} = (0.479\text{ nm})(0.725\text{ nm})(0.978\text{ nm}) = 3.4\times10^{-28}\text{ m}^3$$

$$V_{\text{atom}} = \frac{4}{3}\pi(0.177\text{ nm})^3 = 2.3\times10^{-29}\text{ m}^3$$

$$\text{APF} = \frac{(\#\,atoms)(V_{\text{atom}})}{V_{\text{cell}}} = \frac{(\#)(2.3\times10^{-29}\text{ m}^3)}{3.4\times10^{-28}\text{ m}^3} = 0.547$$

So the numeric of atoms in each unit cell, #, is 8

Theoretical density, $\rho\left(=\frac{\textit{Mass of atoms in unit cell}}{\textit{Total volume of unit cell}}\right)$

$$\rho = \frac{nW}{V_{\text{cell}}N_A} = \frac{8\,atoms\left(126.91\frac{g}{mol}\right)}{3.4\times10^{-28}\text{ m}^3\left(6.022\times10^{23}\text{atom / mol}\right)} = 4.96\times10^6\frac{g}{\text{m}^3}$$

A crystal system is a set of geometric symmetries with Bravais lattices. Face-centered, body-centered, and simple systems do not have to be cubic. They could be other parallelepiped shapes as well. The seven crystal systems that all 3D crystals belong to are cubic, tetragonal, orthorhombic, rhombohedral, monoclinic, triclinic, and hexagonal. Informally, two crystals are in the same crystal system if they have similar symmetries. Some engineering implementations necessitate single crystals such as diamonds for abrasives. Properties of crystalline materials is frequently connected to crystal structure (Figure 6.31).

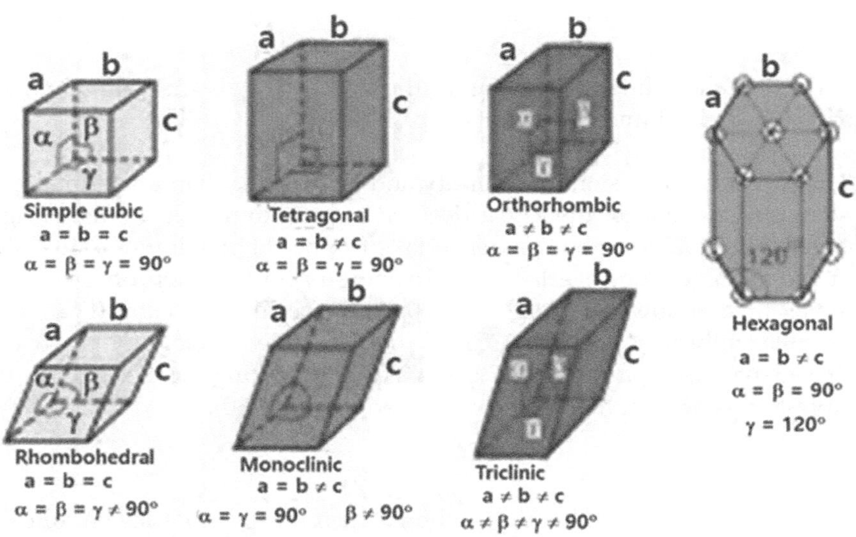

FIGURE 6.31
Crystal structure: seven types of unit cells.

Think about the close-packing of incompressible (hard) spheres. That is, regular close-packing in 2D necessitates a hexagonal array (HCP). Nearly all well-organized methods to pack spheres of single size is six nearest neighbors. In 3D, regular close-packing requires stacking 2D HCP arrays. An HCP crystal is a close-packed structure with the stacking sequence as follows: 1) *first layer*: 2D HCP array (layer A), 2) *second layer*: HCP layer with each sphere placed in alternate interstices in first layer, and 3) *third layer*: HCP layer positioned directly above first layer (repeat of layer A). HCP is the two interpenetrating simple hexagonal lattices displaced by $a_1/3 + a_2/3 + a_3/2$. Orientation occurs in turn repeatedly with each layer. Each sphere is in contact with 12 equidistant neighbors (six in plane, six out-of-plane). Structure has greatest packing fraction feasible for single-sized spheres (APF = 0.74) (Figure 6.32).

Most engineering materials are polycrystals composed of randomly oriented crystalline regions, especially as distinguished from a single crystal. Each "grain" is an individual crystal. Grain sizes generally vary from 1 nm to 2 cm (i.e., from a few to millions of atomic layers). General component properties thus are not directional. The property of single crystals varies with orientation (anisotropic) such as the modulus of elasticity (E) in BCC

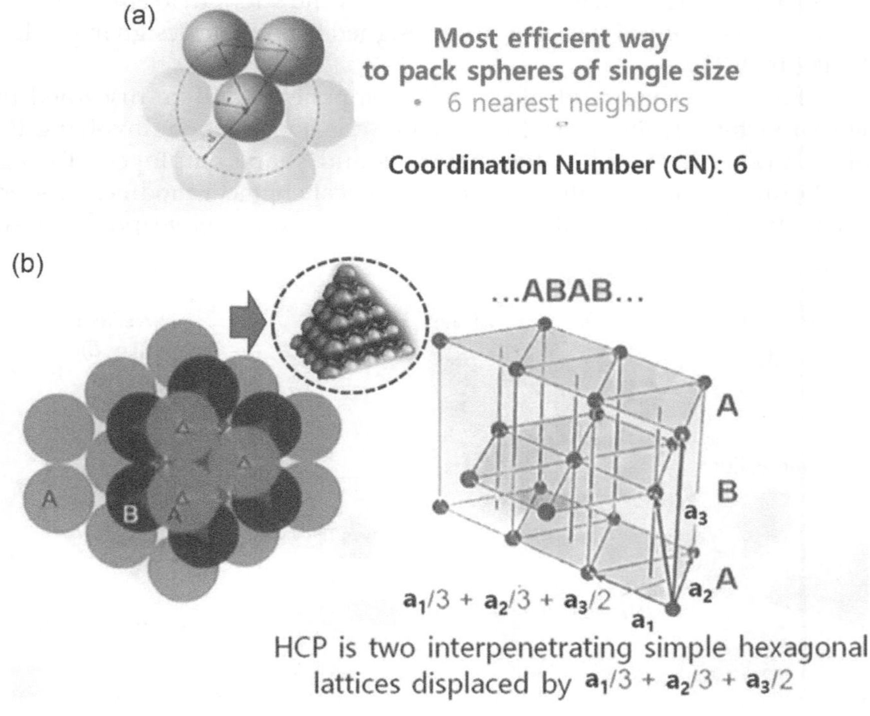

FIGURE 6.32
Close-packing structure: (a) regular close-packing in 2D; (b) close-packing in 3D.

iron. On the other hand, the properties can/cannot vary with direction. As grains are randomly aligned, isotropically ($E_{poly\ iron}$ =210 GPa).

6.4 Failure Mechanics and Design for Mechanical Products

6.4.1 Introduction

Design characteristics in a new mechanical structure must have sufficient strength and stiffness to withstand a number of loads. Through the design of the product, strength, resistance against irreversible deformation, is always required to be great. On the other hand, the product also requires stiffness—the resistance against reversible deformation. It may depend on application and change over a broad span. Because of product design due to inappropriate material and shape, an insufficient strength and stiffness in the product may cause permanent deformation, cracks, fractures, and loss of the product's intended functions. If there is a design defect in the construction where the loads are applied, the product may fracture during its lifetime. Engineers want to make products with sufficient stiffness and strength to endure stress. Finally, the mechanical system will be designed to be a fine design in product life (Figure 6.33).

The failure mechanics of mechanical components will be discerned by the two factors: 1) the stress (or loads) on the construction, involving the failure mechanism, and 2) the materials and shape employed. To prevent failure in the marketplace, the mechanical engineer modifies product designs by choosing an adequate material and form before products are

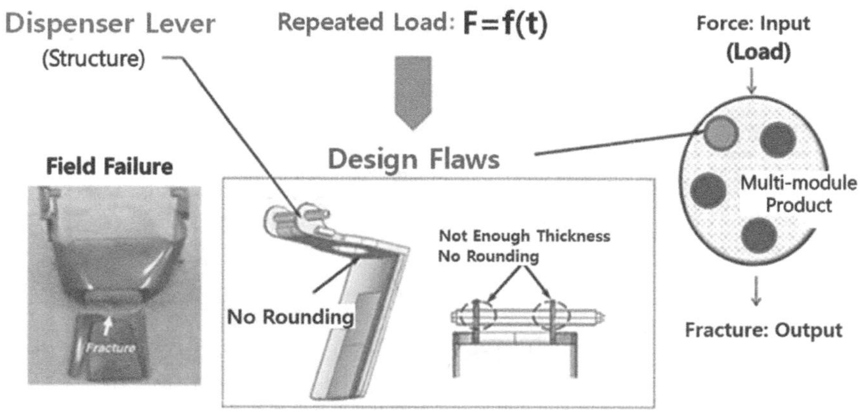

FIGURE 6.33
Failure mechanics produced by a load on a part produced from a particular material.

released. Especially due to the existence of a tiny crack or preexisting defect on the outside of a component in most mechanical systems, fatigue appears. Though the product is optimally designed, it could have design imperfections that will immediately become visible in the field. To stop this issue, the newly designed structure might be confirmed by means of scientific experiment whether or not it has a good-enough quality—enough stiffness and strength. That is, the design issues might be replicated and modified by reliability testing, such as parametric ALTs.

6.4.2 Flux

As seen in Figure 6.34, the flux F of any species is expressed as the $F \equiv$ numeric moving through the unit area in unit time. To obtain a mathematical relationship expressed in symbols for the flux F, consider the example of the motion of charged impurities in a positive manner in a crystal. The atoms of the crystal set up some potential hills that impede the motion of the charged impurities. The maximum of the potential barrier W is normally on the order of electron volts in almost all materials. The interval between following continuously potential barriers a, is on the arrangement of the lattice spacing, which is usually several angstroms.

As a repeated electric field is applied, the potential distribution as a function of distance will be created to proceed into an inclined location, as seen in Figure 6.34(b). It will roam positively charged particles to the right more easily and make their passage to the left harder. Let us now calculate the flux F at position x. This flux will be the mean of the fluxes at position $(x - a/2)$ and at $(x + a/2)$. Immediately one after another, these two fluxes are defined by F_1–F_2 and F_3–F_4, individually.

Think about the component F_1. It may be defined by the product of 1) the density per unit area of impurities at the potential hill at $(x$-$a)$; 2) the

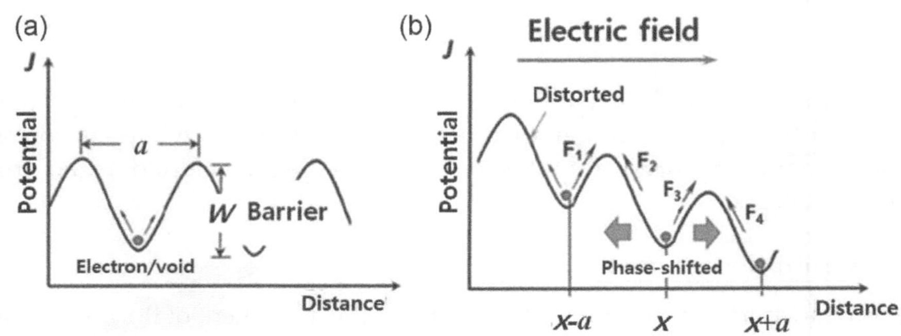

FIGURE 6.34
Model of ionic motion in a crystal: (a) initial state; (b) final state.

probability of a leap of any of these impurities to the valley at x; and 3) the frequency of attempted jumps v. Thus:

$$F_1 = [aC(x-a)] \cdot exp\left[-\frac{q}{kT}\left(W - \frac{1}{2}a\xi\right)\right] \cdot v \tag{6.23}$$

where $[aC(x—a)]$ is the density per unit area of particles situated in the valley at $(x\text{-}a)$, and the exponential factor is the probability of a successful leap from the hill at $(x\text{-}a)$ to the hill at x. Remember the lowering of the barrier due to the electric field ξ.

Similar formulas will be defined for F_1, F_2, and F_3. As these are merged to create a formula for the flux F at position x, with the concentrations $C(x \pm a)$ near to $C(x) \pm a(\partial C / \partial x)$, we obtain

$$F_1 = -\left[a^2 ve^{-qW/kT}\right] \cdot \cosh\frac{qa\xi}{2kT}\frac{\partial C}{\partial x} + \left[2ave^{-qW/kT}\right] C\sinh\frac{qa\xi}{2kT} \tag{6.24}$$

A very main limiting formation of this equation is attained for the instance when the electric field is relatively little, i.e., $\xi = kT / qa$. In this instance, we may enlarge the cosh and sinh words in Equation (6.24). Consider that $\cosh(x) = 1$ and $\sinh(x) = x$ for $x \to 0$, which follows in the limiting shape of the flux equation for a positively charged species:

$$F(x) = -D\frac{\partial C}{\partial x} + \mu\xi C \tag{6.25}$$

where $D \equiv a^2 ve^{-qW/kT}$ and $\propto \equiv a^2 ve^{-qW/kT} / kT / q$

Consider that the mobility $\propto$ and the diffusivity D are attached by

$$D = \frac{kT}{q}\mu \tag{6.26}$$

It is normal to acknowledge the contribution to the flux, which is proportional to the concentration gradient as the diffusion term, while the contribution that is proportional to the concentration itself is referred as the drift expression.

6.4.3 Diffusion

Mass transport in a gas or liquid necessitates the movement of fluid (e.g., convection currents), although atoms also diffuse. However, solids will endure shear stresses and therefore do not move except by diffusion necessitating the leaping of atoms on a secured network of locations. Assume that such jumps will be obtained in the solid state, with a frequency ν with each jump over a distance λ.

For random jumps, the root mean square distance is

$$\overline{x} = \lambda\sqrt{n} = \lambda\sqrt{\nu t}, \text{ where } n \text{ is the number of jumps, and } t \text{ is the time.} \quad (6.27)$$

Concentration of solute C, number/m^3. Each plane has $C\lambda$ atoms/m^2

$$\delta C = \lambda \frac{\partial C}{\partial x} \quad (6.28)$$

Atomic flux J atoms/(m^2 s):

$$J_{L\to R} = \frac{1}{6}\nu C\lambda \quad (6.29)$$

$$J_{R\to L} = \frac{1}{6}\nu(C+\delta C)\lambda \quad (6.30)$$

Thus the net flux along x is stated by

$$J_{\text{net}} = -\frac{1}{6}\nu\delta C\lambda = -\frac{1}{6}\nu\lambda^2\frac{\partial C}{\partial x} \equiv -D\frac{\partial C}{\partial x} \quad (6.31)$$

This is Fick's first law, where the constant of proportionality is defined as the diffusion coefficient in m^2/s. Fick's first law is relevant to steady-state flux in a constant concentration gradient. Thus the equation for the mean diffusion distance will be expressed with regard to the diffusivity:

$$\overline{x} = \lambda\sqrt{\nu t} \text{ with } D = \frac{1}{6}\nu\lambda^2 \text{ giving } \overline{x} = \sqrt{6Dt} \simeq \sqrt{Dt} \quad (6.32)$$

Presume that the concentration gradient is not constant

$$Flux\,in = -D\left(\frac{\partial C}{\partial x}\right)_1 \quad (6.33)$$

$$Flux\,out = -D\left(\frac{\partial C}{\partial x}\right)_2 = -D\left[\left(\frac{\partial C}{\partial x}\right)_1 + \delta x\frac{\partial^2 C}{\partial x^2}\right] \quad (6.34)$$

In the time interval δt, the concentration changes δC:

$$\delta C\delta x = (Flux\,in - Flux\,out)\delta t \quad (6.35)$$

$$\frac{\partial C}{\partial t} = D\frac{\partial^2 C}{\partial x^2} \quad (6.36)$$

Assume that the diffusivity is unrelated to the concentration. This is Fick's second law of diffusion. This is compliant with numerical outcomes for the common instance. However, there are a small number of fascinating structured answers for particular boundary conditions. For an instance where a fastened amount of solution is plated onto a semi-infinite bar.

The boundary circumstances may be expressed:

$$\int_0^{\infty} C(x,t)dx = B \text{ and } C(x,t=0)=0 \tag{6.37}$$

$$C(x,t)=\frac{B}{\sqrt{\pi Dt}}exp\left(\frac{-x^2}{4Dt}\right) \tag{6.38}$$

Presume that we create the diffusion couple by heaping an infinite number of narrow sources on the end of one of the bars. Diffusion will thus be served by taking a whole group of the exponential functions just obtained, each rather displaced along the x-axis, and summing (integrating) up their individual outcomes. The integral is actually the error function.

$$erf(x)=\frac{2}{\sqrt{\pi}}\int_0^x exp\left(-u^2\right)du \tag{6.39}$$

Thus the outcome to the diffusion equation, established on boundary situations:

$$C(x=0,t)=C_s \text{ and } C(x,t=0)=C_0 \tag{6.40}$$

$$C(x,t)=C_s-(C_s-C_0)erf\left(\frac{x}{2\sqrt{Dt}}\right) \tag{6.41}$$

This solution will be utilized in numerous circumstances where the surface concentration is kept constant, for instance, in carburization or decarburization processes. The outcomes stated here also apply to the diffusion of heat. Atoms in the solid-state drift by jumping into vacancies that will be interstitial or in substitutional sites. However, there is a hurdle to the movement of the atoms because the movement is connected to a transient distortion of the lattice.

Presuming that the atom tries to jump at a frequency v_0, the frequency of doing well jumps is expressed:

$$\nu = v_0\exp\left(\frac{-G^*}{kT}\right) \equiv v_0\exp\left(\frac{S^*}{k}\right)\exp\left(\frac{-H^*}{kT}\right) \tag{6.42}$$

where k is the Boltzmann constant, T is the absolute temperature, individually, and H^* and S^* are the activation enthalpy and activation entropy, individually. Since $D \propto \nu$, we find

$$D = D_0 exp\left(\frac{-H^*}{kT}\right) \tag{6.43}$$

6.4.4 Mechanism of Slip

As a mechanical product is exposed to a repetitive stress (or loads), plastic deformation happens at a premature stage. In material science, a dislocation is a linear crystallographic flaw that has an unexpected interchange in the arrangement of atoms. The movement of dislocations (or slip) leads atoms to tilt over each other at low stress levels. Slip occurs on planes that have the highest planar density of atoms and in the direction with the highest linear density of atoms. Specifically, slip occurs in the directions in which the atoms are packed with little space since this requires the lowest quantity of energy. So they will move smoothly past each other with force. The slide motion depends on the repetitive construction of the crystal, which guides the atoms to suddenly change direction from their first neighbors. It therefore goes along the face and joins with the atom of new crystals.

As a result of a simple shearing stress, a slip occurs. The determination of the axial tensile load F takes two loads. Namely, one F_s is the shear load along the slip plane, and the other F_N is a normal tensile load at an angle of 90° to the plane. As stated by the analysis and test, the maximum shear stress occurs at 45°. Figure 6.35 is the packing of atoms on a slip plane. It can be recognized that there are three close-packed ways of the atoms in which these will be the simple slip direction.

Components of the crystal on either side of a specific slip plane go against directions and become responsible for the atoms in nearly equilibrium places, so that there is a minute change in the lattice arrangement. Thus the external

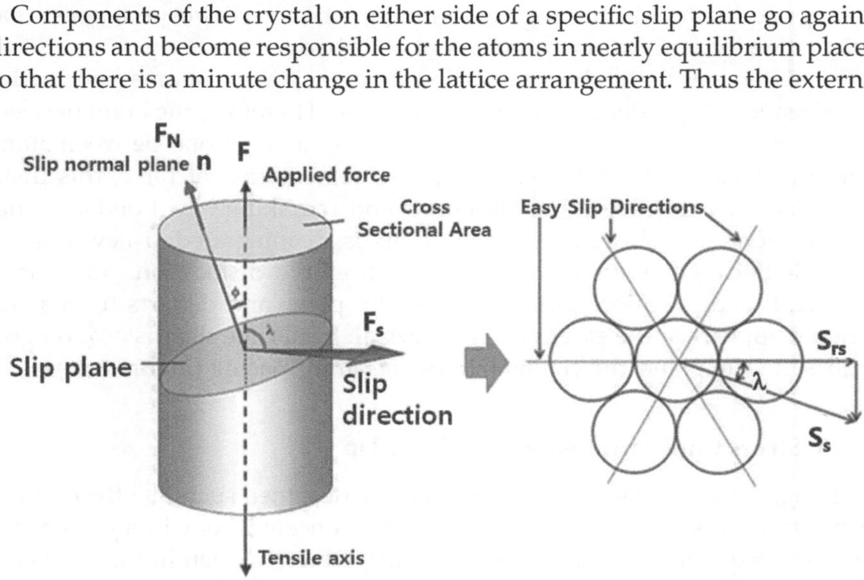

FIGURE 6.35
Components of force on a slip plane.

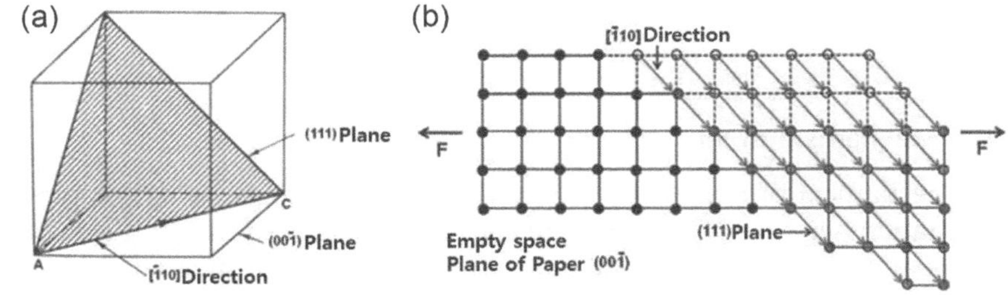

FIGURE 6.36
(a) Clarified picture of the slip plane in the FCC lattice; (b) simplified illustration.

shape of the crystal is modified with not ending it. Straightforwardly, a slip is explained in the face-centered cubic (FCC) lattice. The (111) plane is the slip plane having the maximum number of atoms. It bisects the (001) plane in the line AC (110) direction having the maximum number of atoms on it. A slip is considered a movement along the (111) planes in the close-packed (110) positioning.

From the simple picture of slip in an FCC crystal, one can assume that the atoms slip one after another with no interruption, starting at one location or at a few locations in the slip plane, and thus moving external over the rest of the plane. For instance, if one attempts to move the entire rug smoothly as one piece, the resistance is too great. One makes a fold in the rug and thus slides the whole rug a little at a time by thrusting the fold along. An alike analogy to the fold in the rug is the motion of an earthworm. It goes in an orientation by proceeding a component of its body at a time.

By implementing the shear force, first, an extra plane of atoms (specified as a dislocation) is produced above the slip plane. Therefore, the bond between atoms breaks, which leads to the production of a new bond between atoms and a dislocation. With the continued implementation of force, this dislocation goes by disintegrating old bonds and remaking new bonds. In such a movement, as the bond between atoms is disintegrated, a new bond is re-made between atoms repetitively, resulting in a dislocation. As a consequence, this dislocation goes across the slip plane and departs from a step when it appears at the exterior of the crystal. Each time the dislocation goes across the slip plane, the crystal moves one atom spacing (Figure 6.36).

6.4.5 Stress Concentration at the Crack Tip

Although nearly all mechanical products are designed such that the nominal stress stays in elastic stress ($S_n < \sigma_{ys}$), stress concentrations in a mechanical product frequently produces plastic strains, which happen in the locality of design defects or stress raisers such as holes, fillets, notches, and grooves where the stress is enlarged. The fracture strength in a material is related to the cohesive forces between atoms. It may be approximated that the conceptual cohesive strength of a material might be one-tenth of the elastic modulus (*E*). However, the investigative fracture strength for a brittle material is normally $E/100$–$E/10{,}000$ below this conceptual quantity. This much lower fracture strength originates from the stress concentration due to the existence of microscopic defects or cracks found on the outside of the material. The stress contour along the *x*-axis is focused on an elliptically shaped internal crack (Figure 6.37).

Stress has a greatest value at the crack tip, and it lessens to the theoretical exerted stress with expanding distance away from the crack. Stress raisers have the capacity to grow the stress at a nominated position. The value of amplification depends on the crack direction and geometry. Inglis's solution (1913) utilized not only elliptical coordinates but also complex numbers to find an answer to the elliptical hole problem [2]. He attained the outcomes that the enclosed stresses around a corner or hole in a stressed plate might be many times larger than the mean acted stress. The maximum stress at the end of the ellipse is connected to its dimension and shape by

$$\sigma_{\max} = \sigma_{\infty}\left[1+2\frac{a}{b}\right] \tag{6.44}$$

In a way that is easy to understand, Inglis's elliptical result lessens to the known $\sigma_{\max} = 3\sigma_{\infty}$ for the specific instance of a hole when $a = b$. Namely, a close outcome could be seen in most strength of materials. On the other hand, the greatest stress is predicted to proceed toward infinity as the ellipse makes flat ($b \to 0$).

The radius of curvature ρ at the end of an ellipse is related to its length and width by

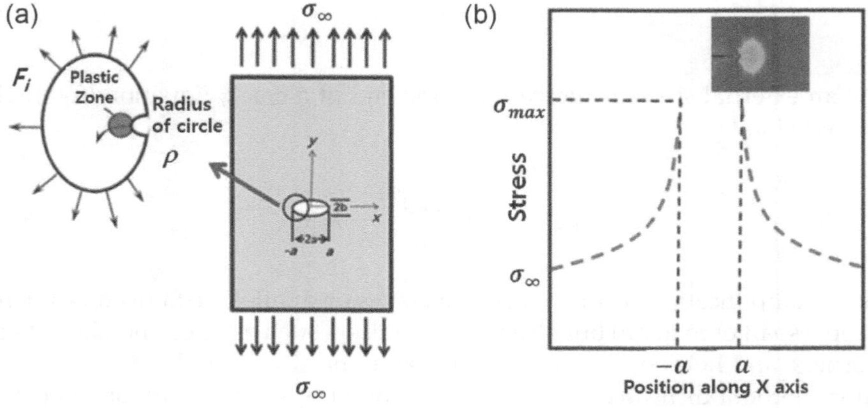

FIGURE 6.37
(a) Stress concentration at crack tip locations; (b) simple stress profile along the *x*-axis at the crack tip.

$$\rho = \frac{b^2}{a} \tag{6.45}$$

Solving this for b and interchanging it into the a/b proportion in Equation (6.44), it may be roughly calculated by Equation (6.46):

$$\sigma_{max} = \sigma_{\infty}\left[1 + 2\sqrt{\frac{a}{\rho}}\right] \tag{6.46}$$

where ρ is the radius of curvature, σ_{∞} is the exerted stress, $\tilde{A}_{max}$ is the stress at the crack tip, and a is the half-length of the inside crack or the complete length for a surface fault.

As the crack is similar to an elliptical hole through the plate and is lined up at an angle of 90° to the applied stress, the maximum stress σ_{max} occurs at a crack tip. The amount of the conceptual acted tensile stress is σ_{∞}; the radius of the curvature of the crack tip is ρ; and a shows the length of a surface crack, or half-length of an interior crack. For a relatively long microcrack, the factor $(a/\rho)^{1/2}$ can be very big. Thus Equation (6.46) may be redefined:

$$\sigma_{max} \cong 2\sigma_{\infty}\left(\frac{a}{\rho}\right)^{1/2} \tag{6.47}$$

The proportion between the greatest stress and the nominal acted tensile stress is expressed as the stress concentration factor K_t. The stress concentration factor is a uncomplicated portion of the degree to which an outer stress is increased at the end of a minute crack and is expressed as

$$K_t = \frac{\sigma_{max}}{\sigma_o} \approx 2\left(\frac{a}{\rho}\right)^{1/2} \tag{6.48}$$

As an external stress is increased at the end of a crack, Equation (6.48) will be redefined:

$$\sigma_{max} = 2\sigma_{\infty}\left(\frac{a}{\rho}\right)^{1/2} = K_t\sigma_{\infty} \tag{6.49}$$

Stress amplification occurs not only at cracks or small-scale faults on a microscopic span of material but also in stress raisers such as fillets, notches, sharp corners, and holes on the macroscopic amount. Cracks with keen tips grow more without difficulty than cracks with blunt tips. As a result of an enlarging acted stress, stress concentration may occur at microscopic defects, sharp corners, notches, interior discontinuities (inclusions/voids), and scratches that are generally expressed as stress raisers. Stress concentrators are generally

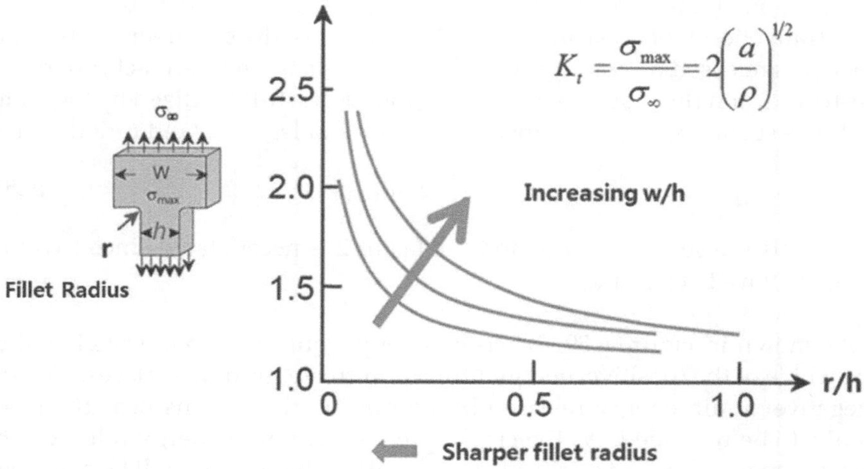

FIGURE 6.38
Stress concentration at sharp corners in accordance with fillet radius [3].

more destructive in brittle materials. Ductile materials have the capacity to plastically deform in the region adjacent to the stress raisers, which in a row uniformly distributes the stress load near the defect. The factor of maximum stress concentration outcomes in a value less than that found for the theoretical value. As brittle materials will not deform in a plastic manner, the stress raisers will create the conceptual stress concentration in certain circumstance, with the amount of amplification dependent on microcrack arrangement, dimensions, and geometry. For example, the stress concentration at sharp-edged corners depends on the fillet radius (Figure 6.38).

6.4.6 Fracture Toughness and Crack Propagation

Cracks with sharp-edged tips increase in size and spread more easily than cracks having blunt-edged tips. In ductile materials, plastic deformation at a crack tip "blunt" develops to the crack. Elastic strain energy is kept in a material as elastically deformed. This energy is freed when the crack increases. The emergence of new surfaces necessitates energy. As a crack increases into a solid to a depth a, an area of material adjacent to the free surfaces is unloaded, and its strain energy is allowed to move. An uncomplicated manner of discerning this energy release is to think about the two triangular areas near the crack flanks, of width a and height πa, as totally unloaded, as the existing material remains in the filled stress σ. The entire strain energy U released is thus the strain energy per unit volume in both triangular areas:

$$U^* = -\frac{\sigma^2}{2E} \cdot \pi a^2 \tag{6.50}$$

In this moment, the area normal to the *x-y* plane is taken to be unity, so U is the strain energy of releasing per unit thickness of the specimen. This strain energy is released by crack growth. However, in shaping the crack, bonds are destroyed, and the required bond energy is successfully utilized by the material. The surface energy S connected to a crack of length a (and unit depth) is

$$S = 2\gamma a \tag{6.51}$$

where γ is the surface energy and the factor 2 is necessitated since two free surfaces have been found.

As shown in Figure 6.39, the entire energy connected to the crack is thus the whole of the (positive) energy utilized to create the new surfaces, plus the (negative) strain energy released by permitting the regions near the crack flanks to be unloaded. As the crack spreads, the strain energy relies on the surface energy. Beyond a critical crack length a_{c}, the system will be at its lowest energy by allowing the crack to grow longer. Up to the point where $a = a_{c}$, the crack will be created only if the stress grows. Beyond the location, crack growth is not constrained and disastrous.

The amount of the critical crack length will be attained by placing the derivative of the entire energy $S + U^*$ to zero:

$$\frac{\partial\left(S+U^*\right)}{\partial a} = 2\gamma - \frac{\sigma_f^2}{E}\cdot \pi a = 0 \tag{6.52}$$

Because fast fracture is approaching when this circumstance is fulfilled, we may resolve Equation (6.52). The required critical stress for crack propagation is defined:

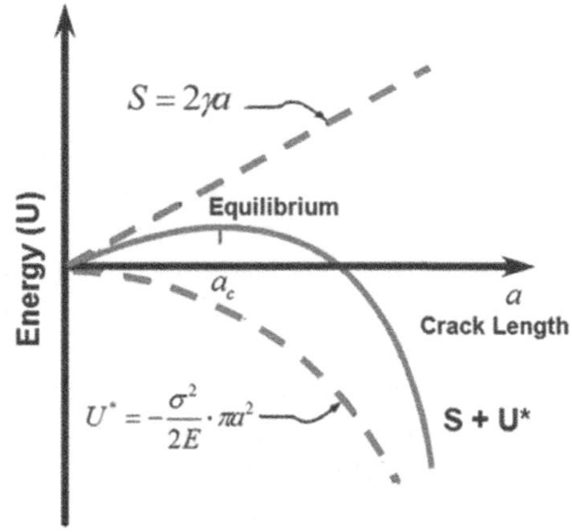

FIGURE 6.39
Entire energy connected to the crack.

$$\sigma_c = \left(\frac{2E\gamma_s}{\pi a}\right)^{1/2} \tag{6.53}$$

where γ_s is the specific surface energy.

When the tensile stress at the point of the crack exceeds the critical stress amount, the crack increases and develops in fracture. Nearly all polymers and metals have plastic deformation. For ductile materials, the particular surface energy γ_s will be exchanged into $\gamma_s + \gamma_p$, where γ_p is the plastic deformation energy. Thus Equation (6.53) will be defined:

$$\sigma_c = \left(\frac{2E\left(\gamma_s + \gamma_p\right)}{\pi a}\right)^{1/2} \tag{6.54}$$

For highly ductile materials, $\gamma_p \gg \gamma_s$ is accurate. Thus Equation (6.54) will be restated as follows:

$$\sigma_c = \left(\frac{2E\gamma_p}{\pi a}\right)^{1/2} \tag{6.55}$$

Nearly all brittle materials have a population of small-scale imperfections that are a number of physical magnitudes. As the dimensions of the tensile stress at the point of the crack exceeds the critical stress amount, the crack increases and develops into fracture. The defect-free metallic and ceramic materials have been created with facture strength that estimates their conceptual amounts. As $a \to 0$ or $w \to \infty$, $Y \to 1$. As the sample thickness grows, the fracture toughness reduces until the plane strain area is obtained. Fracture toughness depends on temperature, strain rate, and microstructure. Its size lessens with growing strain rate and decreasing temperature. As the yield strength is enhanced by alloying and strain hardening, the fracture toughness will increase with a reduction in grain size.

6.4.7 Crack Growth Rates

Metal fatigue begins at a surface (or an inside) imperfection by concentrated stresses and moves at the begin of shear flow along slip planes. Slip may happen in the (111) plane in an FCC lattice because the atoms are nearly all packed in a way that uses little space. Over several (random) loading cycles in the field, this slip generates intrusions and extrusions that start to look like a crack. An accurate crack moving toward the inside from an intrusion region will increase in size at first along one of the first slip flat surfaces but, in the end, turns to proceed transversely to the principal normal stress. If repeated loads are applied, the slip bands will evolve into small shear-driven microcracks. These Stage I cracks will be described as a back-and-forth slip on a series of neighboring crystallographic planes to construct a band. In these slip bands, pore nucleation and the merging of voids to form whole occur. It ultimately leads to microcrack formation. Often, extrusion and intrusions will also become visible, which is a much-localized discontinuity and the results in much faster microcrack emergence.

In Stage II of fatigue, microcracks connect to come out a macrocrack. At this point, the crack is of sufficient length to escape shearing stress control and to be worked by normal stress, which creates a constant growth, cycle by cycle, on a plane and which is no longer more crystallographic but rather normal to outer loads. In front of this macrocrack, two plastic lobes are caused by stress concentration. The cracks increase at an angle of 90° to the ruling stress and increase remarkably due to plastic stresses at the crack tip, as shown in Figure 6.40.

It is critical that engineers may predict the rate of crack growth during load cycling in engineering structures so that the problematic components may be altered before the crack expands to a crucial length. A very large quantity of test verification shows that the crack growth rate will be modified with cycle variation in the stress intensity factor [4]:

$$\frac{da}{dN} = A\Delta K^{m} \tag{6.56}$$

where da/dN is the fatigue crack rate per cycle, $\Delta K = K_{\min} - K_{\max}$ is the stress intensity factor range during the cycle, and A and m are parameters that depend on the material, environment, frequency, temperature, and stress ratio.

During Stage II, the propagation rate of fatigue crack depends on the stress level, crack size, and materials. It is well known as the Paris law, which employs charts like those described in Figure 6.41.

Specified values of constants m and A for some alloys are shown in Table 6.1. The exponent m is frequently near 4 for metallic systems, which could be explained as the damage accumulation connected to the volume V_p of the plastic zone: Since the volume V_p of the zone scales with r_p^2 and $r_p \propto K_I^2$, thus $da/dn \propto \Delta K^4$.

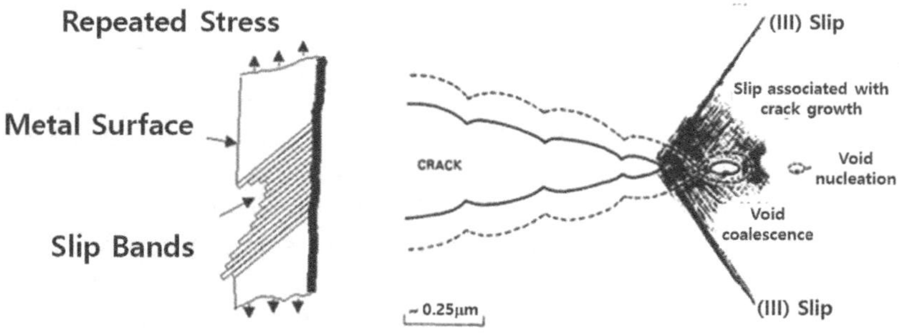

FIGURE 6.40
Simplified illustration of common slip generating nucleation and growth of voids.

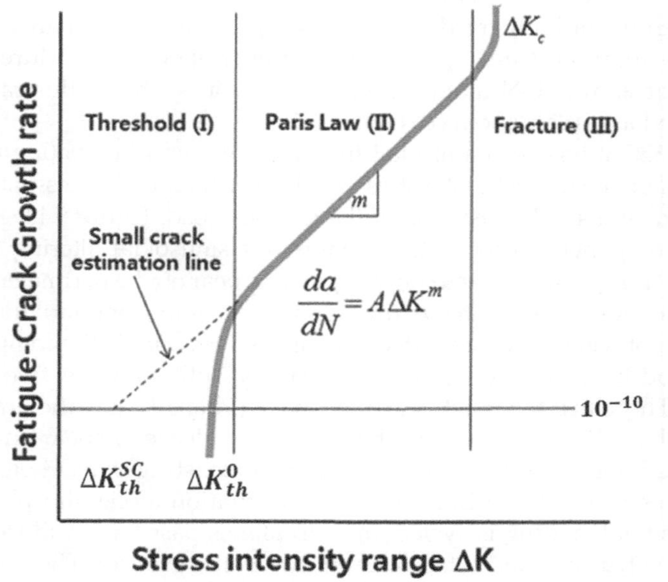

FIGURE 6.41
Paris law.

TABLE 6.1
Numerical Parameters in the Paris Equation

Alloy	m	A
Steel	3	10^{-11}
Aluminum	3	10^{-12}
Nickel	3.3	$4 \cdot 10^{-12}$
Titanium	5	10^{-11}

6.5 Fatigue and Fracture Failure

6.5.1 Introduction

In the middle of the 19th century, Wöhler at first began the contemporary investigation of fatigue. As German engineer on the Lower Silesia–Brandenberg Railroad, he labored for the railroad system and was the head manager of rolling stock. He investigated the sources of fracture in railcar axles after elongated usage. A railcar axle is fundamentally a circular beam with four-point bending, which yields a compressive stress along the peak area and a tensile stress along the bottom. As the axle is revolved a half turn, the bottom begins to be the top and conversely, so that the stresses on a particular area of the material at the surface alter repeatedly from compression to tension. As the metal begins to weaken, fatigue was identified to describe this pattern of damage. This is just now familiar as totally reversed fatigue loading. Some of Wöhler's data are for Krupp axle steel and are plotted with regard to nominal stress (*S*) versus the number of cycles to failure (*N*), which is familiar as the S–N diagram. Each curve on such an illustration is still mentioned as a Wöhler curve (Figure 6.42).

Since 1830, it has been admitted that metal under a fluctuating or repeated load shall be unsuccessful at a stress level lower than that necessitated to guide failure under a single execution of the identical load. Figure 6.43 shows a bar-shaped component exposed to a continuous sinusoidal altering force. As a period of time passes, a crack could begin to appear on the perimeter of the hole. This crack thus increases in size throughout the component until the entire segment cannot withstand the exerted stresses, and eventually the component fails.

The bodily growth of a crack is usually split into the two dissimilar phases. These are the crack starting phase (Phase I) and the crack growth phase (Phase II). Fatigue cracks begin by the release of shear strain energy. Figure 6.43 shows how a minute crack or preexisting defect begins and the shear stresses evolve in inner plastic deformation along slip planes. When the sinusoidal loading is cycled, the slip planes pass back and forth, following little extrusions and intrusions on the crystal exterior. These surface disturbances are roughly 1–10 μm in height and add up to embryonic cracks.

Phase I growth follows the direction of the greatest shear plane, or 45° to the orientation of loading. A crack begins in this way and expands to

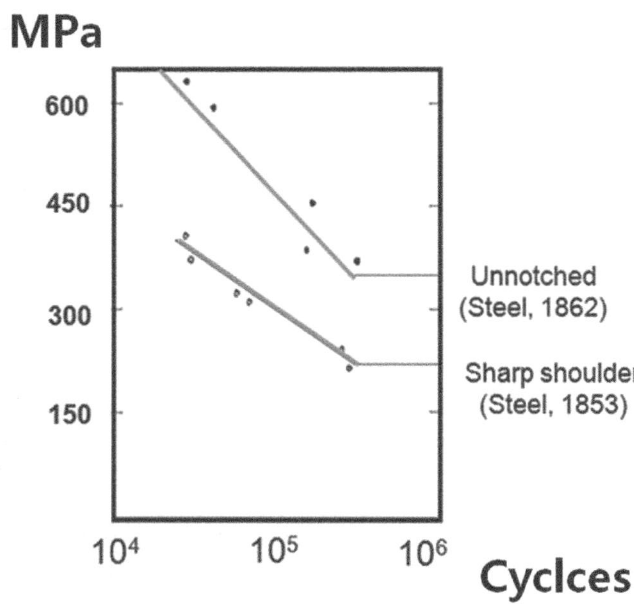

FIGURE 6.42
Some of Wöhler's data for rail car axle steel on the S–N curve [5].

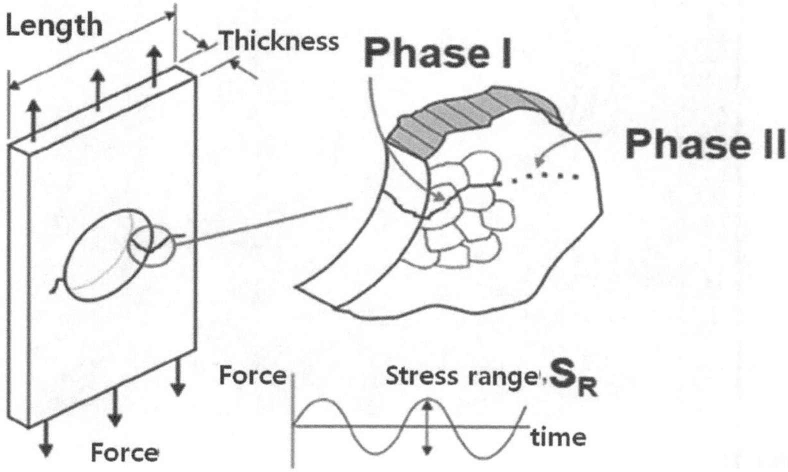

FIGURE 6.43
Bar-shaped part exposed to a continuous sinusoidal changing force.

the grain boundary. The mechanism at this place is moved to the neighboring grain at a slow speed. Once the crack has grown through roughly three grains, it is seen to modify its direction of propagation. The physical mechanism for fatigue moves to Phase II. The crack is sufficiently large to create a geometrical stress concentration. A tensile plastic zone is created at the crack tip, as shown in Figure 6.43. After this phase, the crack proceeds at an angle of 90° to the orientation of the acted load.

Roughly 90% of all structural failures occur through a fatigue mechanism. Fatigue happens after a member is exposed to repetitive cyclic loads and repeated deformations. It relies on the form and material and on how close to the elastic limit it deforms. It also shows itself in the formation of cracks, which are developing at particular locations in various kinds of mechanical structures and which would involve mechanisms such as airplanes, turbines, cranes, machine components, bridges, offshore platforms, reactor vessels, transmission towers, canal lock doors, masts, and chimneys.

If the supplied stress is sufficient large but not too high to produce an unanticipated complete fracture, the crack tip advances by a very little interval in each loading cycle. It cannot show any growing variations in material during the fatigue process, so brittle failures generally occur without care. Rest time with the fatigue stress removed does not lead to any clear recovery or healing. With the naked eye, we may observe a "clam shell"' structure in the crack plane. Under a microscope, striations shall be observed, which distinguish the positions of the crack tip after each single loading cycle. As crack growth is very small-scale in each loading cycle, a large number of cycles are necessitated before the whole failure occurs. Fatigue was first found in the 19th century by noticing the poor service lifetime of railroad axles established on static design limits (Figure 6.44).

FIGURE 6.44
Fracture in train wrecks due to metal fatigue (Wikipedia).

The fatigue lifetime of a construction exposed to repeatedly cyclic loads is defined as the numeric of stress cycles that shall come before failure. The bodily outcome of a repeated load on a matter is different from that of a static load. That is, failure without exception shall be brittle fracture:

> Irrespective of whether the matter is ductile or brittle. Fatigue mostly occurs at stresses well below the static elastic strength of the matter. Relying on the constructional specific shape, its fabrication or the matter utilized, four crucial factors shall influence the fatigue strength: (1) the stress difference called the stress range, (2) the material, (3) the structural shape, and (4) the environment.

Fatigue in ductile metals is the emergence of cracks with high stress concentrations, such as holes and grooves, and spreading [6,7]. Fatigue stresses in mechanical products take the shape of a sinusoidal pattern. In periodic patterns, the summits on both the low side and the high side are critical. The altered Goodman diagram explains the fatigue of components occurring repeatedly with, in turn, stress on the y-axis and mean stress on the x-axis. Therefore:

$$\frac{S_a}{S_e} + \frac{S_m}{S_{ut}} = 1 \tag{6.57}$$

where S_a is the alternating stress, S_m is the mean stress, S_e is the fatigue limit for totally reversed loading, and S_{ut} is the ultimate tensile strength of the material.

The cycles to failure are established as a function of mean stress and range along the lines of constant R-values [8]. The modified Goodman relation indicates a failure envelope such that any alternating stress that falls vertically in the diagram may not cause failure (Figure 6.45).

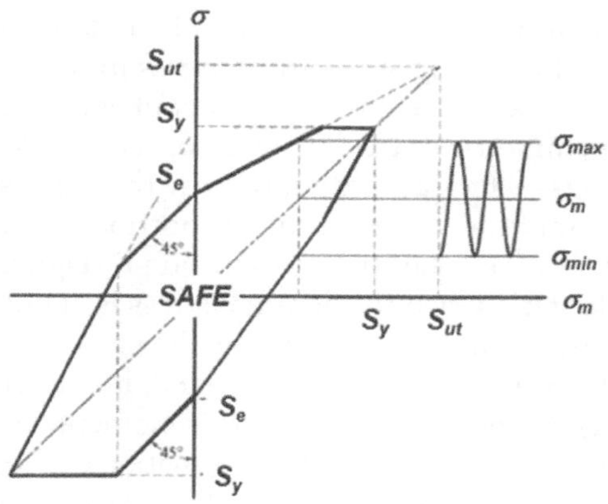

FIGURE 6.45
"Complete" modified Goodman diagram [9].

However, fatigue failure standards, holding the classic Goodman diagram [10–13], have difficulty estimating the life cycles of products because small samples, not modules, are tested and component failures due to design flaws seldom happen in the field. Because these organized procedures are unsuccessful in replicating market failure, they still have been disputed.

If there are design defects where a product is exposed to repeated loads, it will be unsuccessful in its life due to fatigue. It is the weakening of a material conducted by cyclic loading. Especially, the regime of low-cycle fatigue (LCF) is recognized by high cyclic stress levels in excess of the endurance limit of the material and is normally understood to be in the region of 10^4–10^5 cycles. A significant level of awareness is being directed to LCF performance of superalloys, mainly in the region of turbine engine designs [14–18].

Numerous papers have been issued about the LCFs of nickel-base polycrystalline [19–21] and monocrystalline alloys [22]. There are some fatigue tests on different kinds of samples as expressed by specified fatigue testing standards [23–25]. However, it is not easy to get test data of products accessible to directly construct S–N curves for LCFs because complete testing would have been too high-cost due to period and sample size. In Chapter 7 & 8, a structured method will be discussed for reliability tests for increasing the LCF failure of a mechanical system exposed to repetitive loading. It involves (1) an ALT plan established on the BX life, (2) a load examination, (3) a tailored ALT with the modifications, and (4) a judgment of whether the designs of the mechanical system reach the aimed BX life.

6.5.2 Fluctuating Load

Most mechanical structure including mechanisms are exposed to repetitive variable, fluctuating loading. This sort of loading is very intimidating not only because limited stresses are much lower than those received for static loading but also because of the nature of fatigue failure; products are suddenly unsuccessful without symptoms of yielding in their life. There are three different fluctuating stress–time types in fatigue loading: 1) symmetrical about zero stress, 2) asymmetrical about zero stress, and 3) random stress cycle. For asymmetrical stress of approximately zero, a fluctuating load (stress), which brings on fatigue, is discerned by the mean stress σ_m, the range of stress $\Delta\sigma$, the amplitude stress σ_a, the proportion of the mean stress over the amplitude stress χ, and the stress proportion R (Figure 6.46).

The fatigue cycle starts when the nominal stress is at position (a). When the stress grows in tension through position (b), the crack tip is open to produce enclosed plastic deformation. On the other hand, it enlarges into the virginal metal at position (c). As the tensile stress at this moment decreases through (d), the crack tip closes, and the final plastic deformation creates a distinctive saw-tooth profile known as a striation. On realization of the cycle at position (e), the crack has begun through length Δa and has established an additional striation.

As shown in Figures 6.47 and 6.48, extra cyclic stresses that bring on fatigue are 1) periodic and symmetrical about zero stress and 2) random stress fluctuations. In (mechanical) systems such as aircraft, refrigerators, automobiles, machine parts, and bridges, fatigue under fluctuating/cyclic stresses are required:

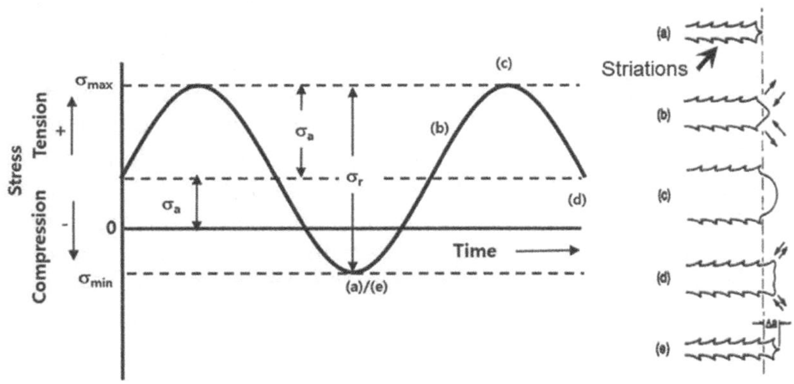

FIGURE 6.46
Fatigue with striations under fluctuating and asymmetrical cyclic stresses at approximately zero stress.

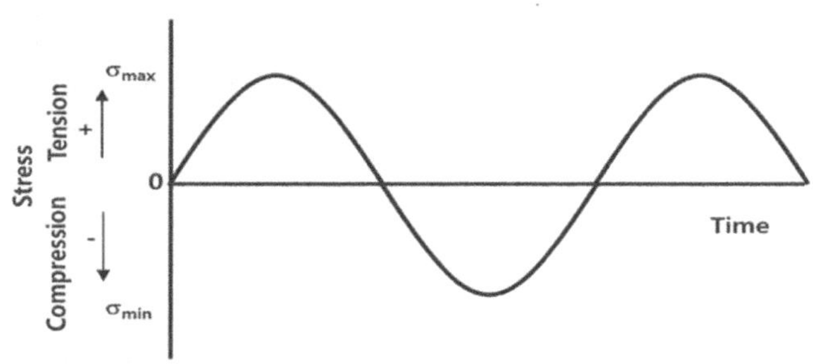

FIGURE 6.47
Periodic and symmetrical about zero stress in fatigue.

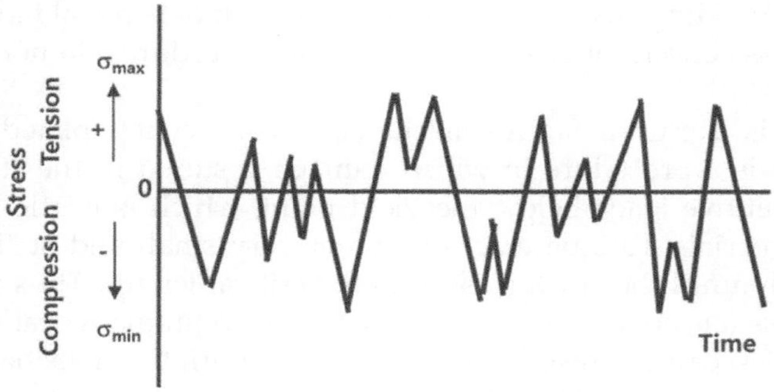

FIGURE 6.48
Random stress fluctuations in fatigue.

- A great enough tensile stress of extremely high value
- A great enough variation or fluctuation in the applied stress
- An indefinite enough quantity of exerted stress cycles

These are constituted in the subsequent equations:

$$\sigma_m = \frac{(\sigma_{max} + \sigma_{min})}{2} \tag{6.58}$$

$$\Delta\sigma = (\sigma_{max} - \sigma_{min}) \tag{6.59}$$

$$\sigma_a = \frac{(\sigma_{max} - \sigma_{min})}{2} \tag{6.60}$$

$$\chi = \frac{\sigma_m}{\sigma_a} \tag{6.61}$$

$$R = \sigma_{min}/\sigma_{max} \tag{6.62}$$

There are the patterns of two classic loads: one with $\chi = 1$, the load is pulsating, and the other with $\chi = 0$, the load is turned toward the direction opposite to that previously stated. The latter is the most intimidating pattern of load difference. The stress bound for fluctuating loading (the endurance limit) is defined as a kind of stress that is fixed for a stated specimen irrespective of the number of load recurrences. It normally remains in a compact connection to the ultimate stress limit.

There are the two procedures to determine when a component is likely to incur metal fatigue: 1) forecasting when a failure may occur due to the shape/iteration/matter/force mix and adding in place of the risky materials before this occurs or 2) performing analysis to recognize the microscopic cracks and carrying out replacements once they occur. Selecting materials that are not likely to be hurt by fatigue in the product lifetime is the optimal outcome. Avoiding shapes with sharp corners hinders metal fatigue by lessening stress concentrations. However, these procedures do not remove the stress.

Fatigue is a crucial failure mechanism to be contemplated in product designs. It is a procedure in which damage is stored in the structure due to the repetitive loads below the yield point, which is brittle-like even in ductile materials. Fatigue cracks begin at very small and at first increase very slowly until the crack is close to the critical length. Thus it is dangerous because it is not easy to initially recognize accumulative fatigue damage with no telescope, microscope, or other device until the crack has developed to near-critical length. The widespread fracture surface is at right angles to the orientation of the acted stress. Fatigue failure has three understandable phases: 1) crack beginning in the areas of stress concentration (near stress raisers), 2) incremental crack propagation, and 3) ultimate disastrous failure.

A very large number of "fatigue" sources have been investigated when the Versailles rail accident, the disaster of the Comet aircraft, etc. became big enough to spread catastrophically (see Chapter 1). Fatigue failure occurs in both metallic and nonmetallic materials, and it accounts for roughly 80–90% of all structural failures such as motor shafts, aircraft landing gear, machine components, bridges, automobile crankshaft, etc. As a result, designing for a maximum stress will not assure actual product life because most fractures generated are related with this categorization. Most engineering stress is asymmetrical near stress raisers—fillets, notches, or holes that concentrate the stress. In drawings, engineers frequently neglect these design flaws, which can create reliability disasters. For instance, the vibration of an aircraft wing in a prolonged flight will result in tens of thousands of loading cycles. Thus unless correctly designed, these structures will fall apart. It is notable to find the design flaws and rectify them. In Chapters 7 and 8, we will write how to explore the design defects by employing the ALT.

The major issue in designing against fracture in high-strength materials is that because of the existence of cracks from an earlier time, the engineer will change the local stresses to such a size that elastic stress done so on purpose by the engineers are insufficient in analyzing it. As a crack enlarges to a definite critical length, it will grow to a very failed size throughout the structure, although the total stress is much less than might generally bring on failure in a tensile specimen. The phrase "fracture mechanics" mentions a crucial specialization in solid mechanics in which the existence of a crack is presumed that involves quantitative relationships between the crack length, the material intrinsic resistance to crack growth, and the stress at which the crack spreads at high speed to bring on constructional failure. A speedy fracture will occur in a few loading cycles. For example, fatigue in the shafts of a motor rotating at 1,200 RPM took less than 12 hours from installation to the final fracture, roughly 830,000 cycles. On the other hand, the crack growth of shafts in steadily revolving equipment has taken many months and more than 10,000,000 cycles to fail.

6.5.3 Fracture Failure

Fracture is the failure of a structure into segments when exposed repeatedly to stress. It happens whenever the acted loads (or stresses) are more than the resisting strength of the body. It starts with a crack that shatters without totally separating. Fracture due to overstress is probably the most common failure mechanism and may be categorized as brittle (cleavage) fracture, ductile (shear) fracture, fatigue fracture, de-adhesion, and crazing. As a crack spreads, a new free surface having a particular surface energy $\gamma$is generated. It is provided by the external load and is also attainable as accumulated elastic energy. However, not all of the available energy is used for the making new crack surfaces. It is also turned into other types of energies, such as dissipative heat or kinetic energy. If enough available energy is used for crack growth, the fracture is designated as brittle. As a considerable quantity of energy is moved into other energies, mainly due to dissipative mechanisms, the fracture is said to be ductile.

As shown in Figure 6.49, brittle fracture spreads rapidly on a crack with minimal energy absorption and plastic deformation. It occurs along crystallographic planes known as cleavage planes. The mechanism of the brittle fracture was initially explained by the Griffith hypothesis [26]. Griffith proposed that there are microcracks in a brittle material, which act on the concentrated stress at their tips. The crack would arise from an unspecified numeric of origins as flow occurred during a surface scratch or solidification. As plastic deformation at the crack tip is prohibited, the crack will proceed through grains by shattering atom bonds in lattice planes. This cleavage fracture will be in materials without close-packed planes, having BCC or HCP crystal structures.

Brittle materials are ceramics, glasses, metals, and some polymers. They have the following features:

- There is no sizable plastic deformation, so crack propagation is too fast.
- The crack spreads almost at a 90° angle to the orientation of the acted stress.

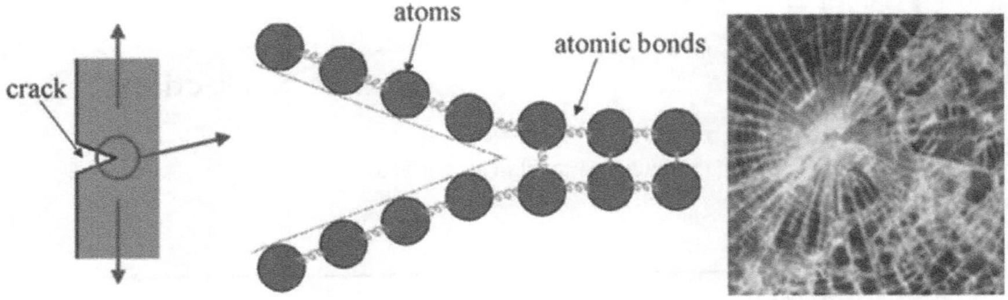

FIGURE 6.49
Brittle fracture mechanism and its instance (glass).

- Cracks are often propagated by cleavage—shattering atomic bonds along specific crystallographic planes (cleavage planes).

For a ductile fracture, as in a crystalline material, dislocations begin in the lattice due to local shear stresses, and numerous dislocations will grow. As the inner structure is modified in a way that cannot be undone, the macroscopic deformation is permanent (plastic). The dislocations join at grain boundaries and gather to establish a void. These voids will increase, and one or more of them will proceed into a macroscopic crack. As the beginning, the growth of cracks is caused by shear stresses, by means of a mechanism referred to as shearing. Plastic deformation is critical, so this mechanism is generally observed in FCC crystals, which have numerous close-packed planes. The fracture surface has a "dough-like" structure with dimples, the shape of which lends its name to the loading of the crack (Figures 6.50 and 6.51).

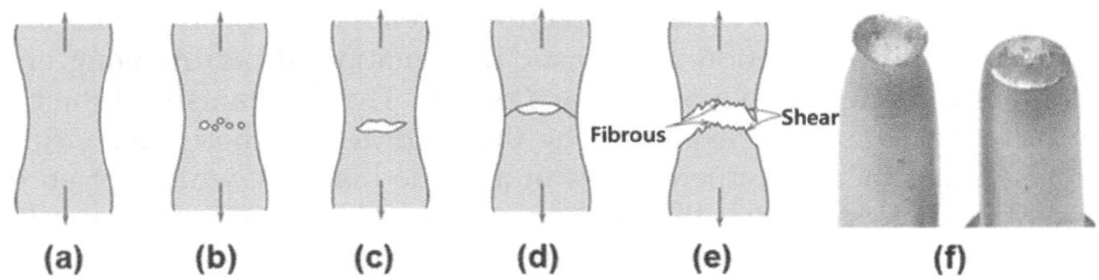

FIGURE 6.50
Ductile fracture failure mechanism: (a) necking, (b) formation of micro-voids, (c) coalescence of micro-voids, (d) crack propagation by shear, (e) fracture, (f) cup and cone.

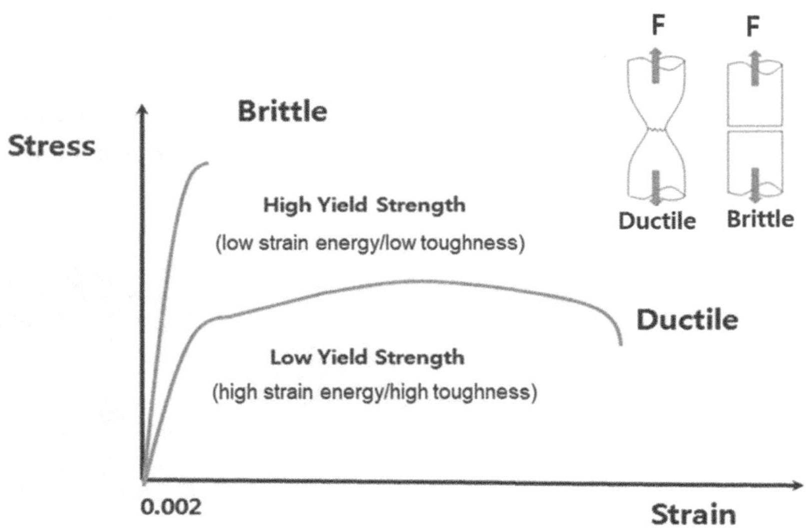

FIGURE 6.51
Brittle versus ductile fracture in matter.

- *Brittle fracture*: Separation along crystallographic planes due to the breakdown of atomic bonds (V-shaped chevron, cleavage, intergranular)
- *Ductile fracture*: Initiation, growth, and coalescence of micro-voids (cup-and-cone, dimple).

In polycrystalline material, however, crystal axes are located at random. As a polycrystalline material is exposed to stress, slip starts first in those grains in which the slip system is nearly all positively situated with respect to the acted stress. As contact at the grain boundaries is carried on, it may be necessary for more than one slip system to run. The rotation into the axis of tension brings other grains, at first less positively located, into a position where they may just now deform. As deformation and rotation proceed, the single grain is inclined to elongate in the direction of flow. As a crystal twists the form, there is some deformation of the lattice structure. This distortion is greatest on the slip planes and grain boundaries and increases with deformation. This is clear by a growth in resistance to further deformation. The material experiences strain hardening or work hardening. As dislocations pile up at grain boundaries, metals will be hardened by increasing the relative extent of the grains.

6.5.4 Ductile–Brittle Transition Temperature (DBTT)

The ductile-to-brittle transition temperature (DBTT) is normally observed in metals whose composition makes the material vulnerable. For some steels, the transition temperature will be roughly 0°C, and in winter, the temperature in some regions of the world is below this criterion. As a result, some steel structures are very likely to fail in winter. At high temperatures, where the impact energy taken in before failure is large, ductile failure occurs with completely elastic and plastic deformation before failure. At low temperatures, where the impact energy absorbed before failure is low, brittle fracture happens with little deformation before failure.

The governing mechanism of this change remains unknown despite the numerous attempts in theoretical and experimental research. All ferrous materials (except the austenitic grades) are visible to a transition from ductile to brittle when tested above and below a definite temperature, expressed as the ductile-to-brittle transition temperature (DBTT). Face-centered cubic (FCC) metals such as Cu and Ni stay ductile down to very low temperatures. For ceramics, this sort of transition happens at much higher temperatures than for metals (Figure 6.52).

Steels having DBTTs just below room temperature were used. At low temperatures, steels can be brittle in a solemn manner. At higher temperatures, the impact energy is big, which is in agreement with a ductile mode of fracture. As the temperature falls, the impact energy lessens immediately and unexpectedly over a relatively narrow temperature span, which is steady with the mode of brittle fracture. Fatigue cracks establish a nucleus at the corners of square hatches and spread swiftly by means of brittle fracture.

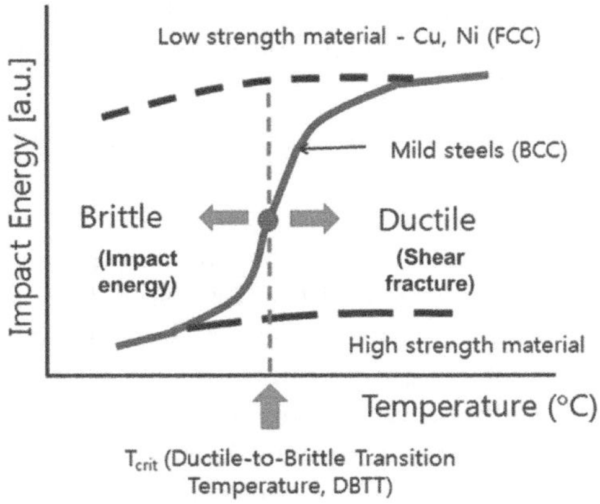

FIGURE 6.52
Ductile-to-brittle transition temperature.

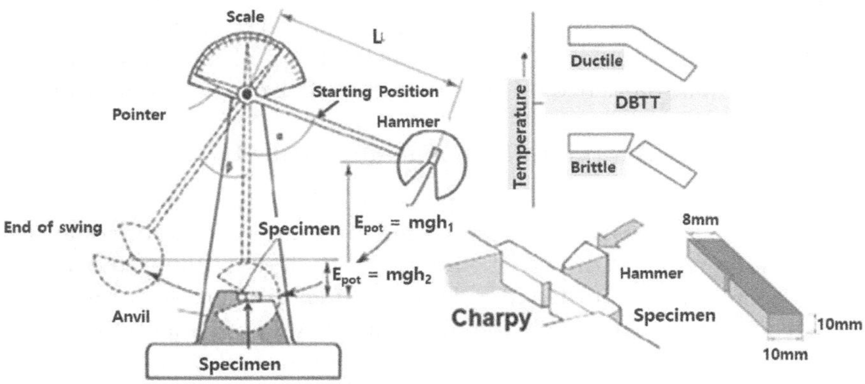

FIGURE 6.53
Schematic of conventional Charpy V-notch testing.

As the familiar weld fractures in numerous U.S. army ships such as Liberty Ships, tankers, etc. in World War II are inspected, the ductile-to-brittle change is calculated by impact testing such as Charpy V-notch testing (Figure 6.53). Though the tensile stress–strain curve provides a sign for ductile/brittle failure, the very typical test to inspect this is the Charpy V-notch test. The principal advantage of this test is that it supplies an uncomplicated measure for the dissipated energy in fast crack propagation.

The sample is a beam with a 2 mm deep V-shaped notch, which owns a 90° angle and a 0.25 mm root radius. It is performed and loaded as in a three-point bending test. The load is provided by the impact of a weight at the end of a pendulum. A crack will start at the tip of the V-notch and move throughout the sample. The material twists at a strain rate of normally 103 s^{-1}. The energy dissipated during fracture will be computed effortlessly from the high point of the pendulum weight before and after impact. The dissipated energy is the impact toughness C_v.

The impact toughness will be determined for various specimen temperatures *T*. For intrinsic brittle materials such as high-strength steel, the dissipated energy will be low for all temperatures *T*. For intrinsic ductile materials such as FCC metals, C_v may be high for all *T*. Numerous materials show a change from brittle to ductile fracture with increasing temperature.

References

1. Duga, J.J., Fisher, W.H., Buxaum, R.W., Rosenfield, A.R., Buhr, A.R., Honton, E.J., McMillan, S.C., 1982, The economic effects of fracture in the United States, *Final Report*, September 30, Battelle Laboratories, Columbus, OH. Available as NBS Special Publication 647–2.
2. Inglis, C.E., 1913, Stresses in a plate due to the presence of cracks and sharp corners, *Transactions of the Royal Institute of Naval Architectes*, 60, 219–241.
3. Neugebauer, G.H., 1943, Stress concentration factors and their effect in design, *Product Engineering*, 14, 82–87.
4. Paris, P.C., Gomez, M.P., Anderson, W.E., 1961, A rational analytic theory of fatigue, *The Trend in Engineering*, 13, 9–14.
5. Wöhler, A., 1870, Über die Festigkeitsversuche mit Eisen und Stahl, *Zeitschrift für Bauwesen*, 20, 73–106.
6. Kim, W.H., Laird, C., 1978, Crack nucleation and stage I propagation in high strain fatigue—II. Mechanism, *Acta Metallurgica*, 26 (5), 789–799.
7. Woo, S., O'Neal, D., Pecht, M., 2010, Reliability design of a reciprocating compressor suction reed valve in a common refrigerator subjected to repetitive pressure loads, *Engineering Failure Analysis*, 17 (4), 979–991.
8. Sutherland, H.J., 1999, *On the Fatigue Analysis of Wind Turbines, SAND99–0089*, Albuquerque, NM: Sandia National Laboratories, p. 132.
9. Nisbett, R.G., Budynas, J.K., 2014, *Shigley's Mechanical Engineering Design*, 10th ed., Boston, MA: McGraw-Hill Higher Education, pp. 308–324.
10. Miner, M.A., 1945, Cumulative damage in fatigue, *Journal of Applied Mechanics*, 12, 149–164.
11. Palmgren, A.G., 1924, Die Lebensdauer von Kugellagern Zeitschrift des Vereines Deutscher Ingenieure, *Scientific Research*, 68 (14), 339–341.
12. Burhan, I., Kim, H., 2018, S–N curve models for composite materials characterization—An evaluative review, *Journal of Composites Science*, 2 (3), 38–66.
13. Sutherland, H.J., Mandell, J.F., 2005, *Optimized Goodman Diagram for the Analysis of Fiberglass Composites Used in Wind Turbines Blades*. 43rd AIAA Aerospace Sciences Meeting and Exhibit, Reno, Nevada.
14. Murakami, Y., Miller, K.J., 2005, What is fatigue damage? A view point from the observation of low cycle fatigue process, *International Journal of Fatigue*, 27 (8), 991–1005.

15. Qiu, B., Kan, Q., Kang, G., Yu, C., Xie, X., 2019, Rate-dependent transformation ratcheting-fatigue interaction of superelastic NiTi alloy under uniaxial and torsional loadings—Experimental observation, *International Journal of Fatigue*, 127, 470–478.
16. Sun, J., Su, A., Wang, T., Chen, W., Guo, W., 2019, Effect of laser shock processing with postmachining and deep cryogenic treatment on fatigue life of GH4169 super alloy, *International Journal of Fatigue*, 119, 261–267.
17. Kumar, N., Goel, S., Jayaganthan, R., Owolabi, G.M., 2018, The influence of metallurgical factors on low cycle fatigue behavior of ultrafine grained 6082 Al alloy, *International Journal of Fatigue*, 110, 130–143.
18. Maier, G., Riedel, H., Somsen, C., 2013, Cyclic deformation and lifetime of Alloy617B during isothermal low cycle fatigue, *International Journal of Fatigue*, 55, 126–135.
19. Sola, J.F., Kelton, R., Meletis, E.I., Huang, H., 2019, Predicting crack initiation site in polycrystalline nickel through surface topography changes, *International Journal of Fatigue*, 124, 70–81.
20. Liu, Y., Kang, M., Wu, Y., Wang, M., Wang, J., 2018, Crack formation and microstructure-sensitive propagation in low cycle fatigue of a polycrystalline nickel-based superalloy with different heat treatments, *International Journal of Fatigue*, 108, 79–89.
21. Wahi, R.P., Auerswald, J., 1997, Damage mechanisms of single and polycrystalline nickel base superalloys SC16 and IN738LC under high temperature LCF loading, *International Journal of Fatigue*, 19 (93), 89–94.
22. Wilhelm, F., Affeldt, E., Fleischmann, E., Glatzel, U., Hammer, J., 2017, Modeling of the deformation behavior of single crystalline Nickel-based superalloys under thermal mechanical loading, *International Journal of Fatigue*, 97, 1–8.
23. American Society for Testing and Materials, 1999a, Standard practice for strain-controlled fatigue testing, In *Annual Book of ASTM Standard, Vol. 03.01*, American Society for Testing and Materials, West Conshohocken, PA, no. E606.
24. American Society for Testing and Materials, 1999b, Standard practice for cycle counting in fatigue analysis, In *Annual Book of ASTM Standard, Vol. 03.01*, American Society for Testing and Materials, West Conshohocken, PA, no. E1049.
25. American Welding Society, 1996, *Structural Welding Code—Steel*, 15th ed., Miami, FL: American Welding Society, prepared by AWS Structural Welding Committee.
26. Griffith, A.A., 1921, The phenomena of rupture and flow in solids, *Philosophical Transactions of the Royal Society of London, A*, 221, 163–198.

7

Design Methodology: Parametric Accelerated Life Testing

7.1 Introduction

Due to the requirements of intense competition in the field, mechanical systems might be designed to attain the required quality and high reliability. As the designs are evaluated before release, new features are rapidly combined into a product and conducted to the marketplace. The implementation of all newly designed characteristics influences a great variety of end user areas where constructional safety is a main concern, such as airplanes, refrigerators, automobiles, nuclear power plants, naval or civil structures, etc. Without either testing or clear understanding of how design attributes will be used by the consumer, product launching with design defects can unsatisfactorily influence the manufacturer's brand due to field recalls [1].

For example, to transfer power such as the multiplication of speed and torque, a gear system of an auger motor used in an ice-maker is a revolving mechanical component possessing cut teeth, which engage with a compatible toothed part. Mounting in a favorable positions of gear drives will supply their precise velocity, high powers, good efficiency, dependable service, and tightly packed layout. On the other hand, manufacturing gears calls for specific tools and apparatus. If some fault in cutting teeth has occurred, it creates vibration and noise during operation. The fundamental working of gears is comparable to the principle of levers. Namely, the auger motor assembly, comprising the gear system, is designed to crush the ice with considerable torque by reducing speed through a plurality of driven gears, which are engaged with a drive pinion gear placed in the input shaft. To prevent a gear system from failing in the marketplace before its anticipated life, a company might demonstrate the new gear merged with suitable ISO standards [2,3] and information sheet [4] and/or correctly carry out reliability testing before the product is launched.

Moreover, the Boeing 737 MAX airplane was banned from flying from March 2019 to December 2020 after a pair of crashes with 346 passengers dead. The plane had the CFM International LEAP-1B engines utilizing the optimized 68-inch fan design; these engines used 12% less fuel and were 7%

DOI: 10.1201/9781003451471-7

lighter [5]. Inspectors, incorporating the Ethiopian Civil Aviation Authority, indefinitely concluded that the crash was generated by the airplane's engine. To ensure that a product will be successful in the marketplace, troublesome components have to be identified and modified by utilizing an organized testing approach—parametric ALT, which can create reliable quantitative (RQ) specifications [6].

When exposed to repetitive (impact) loads, product defects, such as cracks in notches, thin areas, etc., may bring on fatigue pen. Fatigue is the main source of failure in mechanical metallized components, demonstrating about 80–95% of all constructional failures [7]. Namely, fatigue in metals begins to be visible in the shape of cracks, which evolve in regions that stress can concentrate, such as sharp-edges, grooves, holes, etc., and spread it. Constructional failures in any of these districts can have evidently important results with regard to the lives of human beings and environmental disasters. The engineer has to perceive the dissimilar mechanisms affecting crucial and subcritical processes in constructional materials and to develop evaluation skills and management skills for the equivalent structures.

To stop these constructional failures, the ALT, combined with the reliability block diagram, has been employed as a different procedure for resolving product troubles [8]. It includes a test plan for the product, identifying failure mechanics—fatigue and fracture—and employing a sample size equation with raised loads, etc. Elsayed [9] arranged statistical, statistics/physics, and experimental/physics-driven models for the failure. Meeker [10] suggested some observed steps to plan for an ALT. Carrying out a parametric ALT [11,12] requires other ideas, such as the BX life for the structured tests, a straightforward life-stress prototype, sample size formulation, and fracture mechanics, because failure may immediately emerge due to susceptible components in a system [13–15]. The current test method [16–19] may not easily replicate the design defects of components in a multimodule product because the skills assess insufficient samples and testing period.

To robustly obtain the design of a mechanical product, engineers have employed traditional methods such as the strength of materials based on stress–strain concepts [20]. Attention is being paid to the evolution of engineering steps expressing the constructional integrity circumstances of an allowed part. Engineers have also employed quantum mechanics to identify which fatigue began from the atomic and microstructural scales in some metallic alloys or the numerous engineering plastics [21,22]. To acknowledge the fatigue emergence in a mechanical product, a (generalized) life-stress expression can be employed in an organized design approach and related perspectives to identify the failure of electronic/mechanical components due to material defects such as tiny cracks. Strength of material and finite element methods (FEMs) may not identify the source of a breakdown because market failure stochastically happens in regions of locally high stress concentration, not supposed as a continuum in material [23–26]. As another possibility, there are other approaches—such as structural health monitoring (SHM)—that allow the watching of the failure source [27]. Nevertheless, it is not easy to achieve test data for multimodule systems because complete tests should have been required, which will be too costly due to a period of test and compulsory samples.

To verify the accomplishment of recognizing and altering the design defects of a mechanical product, a parametric ALT can be employed as an organized matter to create assignment cycles such as RQ statements. It involves 1) a product BX life created on the ALT steps, 2) a load analysis, 3) a modified ALT with alternations, and 4) an assessment of whether the design achieves the aimed BX lifetime. The time-to-failure model, the sample size equation, and BX life are suggested. To establish the quality of being logically ALT, it might be necessary to check the introduction of the new design in the marketplace to ensure that it reaches the targeted reliability.

7.2 Reliability Design in Mechanical Systems

In a (mechanical) system, generated power is transferred to create mechanical advantages via force and movement by adapting various mechanisms within the system. That is, a mechanical system is a group of physical components that alter an input force and movement into a desired output force and movement. Examples of mechanical products are refrigerator systems, including heating, ventilation, and air conditioning (HVAC) systems, air conditioners in building systems, automobiles, aircraft, inkjet printers, plumbing, elevators, and escalators in building systems. Automobiles have an engine, transmission, drive, electrical components, body parts, accelerometers for airbag deployment and electronic stability control, inertial navigation systems (INSs) for detecting yaw, pitch, and roll. Aircraft carry engine systems, power delivery systems, aviation control systems, door systems, machinery systems, MEMS gyroscopes for spontaneously sensing and balancing flying characteristics. Inkjet printers utilize piezoelectrics or thermal bubble ejection to place ink on paper. Arrangement drawings in mechanical system design include information about the parts, fabrication, dimensions, weight, and lifting points. The drawings should show clear details of all the components and how they are assembled.

7.2.1 Automobiles

An automobile utilized for transport is a motorized vehicle. The word "automobile" comes from the Ancient Greek word *autós,* signifying "self," and the Latin word *mobilis,* or "movable." Since their invention, automobiles have taken on additional attributes and controls, such as air conditioning, navigation systems, and rear reversing cameras, consequently making them more complex but also more reliable and easier to operate. That is, the power of an automobile is caused by an inner combustion engine, which functions in the

Carnot cycle, a conceptually ideal thermodynamic cycle. Thus the power is transferred to each wheel, as a consequence, attaining a mechanical advantage through mechanisms such as transmissions, drive systems, etc. An automobile is made up of some dissimilar modules, such as engine, transmission, drive, body components, and electrical components. The entire number of components in an automobile may be in the range of 20,000. For this reason, the whole failure rate of an automobile in its life is the sum of the failure rates of each module. An automobile's life is expected to be a B40 life of no fewer than 12 years (Figures 7.1 and 7.2).

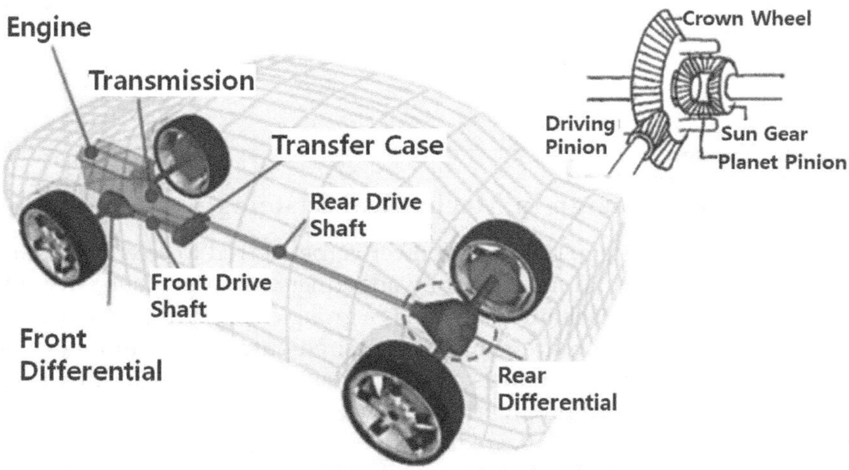

FIGURE 7.1
Automobile structure.

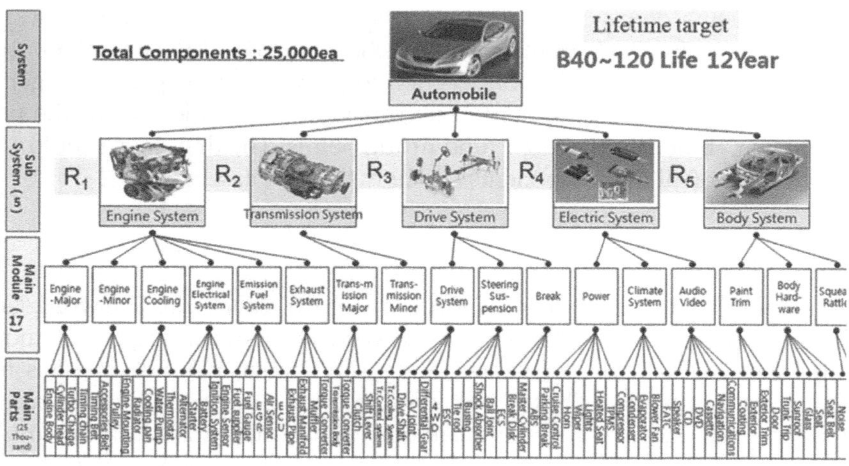

FIGURE 7.2
Breakdown of automobiles with multiple modules.

7.2.2 Airplanes

Airliners are designed to carry passengers or cargo payloads in a specified direction. It is pushed ahead by a thrust force generated from an engine. In particular, as the aircraft goes forward, the wings, depending on a system of wing arrangements, divert the descending air, causing a lifting force to convey aircraft in flight. The word "airplane" originates from the French word *aéroplane*, which arises from the Greek term *ἀήρ* (air) and either the Latin term *planus*, which signifies "level," or the Greek term *πλάνος* (planos), which signifies "wandering" and the wing's movement by air.

An aircraft employs a thrust from the mechanical power created by an aircraft engine, either a jet propulsion or propeller, situated on the aircraft. In particular, jet engines give airplane thrust by sucking in air, squeezing it, introducing fuel into the hot-compressed air mixture in a combustion chamber, and discharging the elevated exhaust rearward through a turbine. Contemporary gas turbine engines work by means of the Brayton cycle, which consists of a gas compressor, a combustion chamber, and an expansion (Figure 7.3).

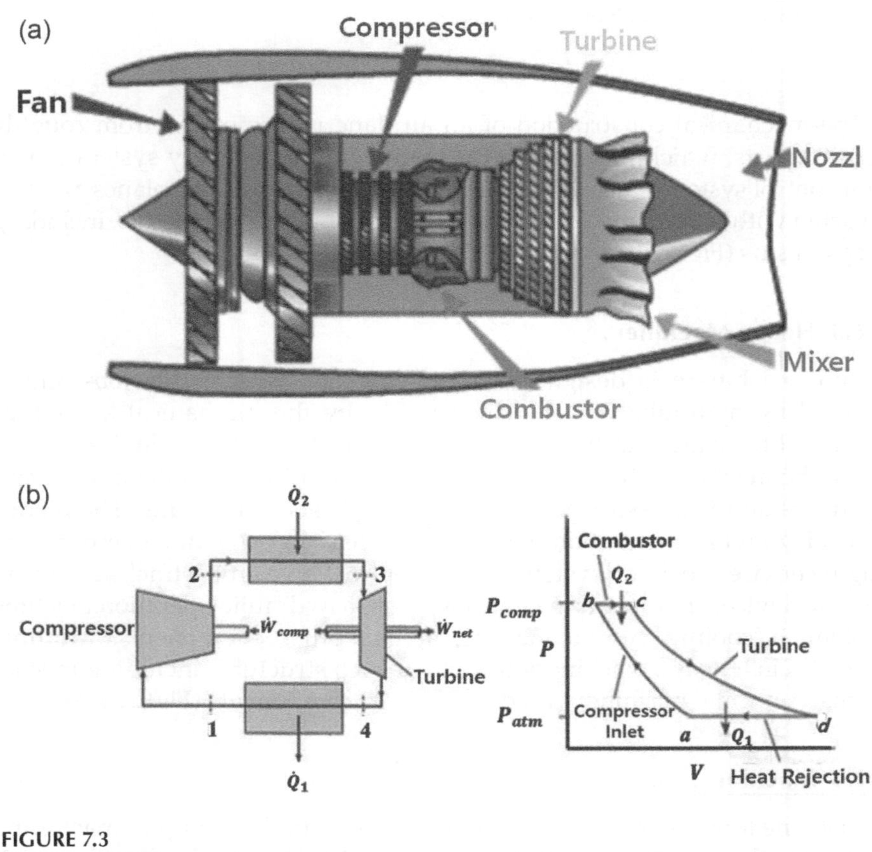

FIGURE 7.3
Modern gas turbine engines: (a) jet-engine structure; (b) Brayton cycle.

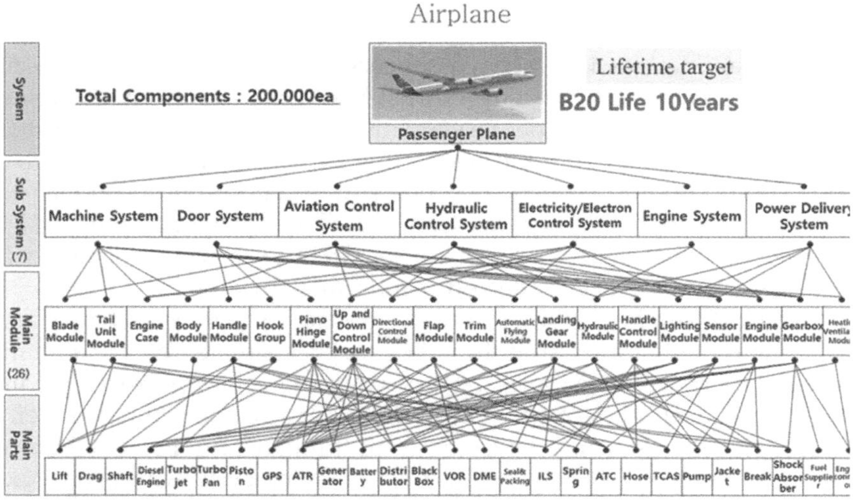

FIGURE 7.4
Structure of an airplane with multiple modules.

The mechanical construction of an airplane is formulated from roughly 200,000 parts, which incorporate the engine, power delivery systems, aviation control systems, machine systems, door systems, etc. Airplanes are constructed with several modules, which have their own components, including mechanisms (Figure 7.4).

7.2.3 Heavy Machinery

Heavy machinery is designed for carrying out construction jobs and is worked by hydraulic power, which works by the mechanical advantage of straightforward machinery. Through correctly utilizing hydraulic circuits, the machinery transfers the hydraulic fluid by hydraulic motors and cylinders and is pressurized in the hoses and tubes of circuits. The hierarchical layout of construction machinery, such as an excavator, comprises an engine device, a driving system, a main control valve unit, a track system, an electric device, an upper appearance system, a hydraulic operation machine system, a cooling system, and numerous other components. Multiple modules in a heavy machinery have their own structures including mechanisms. A normal machinery holds over 5,000 components (Figure 7.5).

7.2.4 Machine Tools

A machine tool is normally a type of machinery for setting up or machining metal or other rigid materials by cutting, boring, grinding, shearing, etc. It

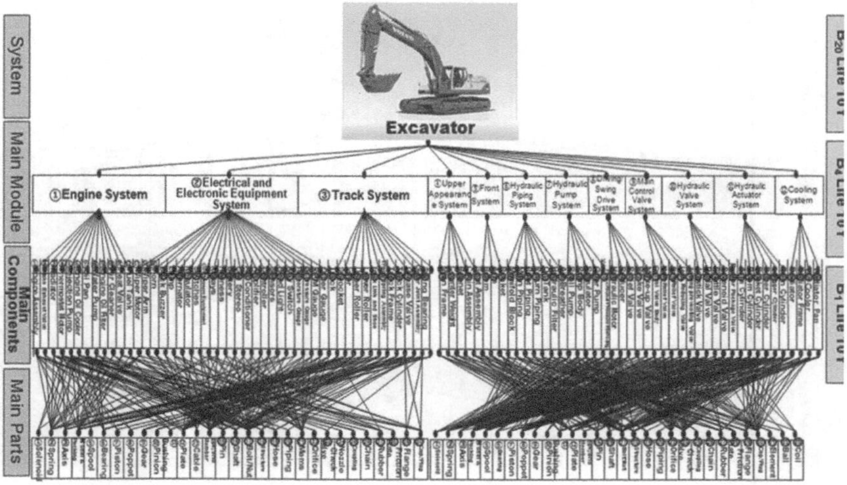

FIGURE 7.5
Structure of excavator with multiple modules.

uses various kinds of cutting or shaping tools. All machine tools have methods of restricting the workpiece and operating a conducted motion of the machinery components. Thus the relative motion between the cutting tool (that is stated as the tool path) and the workpiece is restricted or managed by the machinery to some minimum amount, rather than being completely freehand or offhand.

The arrangement of machine tools hierarchically consists of an automatic tool or pallet changing device, a drive unit, a tilting index table, a spindle unit, a hydropower unit, a computer numerical control (CNC) controller, a turret head, a cooler unit, etc. The machine tools have over 1,000 components. Its reliability design may concentrate on the modules. The machine tools may easily calculate the module reliability if the serial connection system is assumed (Figure 7.6).

7.3 Product Connection in Reliability Block Diagram

7.3.1 Introduction

For instance, a refrigerator through the vapor–compression refrigeration cycle gives cold air to the refrigerator and freezing sections where food is kept fresh. This product includes just about 2,000 parts, comprising doors, cabinets, fixtures, condenser and evaporator, controls and sensors, compressor,

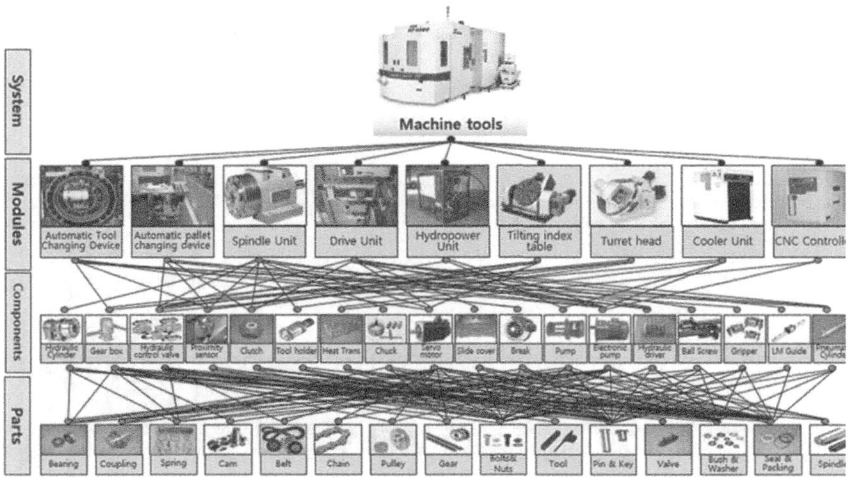

FIGURE 7.6
Structure of machine tools with multiple modules.

water-supplying equipment, etc. By employing an all-inclusive parametric accelerated life testing plan, a reliability design of the system may be attained by acquiring its aimed reliability—lifetime L_B and failure rate λ—after discovering the design defects and altering them. The product (or module) with the altered design should satisfy the allocated reliability aim. As the product (or module) performs testing for notably longer periods, the parametric ALTs could ascertain the defects of the mechanical system in the design phase. Under customer use circumstances, these new structured reliability methods in the developing process may supply the reliability quantitative (RQ) test statements of a mechanical construction, which establish correctness in terms of the reliability aim.

As shown in Figure 7.7, a product may be made up of some dissimilar modules. For instance, automobiles are made up in modules, such as the engine, transmission, driving system, electrical, and body components. The life L_B and failure rate λ_s in a multi-module product might be decided for each module. If there are no inceptive failures in a product, its life might be represented by the module #3 in Figure 7.7(b) that has a design defect. On the other hand, the failure rate of a product is simply defined as the sum total of each module.

In building a reliability block diagram, real arrangements of a product made in parallel or series cannot specify the same logic relationships from a viewpoint of reliability. A system can be defined as a series system if the failure of one or more modules in the product causes the failure of the whole system to happen. Thus a number of mechanical products are the serial system at the two hierarchical levels that are formed of multiple modules.

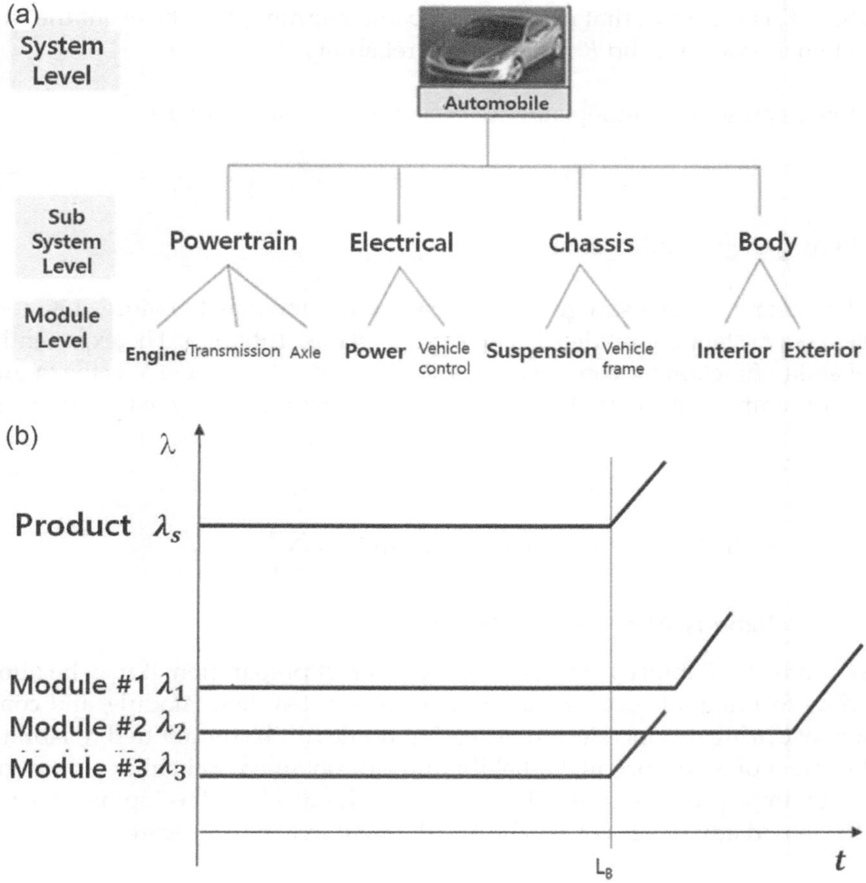

FIGURE 7.7
Product lifetime L_B and failure rate λ_s with multi-modules: (a) breakdown of multi-module automobile; (b) product life L_B and failure rate λ with multi-modules.

For instance, six cylinders in an automobile engine are in series because the engine is said to have failed if one or more cylinders attached in parallel mechanically fail. In a very similar way, an automobile is serially attached by modules, such as power train, engine, electrical and control system, chassis, and body structure. Assume that a mechanical series system is formed of n mutually independent modules. Mutual independence signifies that the failure of one module does not influence the lifetime of other dissimilar modules. Definitionally, accomplishing the operation of a system necessitates all components to be functional. From probability theory, the reliability is

$$R = Pr(E) = Pr(E_1 \cdot E_2 \cdots E_n) \tag{7.1}$$

where E_i is the event that module i is up and running, E is the event that the system is working, and R is the system reliability.

On account of the independence presumption, this becomes

$$R = Pr(E) = Pr(E_1) \cdot Pr(E_2) \cdots Pr(E_n) = \prod_{i=1}^{n} R_i \tag{7.2}$$

where R_i the reliability of module i.

Let's think about a straightforward case where the time to failure of n modules in a system are modeled with exponential distribution. The exponential reliability function for component i is $R_i(t) = \exp(-\lambda t)$, where λ_i is the failure rate of component i. From Equation (7.2), the system reliability may be defined:

$$R(t) = \exp\left(-t\sum_{i=1}^{n} \lambda_i\right) = exp(-\lambda t) \tag{7.3}$$

where λ is the failure rate of the system, and $\lambda = \sum_{i=1}^{n} \lambda_i$.

7.3.2 Reliability Allocation of Product

As product reliability is targeted in the product preparation, it may be allotted by forming a logical sequence to a single subsystem, module, and component during product design. As each module fulfills the allotted reliability, the target of whole product reliability may be obtained. Reliability allocation is a main step of new reliability testing in the product developing process. The gained advantages of reliability allotment may be abridged.

- Reliability allocation implies a reliability aim for each module. The product consists of several module or subsystem, which are produced by inner divisions or suppliers. It is significant that manufacturers share all connected parties with the reliability aim before bringing the product to the end user.
- Quantitative reliability target aims for modules urging parties in charge to upgrade present reliability through usage of reliability skills.
- Required reliability needs are tightly related with engineering pursuits targeted at meeting other consumer prospects in the product developing process.
- Reliability allocation guides a sound comprehension of product hierarchical structure. The process can lead to recognize the component of design fault and afterward upgrade it.
- As a consequence, reliability assignment may function on what is put in other reliability tasks. For instance, reliability allocated to a module may be utilized to the overall verification of design reliability.

Reliability allocation is basically a process characterized by repetition in the product development. It is organized and carried out in the untimely design phase to aid concept design when obtainable information is limited. As the design process begins, the all-inclusive reliability target could be allocated in a different way to lessen the cost of attaining the reliability target. The allocation procedure can be invoked by the failure of one or more modules to achieve the allocated reliability relating to technology restrictions. The process is also replicated whenever a crucial design change happens.

As some components are allocated to very high reliability targets, it can be too difficult to be achieved at all. On the other hand, although there are crucial parts whose failure have safety-related, environmental, or legal effects, low-reliability targets might still be assigned. In this case, it is of great significance to begin some basis for what might be contemplated in reliability allotment. The job of reliability allotment is to choose component reliability aims, $R_1^*, R_2^*, \ldots, R_n^*$ that fulfill the subsequent equality equation:

$$R_S^* \leq g\left(R_1^*, R_2^* R_2, \ldots\ldots, R_n^*\right) \tag{7.4}$$

By means of mathematics, there are a boundless numerics for such sets. Obviously, they are not fine in the same manner, and even some of them are inconvenient. The typical criteria are outlined:

- *Failure possibility*: Components with a high likelihood of failure formerly might be allocated a low-reliability target because of the thorough vigorous effort necessitated to enhance reliability. On the other hand, reliable components can be entrusted high-reliability targets.
- *Complexity*: The numeric of integral components (or modules) in a subsystem reflects the difficulty of the subsystem. A higher complexity tends to lower reliability. It is like the motivation of failure possibility.
- *Criticality*: The failure of some components can bring about intense effects, involving, for instance, loss of life and enduring environmental harm. The circumstance may be very great when such components have a high likelihood of failure. Evidently, criticality is a multiplication of severity and failure probability, as expressed in the FMEA skill. If a design may not eliminate very bad failure modes, the components might have the lowest likelihood of failure. As a consequence, high-reliability targets might be allotted to them.
- *Cost*: Cost is a critical basis that is a target minimized in the industry. The cost outcomes for fulfilling reliability depend on components. Some components bring about a high cost to enhance reliability a little because of the problem in design, verification, and production. So it can be favorable to allocate a higher reliability goal to the components that cost less to improve reliability.

TABLE 7.1

Reliability Target for Mechanical Product (Part Count Method)

Level	Quantity	Target	Remark
System	1 system	B20 life of 10 years	Refrigerator
Module	5 units	B4 life of 10 years	Compressor
Component	500 components	B0.04 life of 10 years	—

Though some methods for reliability allocation have been evolved, the straightforward way here is the identical allocation way. It may only be applicable when the system reliability arrangement is in series. The product reliability is calculated by:

$$R_S^* = \prod_{i=1}^{n} R_i \tag{7.5}$$

The allotted reliability for each subsystem is:

$$R_i = \left(R_S^*\right)^{1/k} \tag{7.6}$$

Products included in a mechanical system may be defined as an airplane, automobile, household appliance, heavy construction equipment, machine tools, and agricultural machine. Based on the hierarchy in product and field data, the reliability target might be allocated to the modules as in Table 7.1. Consequently, they can break down several modules to the single components. The aimed reliability of each module that will center on reliability testing may be quantified as the product of anticipated product life L_B and failure rate λ. For instance, if the aimed reliability of a refrigerator is assigned as B20 life of 10 years, the reliability for the compressor may be B4 life of 10 years.

It is practical to perform the reliability testing per module because test cost for system or component is higher than that of module.

7.4 Parametric ALT for Mechanical Product

7.4.1 BX Lifetime in a Product

In (mechanical) products, (generated) power can be used to a mechanical advantage to provide a task involving forces and movement by correctly adopting mechanisms. In this operation, they may be exposed to repeated loading. They are required to have a robust design that has the strength and stiffness for the anticipated loads. An instance for this is a household refrigerator that is devised to distribute cooled air from an evaporator to each section and maintain the freshness of stocked foods in freezer and refrigerator. One of the modules in a refrigerator is the ice-maker, involving the gear system that lessens the speed and grows the torque. As a consequence, the ice-maker may attain adequate torque to crush the ice. A refrigerator may also embrace a dissimilar span of modules (or subsystems)—cabinet, door, drawers and shelves, compressor, evaporator and condenser, controlled instruments, motor and electronics, water supply equipment, and some other components. The refrigerator life is decided by the random (or wear-out) failure of a newly designed module such as the ice-maker, embracing an auger motor with a gear system that has design faults (Figure 7.8).

To perform a parametric ALT, the BX life (or "Bearing Life") L_B, should be decided as an index of the product life. The BX life, arisen from bearing industry, has become an index for product life that may be utilized in numerous industries. BX life is a period when x% of a number of product samples being thought about and discussed fail. A "BX life of y years" is a good way for manifesting the product lifetime. For example, if a product is a B20 life of 10 years, it means that 20% of a group of samples may be unsuccessful during ten years of functioning. By utilizing this life index, the ALT can identify and modify the defect that is the basic cause of accumulative failure rate and short life in the marketplace.

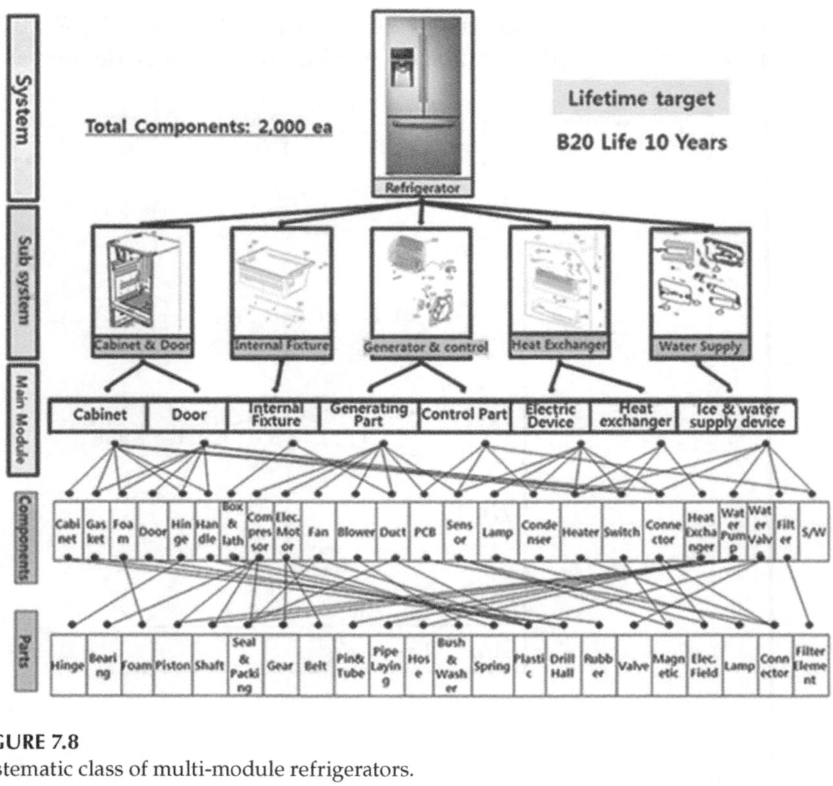

FIGURE 7.8
Systematic class of multi-module refrigerators.

For example, a refrigerator has approximately 2000 components, involving some module such as ice-makers. As an ice-maker has design flaws, it can influence the feasible life of a refrigerator. As a household refrigerator's life is targeted to have a B20 life of 10 years, the life of each module including ice-maker will be a B1 life of 10 years because the refrigerator is made up of as many as 20 modules, and each module has 100 components.

7.4.2 Placing a Whole Parametric ALT Procedure

The reliability can be defined as the ability necessitated to perform the organized function under the stated environmental/operational situations for a required period of time [28]. Product reliability is described by using a "bathtub curve," which has three segments: 1) in the first segment, during the early system life, there is some decreasing in the failure rate after having big failure; 2) in the second segment, during its central lifetime, there is a constant failure rate; and 3) in the third segment, there is an increasing failure rate until the system life is extended.

As T is a random variable meaning the period to failure, the ratio of outlasting at period t can be expressed:

$$R(t) = P(T > t) \tag{7.7}$$

The cumulative distribution function (CDF) $F(t)$ is $F(t)(or\, X) = 1 - R(t)$. The failure rate λ on the (slanted) bathtub curve in Figure 7.9 may be defined as:

$$\lambda(t) = \frac{f(t)}{R(t)} = \frac{dF(t)/dt}{R(t)} = \frac{-R'(t)}{R(t)} \tag{7.8}$$

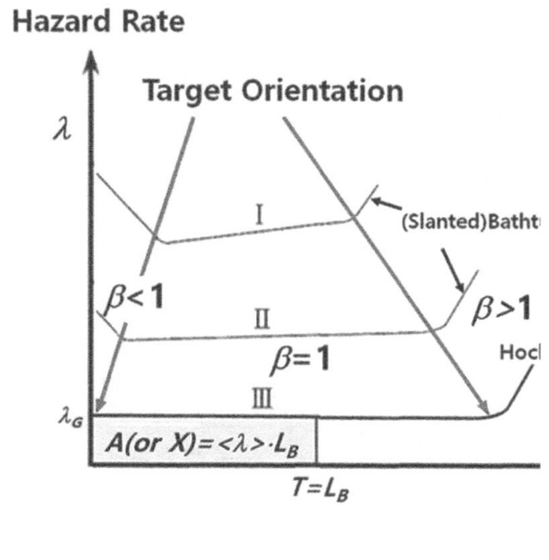

FIGURE 7.9
Life measure: BX Life (L_{BX}) on the bathtub.

where f is the failure density function.

As the failure rate in Equation (7.8) puts the integral, the X% accumulative failure, $F(L_B)$, at $T = L_B$ may be obtained:

$$F(=X) = \int_0^T \lambda(t) \cdot dt = -\ln R(L_B) \cong \lambda \cdot L_B \tag{7.9}$$

Assuming that T_1 is the time of the first failure in the second segment of the bathtub, the reliability function $R(t)$ will also be attained as:

$$R(t) = P(T_1 > t) = P(\text{no failure} \quad \text{in } (0,t]) = \frac{(m)^0 \, e^{-m}}{0!} = e^{-m} = e^{-\lambda t} \tag{7.10}$$

As product design is improved, the failure rate in the field will lessen. On the other hand, the system life expands. For such situations, the reliability of a mechanical product can be defined:

$$R(L_B) = 1 - F(L_B) = e^{-\lambda L_B} \cong 1\text{-}\ \lambda L_B \tag{7.11}$$

Equation (7.11) may be manifested to be below roughly 20% of the accumulative failure [29]. It is also known that the product reliability may be attained from the product of lifetime and failure rate.

As a case investigation, an ice-maker, involving an auger motor with a gear system, continuously necessitates a simple mechanical functioning: 1) water supplies the ice tray, 2) water freezes into ice by moving created cool air by the heat exchanger (evaporator), and 3) the ice produces until the ice bucket is filled. As the consumer pushes the lever, crushed (or cubed) ice may be distributed. The products failed in the field were crucial for understanding the repetitive use procedures of customers and acknowledging design flaws in the constructional system. Developed on the basis of the field data, the root cause(s) of the troublesome ice-maker, involving the gear system, were recognized. After setting the aimed life L_B in the scheme of a parametric ALT, the mechanical system can be modified by identifying the troublesome components and altering them (Figure 7.10).

To set the aimed lifetime of a mechanical product such as an ice-maker in a refrigerator consisting of nine subsystems (or modules) (see Figure 7.8), the subsequent attainable modules are 1) modified independent and self-reliant units, 2) new independent and self-reliant units, and 3) independent and self-reliant units similar to the earlier design based on a need in the field. The ice-maker in the refrigerator case study here was utilized. It was originally modified to enhance the market quality of an ice-maker, involving the auger motor with a gear system. However, the altered ice-maker had a high failure rate and short life because of design flaws that were in need of being altered.

The altered module E from the field data had a failure rate of 0.31% yearly and a B1 life of 3.2 years (Table 7.2). Established on market statistics, the life

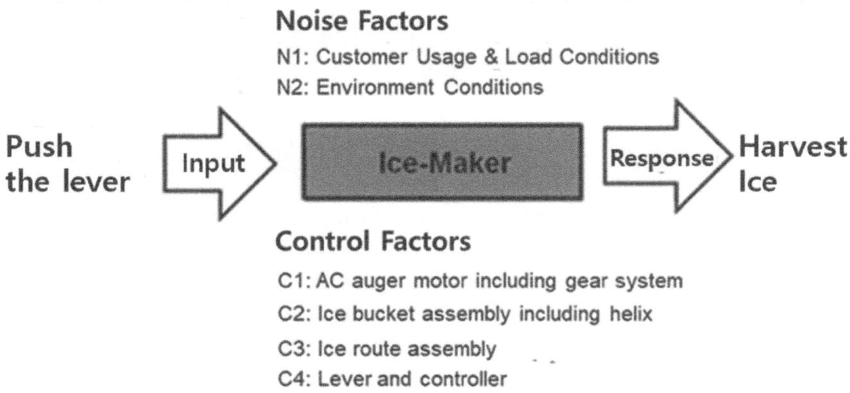

FIGURE 7.10
Intended function of ice-maker with gear system (example).

of an ice-maker involving the auger motor with gear system was expected to be B1 life of 1.61 years because they had a failure rate of 0.62 % yearly. To satisfy the consumer needs, the newly designed life of a module, such as an ice-maker, involving the auger motor with a gear system, was targeted to fulfill the B1 life of ten years.

7.4.3 Failure Model and Sample Size Formulation

As mentioned, mechanical products normally transfer generated power from one place to another by selecting an acceptable mechanism, such as the gear system in the auger motor of an ice-maker. As the end user hopes to use ice daily, the ice-maker is attached in a refrigerator for this new purpose. The chief components in an ice-maker are formed of the ice-bucket, auger motor, bucket case, blade dispenser, helix upper dispenser, helix support, helix dispenser clamp, blade, etc.

A gear train in an auger motor has two or more gears functioning with each other by becoming engaged in their teeth and revolving in a system, creating sufficient torque to crush ice. Because electric motors are employed, the gear systems lessen speed and increase torque. An auger motor operated on alternating current (AC) grows the torque by the gearbox to convey ice and crush it at the end of an ice-maker. Consequentially, the product is exposed to repeated continuous stresses (or impact loading) due to having to crush ice in a seriously freezing temperature (–20°C ↓). If a constructional defect conducts inadequate strength when the (impact) loads are exerted, the product will suddenly fail in its anticipated lifetime. Thus the failed parts need to be found and modified (Figure 7.11).

TABLE 7.2

Complete ALT Strategy of Mechanical Modules in a Product

	Market Data		Predicted Reliability				Intended Reliability	
Modules	**Failure Rate per Year, λ (%/year)**	**BX Life, L_B (year)**	**Failure Rate per Year, λ (%/year)**			**BX Life, L_B (year)**	**Failure Rate per Year, λ (%/year)**	**BX Life, L_B (year)**
A	0.30	3.3	The same	×1	0.30	3.33	0.10	10(BX = 1.0)
B	0.35	2.9	The same	×1	0.35	2.9	0.10	10(BX = 1.0)
C	0.24	4.2	New	×5	1.20	0.83	0.10	10(BX = 1.0)
D	0.15	6.7	Adjusted	×2	0.30	3.33	0.10	10(BX = 1.0)
E	0.31	3.2	Adjusted	×2	0.62	1.61	0.10	10(BX = 1.0)
Others (F/G/H)	0.50	10.0	The same	×1	0.50	10.0	0.50	10(BX = 5.0)
System	1.9	2.9	—	—	3.27	0.83	1.00	10(BX = 10)

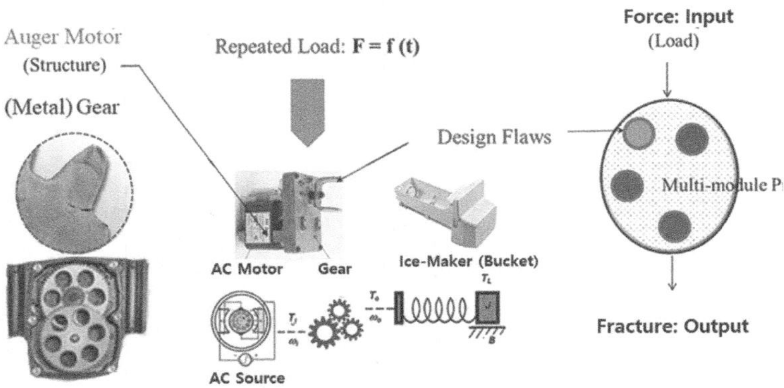

FIGURE 7.11
Fatigue failure on the constructional component produced by repetitive (impact) load and design flaws.

7.4.3.1 Schrodinger Equation for Establishing the Failure Model

First, we employ the Schrodinger equation to find the failure model and understand it. Our starting point is the traditional formulation not based on the theory of relativity for the energy of a particle, which is the entirety of the potential and kinetic energies. We generally assume that the potential is a function of only x. We have a particle energy as follows:

$$E = K + V = \frac{1}{2}mv^2 + V(x) = \frac{p^2}{2m} + V(x) \tag{7.12}$$

We call upon de Broglie's declaration that particles are defined as waves with frequency ω and wavenumber k. Namely, $\mathrm{E} = \hbar\omega$ and $\mathrm{p} = \hbar \mathrm{k}$. It can be redefined as

$$\hbar\omega = \frac{\hbar^2 k^2}{2m} + V(x) \tag{7.13}$$

A wave with frequency ω and wavenumber k will be usually defined as $\psi(x,t) = Ae^{i(kx-\omega t)}$ (the agreement is to insert a minus sign before ωt). We consider that

$$\frac{\partial \psi}{\partial t} = -i\omega\psi \Rightarrow \omega\psi = i\frac{\partial \psi}{\partial t} \text{ and } \frac{\partial^2 \psi}{\partial x^2} = -k^2\psi \Rightarrow k^2\psi = -\frac{\partial^2 \psi}{\partial x^2} \tag{7.14}$$

After the energy formulation in Equation (7.13) is multiplied by ψ:

$$\hbar(\omega\psi) = \frac{\hbar^2}{2m}(k^2\psi) + V(x)\psi \Rightarrow i\hbar\frac{\partial \psi}{\partial t} = \frac{-\hbar^2}{2m}\frac{\partial^2 \psi}{\partial x^2} + V\psi \tag{7.15}$$

In 3D, the x dependence starts to be entirely reliant on three coordinates (x, y, z), and the word $\frac{\partial^2 \psi}{\partial x^2}$ becomes $\nabla^2 \psi$ (the whole of the second derivative). Born's clarification of $\psi(x)$ shows that $|\psi(x)|^2$ is the probability of finding the particle at location x.

7.4.3.2 Infinite Square Well as Typical Solution for Schrodinger Equation

The main problem for parametric ALT is to decide how rapid the potential failure can be identified. To fulfill this motivation, a straightforward failure model is arranged in a systematic manner. Namely, a life-stress (LS) model that has stress and reaction parameters might be evolved. It thus can be explained to contains many failures—fatigue (or fracture). Fatigue failures on the outside of a component happen due not only to incomplete stresses but also to defects such as cracks. Thus fatigue may begin from matter flaws such as void, which appear on a macro, microscopic, or nano span. From such a conceptual view-point, failure can start to be expressed as transport procedures, such as the diffusion of shallow-level dopants of silicon in semiconductors.

First, consider an electric particle that goes only in the x direction from $x = 0$ to $x = a$. The time-independent Schrodinger wave equation in operator configuration is:

$$\hat{H}\grave{\mathrm{E}} = E\psi \tag{7.16}$$

If $\hat{H} = -\frac{h^2}{8\pi^2 m}\frac{d^2}{dx^2} + V$, we may place this in Equation (7.16).

$$-\frac{h^2}{8\pi^2 m}\frac{d^2\psi}{dx^2} + V\psi = E\psi \text{ or } \frac{d^2\psi}{dx^2} + \frac{8\pi^2 m}{h^2}(E - V)\psi = 0 \tag{7.17}$$

where m is the electron mass, h is the Planck constant, and V is the potential energy.

As $V = \infty$ outside the barriers, that can be only when $\psi = 0$. The particle is not outside the walls. As $V = 0$ within the barriers, Equation (7.17) may be expressed as:

$$\frac{d^2\psi}{dx^2} + \frac{8\pi^2 m}{h^2}(E - 0)\psi = 0 \text{ or } \frac{d^2\psi}{dx^2} + K^2\psi = 0 \tag{7.18}$$

where $K^2 = \frac{8\pi^2 mE}{h^2}$.

The solution of Equation (7.18) can be presumed:

$$\psi(x) = A\sin Kx + B\cos K \tag{7.19}$$

where A, B = constants.

As $x = 0$ or $x = a$ at the walls, $\psi(0) = \psi(a) = 0$, $B = 0$, $K = \dfrac{n}{a}$, and $E = \dfrac{n^2h^2}{8ma^2}$, $n = 1,2,3,4$.

Thus we can express Equation (7.19):

$$\psi(x) = A\sin\left(\frac{n}{a}\right)x \tag{7.20}$$

The probability of discovering the particle in a small-scale room between x and $x + dx$ is expressed as:

$$\int_0^a \psi^2(x)dx = 1 \text{ or } \int_0^a \left(A\sin\left(\frac{n}{a}\right)x\right)^2 dx = 1 \tag{7.21}$$

Thus we may attain the solution of Equation (7.19):

$$\psi(x) = \sqrt{\frac{2}{a}}\sin\left(\frac{n\pi}{a}\right)x \tag{7.22}$$

where $\psi(x+a) = \psi(x)$, a is the (periodic) interval, and n is the principal quantum number.

7.4.3.3 Transport Processes in Material

The atoms of the crystal in a material successively form potential walls, which create difficulties for the motion of the charged impurities. As an electromagnetic field ξ is exerted, the obstacles of potential junction energy as a function of interval may be lessened and distorted/phase-shifted. The impurities in a matter, created through electronic movement, are effortlessly drifted to the right because the movement to the left starts to be hard (Figure 7.12).

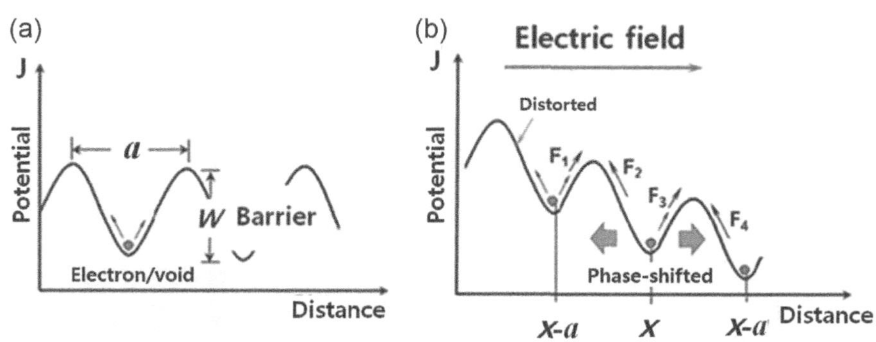

FIGURE 7.12
Potential difference in a material—silicon—when an electromagnetic field is exerted: (a) initial stage; (b) final stage.

TABLE 7.3

Linear Transport Phenomena

Ohm's Law of electrical conduction: $j = -\sigma \nabla V$		
J = electric current density, j (units: A/cm²)	X = electric field, $-\nabla V$ (units: V/cm, V = electrical potential)	L = conductivity, $\sigma = 1/\rho$ (units: ρ = resistivity (Ω cm))
Fourier's Law of heat transport: $q = -\kappa \nabla T$		
J = heat flux, q (units: W/cm²)	X = thermal force, $-\nabla T$ (units: °K/cm, T = temperature)	L = thermal conductivity, κ (units: W/°K cm)
Fick's Law of diffusion: $F = -D\nabla C$		
J = material flux, F (units:/sec cm²)	X = diffusion force, $-\nabla C$ (units:/cm⁴, C = concentration)	L = diffusivity, D (units: cm²/sec)
Newton's Law of viscous fluid flow: $F_u = -\mu \nabla u$		
J = fluid velocity flux, F_u (units:/sec² cm)	X = viscous force, $-\nabla u$ (units:/sec, u = fluid velocity)	L = viscosity, μ (units:/sec cm)

Transport processes are thus expressed (see Table 7.3):

$$J = LX \tag{7.23}$$

where J is a flux vector that is recognized as the transport characteristics. X is expressed as a driving force, recognized as the slope of the electrical potential, fluid velocity, concentration, temperature, etc. L is a transport numerical quantity.

For instance, solid-state diffusion in semiconductor technology is almost totally feasible to control the kind and concentration of impurities in particular areas of a crystal. It involves electromigration-caused voiding, growth of chloride ions, and capturing of electrons or holes.

The solid-state diffusion of impurities doped in silicon J is the amount/area per time. Thus it can be expressed as [30,31]:

$$J = \left[aC(x-a)\right] \cdot \exp\left[-\frac{q}{kT}\left(W - \frac{1}{2}a\xi\right)\right] \cdot v \tag{7.24}$$

[Density/area]·[Jump probability]·[Jump frequency]

$$= -\left[a^2 ve^{-qw/kT}\right] \cdot \cos h \frac{qa\xi}{2kT} \frac{\partial C}{\partial x} + \left[2ave^{-qw/kT}\right] C \sin h \frac{qa\xi}{2kT}$$

$$= \Phi(x,t,T) \sin h(a\xi) \exp\left(-\frac{Q}{kT}\right) = B \sin h(a\xi) \exp\left(-\frac{Q}{kT}\right)$$

where C is the concentration, q is the extent of electric charge, ν is the frequency rate of the attempted jump, a is the distance between atoms, ξ is the applied electric field, k is Boltzmann's constant, T is the absolute temperature, Q is the energy, and B is a constant.

On the other hand, chemical reactions dependent on speed are defined as

$$K = K^{+} - K^{-} = a\frac{kT}{h}e^{-\frac{\Delta E - aS}{kT}} - a\frac{kT}{h}e^{-\frac{\Delta E + aS}{kT}} = B\sin h(aS)exp\left(-\frac{\Delta E}{kT}\right) \quad (7.25)$$

where K is the reaction rate, S is the (chemical) field effect, T is the absolute temperature, k is Boltzmann's parameter, E is the activation energy, and Δ is the difference.

The junction function J from Equations (7.24) and (7.25) may be defined as

$$J = B\sin \text{h}(aS)\text{exp}\left(-\frac{E_a}{kT}\right) \quad (7.26)$$

7.4.3.4 Life-Stress (LS) Prototype

If Equation (7.26) takes a reverse function, the life-stress (LS) model might be redefined as

$$TF = A\left[\sin h(aS)\right]^{-1}\exp\left(\frac{E_a}{kT}\right) \quad (7.27)$$

The sine hyperbolic term $\left[\sin\text{h}(aS)\right]^{-1}$ in Equation (7.27) has the subsequent attributes: 1) $(S)^{-1}$ at the start has some linear effect; 2) $(S)^{-n}$ has what is established as a mid-effect; and 3) $\left(\text{e}^{aS}\right)^{-1}$ in the end is high (Figure 7.13).

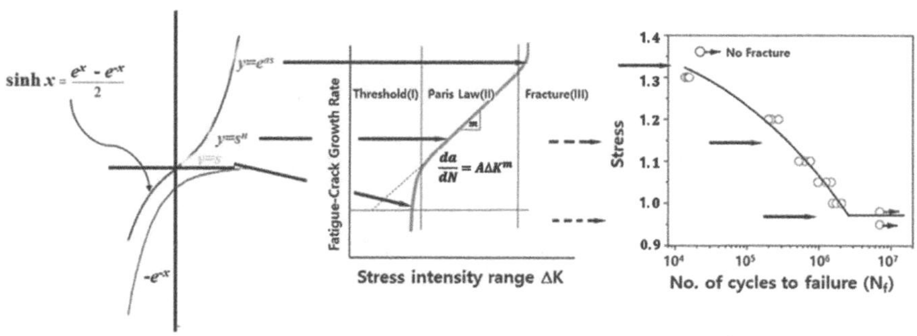

FIGURE 7.13
Explanation of hyperbolic sine stress term on Paris law and S–N curve.

As a parametric ALT in the medium range is generally performed, Equation (7.27) might be defined as:

$$TF = A(S)^{-n} \exp\left(\frac{E_a}{kT}\right) \tag{7.28}$$

Stress is a material amount that expresses the internal forces connecting a small part of a continuum material to apply to withstand each other. As the stress term in the constructional components is hard to compute in parametric ALT, Equation (7.28) might need to be redefined. If the power for some energy district can be expressed as the product of effort and flows, some stresses begin from effort in an energy transport system (Table 7.4) [32].

Equation (7.28) might be redefined as

$$TF = A(S)^{-n} \exp\left(\frac{E_a}{kT}\right) = B(e)^{-\lambda} \exp\left(\frac{E_a}{kT}\right) \tag{7.29}$$

where A and B are quantities that do not change their values.

To attain the acceleration factor (AF), which could be defined as the relationship between the raised stress and normal working conditions, it might be changed to join with the effort concept:

$$AF = \left(\frac{S_1}{S_0}\right)^n \left[\frac{E_a}{k}\left(\frac{1}{T_0} - \frac{1}{T_1}\right)\right] = \left(\frac{e_1}{e_0}\right)^{\lambda} \left[\frac{E_a}{k}\left(\frac{1}{T_0} - \frac{1}{T_1}\right)\right] \tag{7.30}$$

As the greater number of raised tests is carried out at room temperatures, Equation (7.30) can be defined as

$$AF = \left(\frac{S_1}{S_0}\right)^n = \left(\frac{e_1}{e_0}\right)^{\lambda} \tag{7.31}$$

TABLE 7.4

Energy Flow in the Multiport Physical System

Systems	**Effort,** $e(t)$	**Flow,** $f(t)$
Mechanical translation	Force, $F(t)$	Velocity, $V(t)$
Mechanical rotation	Torque, $\tau(t)$	Angular velocity, $\omega(t)$
Compressor, Pump	Pressure difference, $\Delta P(t)$	Volume flow rate, $Q(t)$
Electric	Voltage, $V(t)$	Current, $i(t)$
Magnetic	Magneto-motive force, e_m	Magnetic flux, ϕ

7.4.4 Derivation of Sample Size Equation—Version I

Due to the cost and period restriction for a (mechanical) system, it is hard to carry out a reliability test with large samples. If fewer samples are carried out, the reliability test in statistical analysis will be more ambiguous because this testing will need a substantial period. For a more precise outcome, sufficient samples might be tested. Thus it is important to utilize the sample size equation with *AF* in Equations (7.30) and (7.31) and the basic statistics such as confidence level for raised testing. In statistical testing, the first step of parametric ALT needs to determine how many of the sample could be randomly selected and drawn from the population.

In selecting the set of values of the model parameters that make the likelihood function as great as possible, the maximum likelihood estimation (MLE) in statistics is a way of estimating the parameters of a statistical model from a stated data set. The characteristic life, η_{MLE}, is defined:

$$\eta_{\mathrm{MLE}}^{\beta} = \sum_{i=1}^{n} \frac{t_i^{\beta}}{r} \tag{7.32}$$

Some numerous procedures have been evolved to determine the characteristic life. Weibull analysis is a well-known way of inspecting reliability data in a way that can cover a lot of area. The Weibayes model that is established on Weibull examination may be used to decide the sample size. However, because it is hard to apply it directly due to its mathematical difficulty, it is possible that we will recognize the occasions of failures ($r \geq 1$) from the case of no failures ($r = 0$). In that case, an elucidated equation is thus evolved. As the failure number is $r \geq 1$ and the confidence level is $100(1 - \alpha)$, the characteristic life η_α from Equation (7.32) can be approximated:

$$\eta_{\alpha}^{\beta} = \frac{2r}{\chi_{\alpha}^{2}(2r+2)} \cdot \eta_{\mathrm{MLE}}^{\beta} = \frac{2}{\chi_{\alpha}^{2}(2r+2)} \cdot \sum_{i=1}^{n} t_i^{\beta} \text{ for } r \geq 1 \tag{7.33}$$

Contrarily, assuming there are no failures, the p value is α, and $\ln(1/\alpha)$ is similar to the chi-squared value $\frac{\chi_{\alpha}^{2}(2)}{2}$ in mathematical terms [33]:

$$p-\text{value}: \alpha = \int_{\chi_{\alpha}^{2}(2)}^{\infty} \left(\frac{e^{-\frac{x}{2}} x^{\frac{\nu}{2}-1}}{2^{\frac{\nu}{2}} \Gamma\left(\frac{\nu}{2}\right)} \right) dx = \int_{2\ln\alpha^{-1}}^{\infty} \left(\frac{e^{-\frac{x}{2}} x^{\frac{\nu}{2}-1}}{2^{\frac{\nu}{2}} \Gamma\left(\frac{\nu}{2}\right)} \right) dx \quad \text{for } x \geq 0 \tag{7.34}$$

where ν is the shape parameter, and Γ is the gamma function.

For $r = 0$, the characteristic life, η_α, from Equation (7.33) is defined:

$$\eta_{\alpha}^{\beta} == \frac{2}{\chi_{\alpha}^{2}(2)} \cdot \sum_{i=1}^{n} t_i^{\beta} = \frac{1}{\ln\frac{1}{\alpha}} \cdot \sum_{i=1}^{n} t_i^{\beta} \quad \text{for } r = 0 \tag{7.35}$$

Because Equation (7.33) is correct for all cases $r \geq 0$ from Equation (7.35), the characteristic life η_α is redefined:

$$\eta_\alpha^\beta = \frac{2}{\chi_\alpha^2(2r+2)} \cdot \sum_{i=1}^{n} t_i^\beta \quad \text{for } r \geq 0 \tag{7.36}$$

From CDF in the Weibull distribution from Equation (2.38), the interrelation between characteristic life and BX life may be derived. If taking the logarithm, it will be defined:

$$L_{BX}^\beta = \left(\ln \frac{1}{1-x}\right) \cdot \eta^\beta \tag{7.37}$$

As the roughly computed characteristic life of the p value α, η_α in Equation (7.36), is inserted into Equation (7.37), the BX life is obtained:

$$L_{BX}^\beta = \left(\ln \frac{1}{1-x}\right) \cdot \frac{2}{\chi_\alpha^2(2r+2)} \cdot \sum_{i=1}^{n} t_i^\beta \tag{7.38}$$

As most life testing generally has insufficient samples, the schemed testing time may be defined:

$$n \cdot h^\beta \geq \sum t_i^\beta \geq (n-r) \cdot h^\beta \tag{7.39}$$

If Equation (7.39) is exchanged into Equation (7.38), the BX life formulation may be defined:

$$L_{BX}^\beta \cong \left(\ln \frac{1}{1-x}\right) \cdot \frac{2}{\chi_\alpha^2(2r+2)} \cdot nh^\beta \geq \left(\ln \frac{1}{1-x}\right) \cdot \frac{2}{\chi_\alpha^2(2r+2)} \cdot (n-r)h^\beta \geq L_{BX}^{*\beta} \tag{7.40}$$

If Equation (7.40) is arranged differently, the sample size is attained:

$$n \geq \frac{\chi_\alpha^2(2r+2)}{2} \cdot \frac{1}{\left(\ln \frac{1}{1-x}\right)} \cdot \left(\frac{L_{BX}^*}{h}\right)^\beta + r \tag{7.41}$$

For a 60% confidence level, the first word $\frac{\chi_\alpha^2(2r+2)}{2}$ is estimated to (r + 1). Additionally, if the accumulative failure rate x is below roughly 20%, the denominator of the second word $\ln \frac{1}{1-x}$ is approximated to x by the Taylor expansion (Table 7.5).

Thus the approximated sample size equation (7.41) might be approached:

$$n \geq (r+1) \cdot \frac{1}{x} \cdot \left(\frac{L_{BX}^*}{h}\right)^\beta + r \tag{7.42}$$

TABLE 7.5

Characteristics of $\frac{\chi^2_{0.4}(2r+2)}{2}$ at the α = 60% Confidence Level

r	$1-\alpha$	$\frac{\chi^2_{0.4}(2r+2)}{2}$	$\frac{\chi^2_{\alpha}(2r+2)}{2} \approx r+1$	$1-\alpha$
0	0.4	0.92	1	0.63
1	0.4	2.02	2	0.59
2	0.4	3.11	3	0.58
3	0.4	4.18	4	0.57

Where the sample size equation $n \approx$ (failure numbers + 1)·(1/accumulative failure rate) ((target lifetime/(test hours))^$\beta + r$.

As Equation (7.30) or (7.31) are fastened into the schemed testing period h, Equation (7.42) may be redefined as:

$$n \geq (r+1) \times \frac{1}{x} \times \left(\frac{L_B^*}{AF \cdot h_a} \right)^{\beta} + r \tag{7.43}$$

As the lifetime of a mechanical system such as a compressor is aimed to be a B1 life of 10 years, the allocated mission cycles may be obtained for a defined set of samples exposed to raised loads.

In each ALT, the approximated life L_B is approximated:

$$L_B^{\beta} \cong x \cdot \frac{n \cdot (h_a \cdot AF)^{\beta}}{r+1} \tag{7.44}$$

Let $x = \lambda \cdot L_B$. The roughly calculated failure rate of the design samples λ is defined:

$$\lambda \cong \frac{1}{L_{BX}} \cdot (r+1) \cdot \frac{L_{BX}^{\beta}}{n \cdot (h_a \cdot AF)^{\beta}} \tag{7.45}$$

In each ALT phase, reliability may be calculated from the multiplication of the evaluated BX life L_B and failure rate λ. The functioning cycles of a product life are calculated under the expected customer usage situations. As the failed number, targeted life, accelerated factor (AF), and accumulative failure rate are specified, the compulsory test cycles under the raised situations are obtained from Equation (7.43).

The other form of sample size equation, which is attained by Wasserman [33–35], is expressed:

$$n = -\frac{\chi_\alpha^2(2r+2)}{2m^\beta \ln R_L} = \frac{\chi_\alpha^2(2r+2)}{2m^\beta \ln R_L^{-1}} = \frac{\chi_\alpha^2(2r+2)}{2m^\beta \ln\left(1-F_L^{-1}\right)}$$
$$= \frac{\chi_\alpha^2(2r+2)}{2} \cdot \frac{1}{\ln\left(1-F_L^{-1}\right)} \cdot \left(\frac{L_{BX}}{h}\right)^\beta \tag{7.46}$$

where $m \cong h/L_{BX}$.

As $r = 0$, the sample size equation will be attained:

$$n = \frac{\ln(1-C)}{m^\beta \ln R_L} = \frac{-\ln(1-C)}{-m^\beta \ln R_L} = \frac{\ln(1-C)^{-1}}{m^\beta \ln R_L^{-1}} = \frac{\ln \alpha^{-1}}{m^\beta \ln R_L^{-1}}$$
$$= \frac{\chi_\alpha^2(2)}{2} \cdot \frac{1}{\ln\left(1-F_L^{-1}\right)} \cdot \left(\frac{L_{BX}}{h}\right)^\beta \tag{7.47}$$

7.4.5 Derivation of Sample Size Equation—Version II

To attain the numeric of allocated mission cycles from the aimed BX life on the test plan, the sample size formulation combined with AF should be derived. First, each Bernoulli trial has one of the two outcomes such as failure or success. The accumulative probability that pursues a binomial distribution may be defined:

$$L(p) = \sum_{r=0}^{c} \binom{n}{r} p^r \cdot (1-p)^{n-r} \le \alpha \tag{7.48}$$

where n is the numeric of test samples, and c is the presumed number of failures.

As probability p is small-scale, and n is sufficient, Equation (7.48), which pursues a Poisson distribution, might be restated:

$$L(n \cdot p) = \sum_{r=0}^{c} \frac{1}{r!}(n \cdot p)^r \cdot e^{-(np)} = \sum_{r=0}^{c} \frac{1}{r!} m^r \cdot e^{-m} \le \alpha \tag{7.49}$$

where $m = \text{parameter} = n \cdot p$.

From Equation (7.49), it can be presumed that parameter m pursues the chi-square distribution, $\chi_\alpha^2()$, when the p value is α:

$$m = n \cdot p \sim \frac{\chi_\alpha^2(2r+2)}{2} \tag{7.50}$$

As the product pursues the Weibull distribution, the accumulated failure rate *F(t)* is expressed:

$$F(t) = 1 - e^{-\left(\frac{t}{\eta}\right)^{\beta}} \tag{7.51}$$

where t is time, η is characteristic life, and β is shape parameter.

In the case of unreliability, $p = F(t)$, and reliability, $1 - p = R(t)$, Equation (7.51) may be put into Equation (7.49):

$$L(p) = \sum_{r=0}^{c} \binom{n}{r} \left(1 - e^{-\left(\frac{t}{\eta}\right)^{\beta}}\right)^{r} \cdot \left(e^{-\left(\frac{t}{\eta}\right)^{\beta}}\right)^{n-r} \leq \alpha \tag{7.52}$$

Because $e^{-\left(\frac{t}{\eta}\right)^{\beta}} \cong 1 - \left(\frac{t}{\eta}\right)^{\beta}$, Equation (7.52) can be approximated:

$$L(p) \cong \sum_{r=0}^{c} \frac{1}{r!} \left(\frac{t}{\eta}\right)^{\beta r} \cdot \left(1 - \left(\frac{t}{\eta}\right)^{\beta}\right)^{n-r} \leq \alpha \tag{7.53}$$

Because Equations (7.52) and (7.53) have similar formations, the characteristic life with the confidence level 100 (1–α) will be defined:

$$m = n \cdot p = n \cdot \left(\frac{t}{\eta}\right)^{\beta} \sim \frac{\chi^2_{\alpha}(2r+2)}{2} \text{ or } \eta_{\alpha}^{\beta} = \frac{2}{\chi^2_{\alpha}(2r+2)} \cdot n \cdot t^{\beta} \tag{7.54}$$

At BX life L_B in Equation (7.51), the test period t is h.

$$L_B^{\beta} \cong x \cdot \eta_{\alpha}^{\beta} = x \cdot \frac{2}{\chi^2_{\alpha}(2r+2)} \cdot n \cdot t^{\beta} = x \cdot \frac{2}{\chi^2_{\alpha}(2r+2)} \cdot n \cdot h^{\beta} \geq L_B^{*\beta} \text{ for } x \text{ δ } 0.2 \tag{7.55}$$

where $x = 0.01F(t)$.

As Equation (7.55) is arranged differently, the sample size equation is defined:

$$n \geq \frac{\chi^2_{\alpha}(2r+2)}{2} \times \frac{1}{x} \times \left(\frac{L_B^{*}}{h}\right)^{\beta} \tag{7.56}$$

For a 60% confidence level, because the first word $\frac{\chi^2_{\alpha}(2r+2)}{2}$ in Equation (7.56) may be approximated as $(r + 1)$, Equation (7.56) can be redefined:

$$n \geq (r+1) \times \frac{1}{x} \times \left(\frac{L_B^*}{h} \right)^{\beta} \tag{7.57}$$

As the acceleration factors in Equations (7.30) and (7.31) are inserted to the schemed test period h, Equation (7.57) is

$$n \geq (r+1) \times \frac{1}{x} \times \left(\frac{L_B^*}{AF \cdot h_a} \right)^{\beta} + r \tag{7.58}$$

References

1. Bigg, G.; Billings, S. The iceberg risk in the *Titanic* year of 1912: Was it exceptional? *Significance*. **2014**, *11*(3), 6–10.
2. ISO 1328–2:2020. *Cylindrical gears—ISO system of flank tolerance classification—Part 2: Definitions and allowable values of double flank radial composite deviations: General procedures*. ISO: Geneva, Switzerland, 2020.
3. ISO 17804:2020. *Founding—Ausferritic spheroidal graphite cast irons—classification cylindrical gears*. ISO: Geneva, Switzerland, 2020.
4. AGMA 939-A07. *Austempered ductile iron for gears*. AGMA Information Sheet, 2018.
5. DVB Bank SE Aviation Research (AR). *An overview of commercial Jet aircraft 2013*, 2014, p. 20.
6. Woo, S.; Pecht, M.; O'Neal, D. Reliability design and case study of the domestic compressor subjected to repetitive internal stresses. *Reliability Engineering & System Safety*. **2019**, *193*, 106604.
7. Duga, J.J.; Fisher, W.H.; Buxaum, R.W.; Rosenfield, A.R.; Buhr, A.R.; Honton, E.J.; McMillan, S.C. *The economic effects of fracture in the United States*. Final Report. Available as NBS Special Publication 647–2; Battelle Laboratories: Columbus, OH, 1982.
8. Modarres, M.; Kaminskiy, M.; Krivtsov, V. *Reliability engineering and risk analysis: A practical guide*, 3rd ed. CRC Press: Boca Raton, FL, 2016.
9. Elsayed, E.A. *Reliability engineering*. John Wiley & Sons: Hoboken, NJ, 2012.
10. Hahn, G.J.; Meeker, W.Q. *How to plan an accelerated life test (E-Book)*. ASQ Quality Press: Milwaukee, WI, 2004.
11. Moura, E.C. *How to determine sample size and estimate failure rate in life testing*. ASQ Quality Press: Milwaukee, WI, 2004.
12. McPherson, J. *Reliability physics and engineering: Time-to-failure modeling*. Springer: New York, 2010.
13. Griffith, A.A. The phenomena of rupture and flow in solids. *Philosophical Transactions of the Royal Society of London. Series A*. **1921**, *221*, 163–198.
14. Wang, Y.; He, Z. Experimental and statistical study of the fracture mechanism of Sn96.5Ag3Cu0.5 solder joints via Ball Shear test. *Materials*. 2022, *15*, 2455.
15. Anderson, T.L. *Fracture mechanics—fundamentals and applications*, 3rd ed. CRC: Boca Raton, FL, 2017.
16. ASTM E606/E606 M. *Standard test method for strain-controlled fatigue testing*. ASTM International: West Conshohocken, PA, 2019.

17. ASTM E399. *Standard test method for linear-elastic plane-strain fracture toughness of metallic materials.* ASTM International: West Conshohocken, PA, 2020.
18. ASTM E647. *Standard test method for measurement of fatigue crack growth rates.* ASTM International: West Conshohocken, PA, 2015.
19. ASTM E739–10. *Standard practice for statistical analysis of linear or linearized stress-life (S–N) and strain-life (ε-N) fatigue data.* ASTM International: West Conshohocken, PA, 2015.
20. Elishakoff, I. Stepan Prokofievich Timoshenko and America. *ZAMM—Journal of Applied Mathematics and Mechanics.* **2019**, *99*(3).
21. David, J.G. *Introduction to quantum mechanics,* 3rd ed. Cambridge University Press: Cambridge, 2018.
22. Cemal, B. *Introduction to unified mechanics theory with applications,* 1st ed. Springer Nature: Berlin, Germany, 2021.
23. Weingart, R.G.; Stephen, P. Timoshenko: Father of engineering mechanics in the U.S. *Structure Magazine,* 1 August 2007.
24. Reddy, J.N. *An introduction to the finite element method,* 4th ed. McGraw-Hill: New York, 2020.
25. Dyniewicz, B.; Bajkowski, J.M.; Bajer, C.I. Effective viscoplastic-softening model suitable for brain impact modeling. *Materials.* **2022**, *15*, 2270.
26. Vijaya Kumar, S.D.; Karuppanan, S.; Ovinis, M. Artificial neural network-based failure pressure prediction of API 5 L X80 pipeline with circumferentially aligned interacting corrosion defects subjected to combined loadings. *Materials.* **2022**, *15*, 2259.
27. Vega, M.A.; Todd, M.D. A variational Bayesian neural network for structural health monitoring and cost-informed decision-making in miter gates. *Structural Health Monitoring.* **2022**, *21*(1), 4–18.
28. IEEE Standard Glossary of Software Engineering Terminology. IEEE STD 610.12–1990. *Standards coordinating committee of the computer society of IEEE* (reaffirmed September 2002). Available online: https://ieeexplore.ieee.org/document/159342 (accessed on 31 December 2020).
29. Kreyszig, E. *Advanced engineering mathematics,* 10th ed. John Wiley and Son: Hoboken, NJ, 2011, p. 683.
30. Grove, A. *Physics and technology of semiconductor device,* 1st ed. Wiley International Edition, New York, 1967, p. 37.
31. Minges, M.L. *Electronic materials handbook.* ASM International: Cleveland, OH, 1989; Volume 1, p. 888.
32. Karnopp, D.C.; Margolis, D.L.; Rosenberg, R.C. *System dynamics: Modeling, simulation, and control of mechatronic systems,* 6th ed. John Wiley & Sons: New York, 2012.
33. Wasserman, G. *Reliability verification, testing, and analysis in engineering design.* Marcel Dekker: New York, 2003, p. 228.
34. Tang, L.C. Multiple-steps step-stress accelerated life tests: A model and its spreadsheet analysis. *International Journal of Materials and Product Technology.* **2004**, *21*(5), 423–434.
35. William, E.L.; David, M.; Christoper, N.M.; Stephen, W.B.; Richard, J.F.; Timothy, F.; Thomas, A.S.; Frank, W.G. September 2005. *Mechanical properties of structural steels.* NIST NCSTAR 1–3D.

8

Case Investigations of Parametric Accelerated Life Testing (ALT)

8.1 Redesign of the HKS Returned from the Marketplace

Because the end user wants to close the door gently, a hinge kit system is attached. Its chief parts are torsional shaft, kit cover, support, spring, HKS housing, etc. (Figure 8.1).

In the marketplace, the HKS in a household refrigerator did not function under undescribed customer utilization conditions. Market data designated that the failed component could have design defects (Figure 8.2).

As shown in Figure 8.3, the mechanical load in an HKS may be expressed:

$$M_1 = M_0 + M_A = W_{door} \times b + W_A \times a \tag{8.1}$$

$$(8.1) = T_1 = F_1 \times R \tag{8.2}$$

As the time to failure relied on the impact force due to moment in Equations (8.1) and (8.2), the life-stress prototype (LS model) in Equation (7.29) can be restated:

$$TF = A(S)^{-n} = AT^{-\lambda} = A(F \times R)^{-\lambda} = B(F)^{-\lambda} \tag{8.3}$$

Thus the AF in Equations (7.30) and (7.31) can be expressed:

$$AF = \left(\frac{S_1}{S_0}\right)^n = \left(\frac{T_1}{T_0}\right)^\lambda = \left(\frac{F_1 \times R}{F_0 \times R}\right)^\lambda = \left(\frac{F_1}{F_0}\right)^\lambda \tag{8.4}$$

Based on customer use, the number of daily openings/closings HKS changed from three to ten cycles. Under maximum use for ten years, the operation of an ice dispenser occurs roughly 36,500 use cycles. Manufacturer data shows that the normal door weight is 1.1 kN and that the door with an elevated weight is 2.76 kN. Assuming the damage factor λ is 2, AF is approximately 6.3 in Equation (8.4).

DOI: 10.1201/9781003451471-8

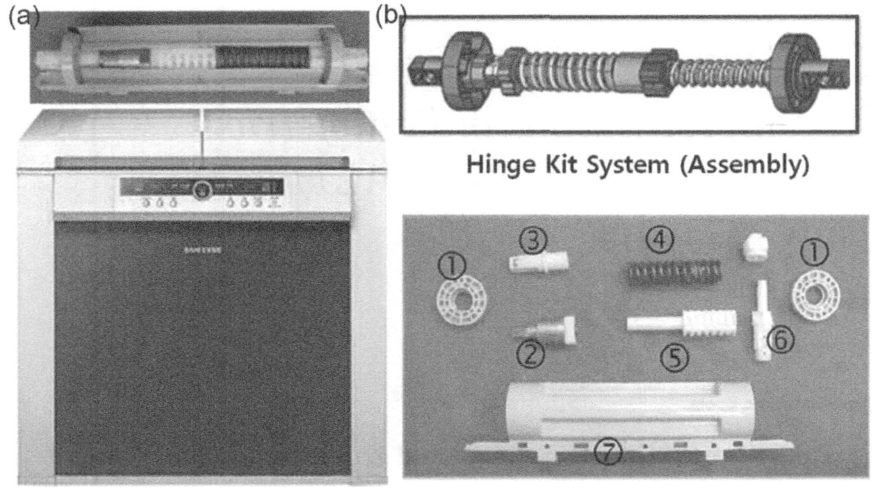

FIGURE 8.1
Household refrigerator and HKS failed from the market: (a) refrigerator; (b) HKS components: ①kit cover, ②oil damper, ④ spring, and ⑦ kit housing.

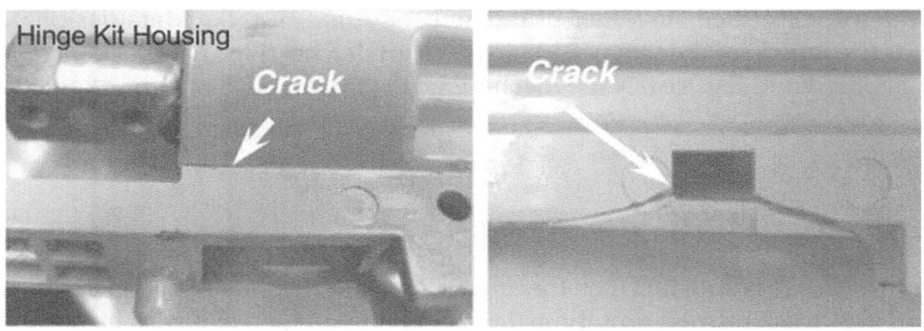

FIGURE 8.2
HKS housing failed from the market.

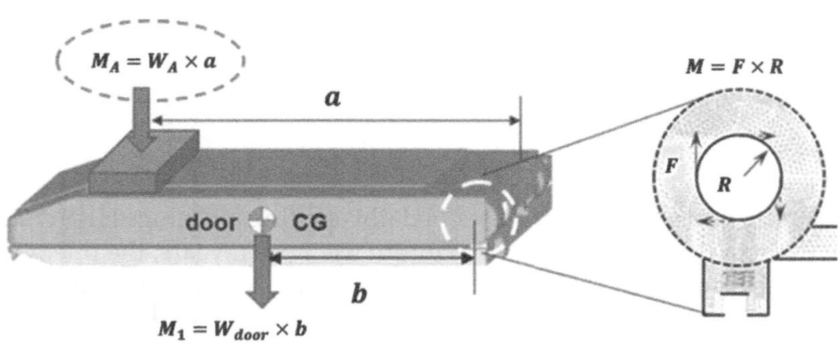

FIGURE 8.3
Functional design idea of a mechanical HKS.

For B1 life of 10 years, with the shape parameter assumed to be 2.0, the allocated mission cycles for six samples computed from Equation (7.43) were approximately 23,000 cycles. This parametric ALT is designed to ensure a B1 life of 10 years if HKS is unsuccessful less than once in 23,000 cycles.

8.1.1 Results and Conclusions

In the 1st ALT, the HKS was unsuccessful at 3,000 cycles and 15,000 cycles. Figure 8.4 indicates a photo showing the fractured components from the field and that from the 1st ALT. Because they are similar, the HKS failed from the field might be replicated by a parametric ALT. An easily broken structure originated from design defects. Thus to have adequate strength against repetitive impact loads, the construction of HKS was strengthened as follows: 1) the notch of housing was eliminated and the inside and outside of the HKS was shaped approximately like a circle, 2) strengthened ribs were also fastened to the deck and housing (Figure 8.5).

When taking apart the problematic HKS, the leaked oil damper failed at 15,000 cycles. The basic cause for this arose from the structure of the sealing design in an oil damper that had held an O-ring, Teflon, and O-ring with a space of 0.5 mm. We came to the conclusion that something might be intruded between the O-ring and Teflon. To clasp the O-ring firmly and closely, the sealing design was modified as indicated in Figure 8.6.

In the 2nd ALT, the fractured HKS cover failed at 7,800 and 9,200 cycles. The material was modified from plastic to aluminum die casting (Figure 8.7).

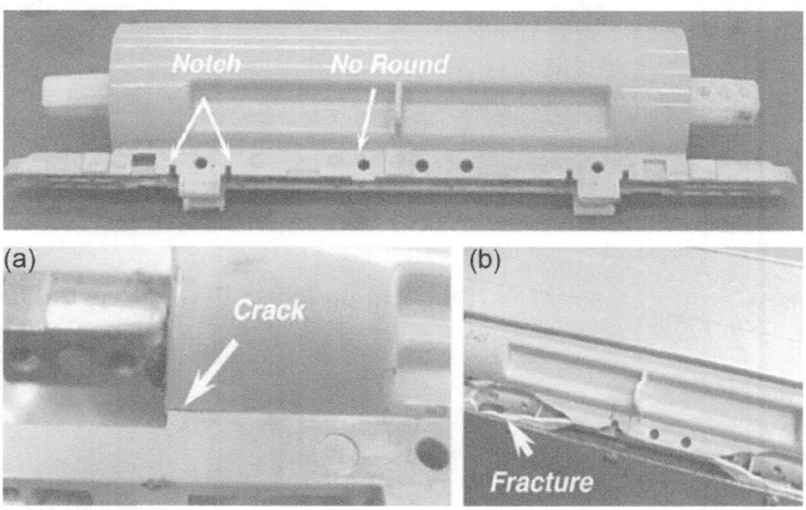

FIGURE 8.4
Failed components from the market and ALT: (a) products failed from the field; (b) failed product after 1st ALT.

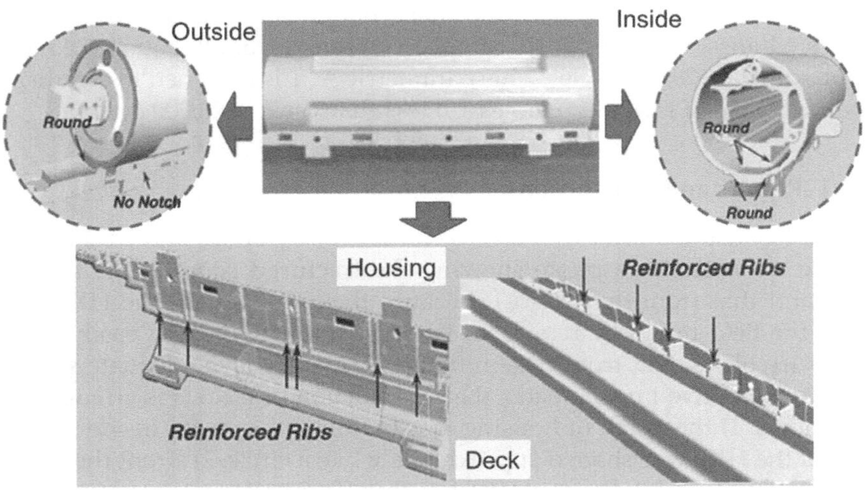

FIGURE 8.5
Modified hinge kit housing.

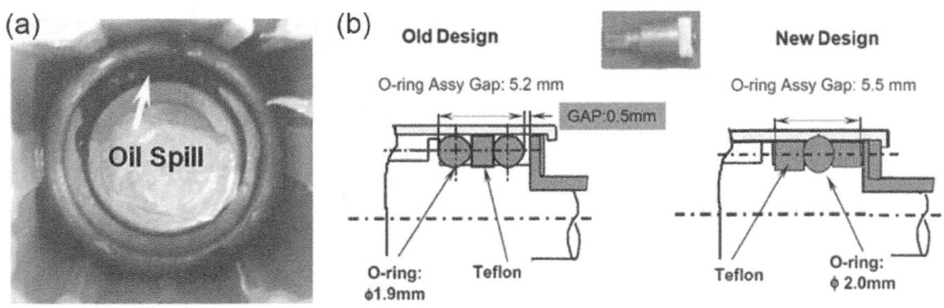

FIGURE 8.6
Altered oil damper: (a) products failed from the 1st ALT; (b) design modifications.

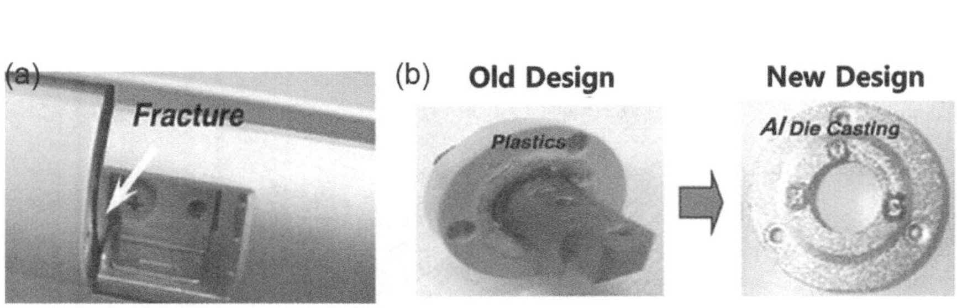

FIGURE 8.7
Altered housing cover: (a) unsuccessful products from the 2nd ALT; (b) design alterations.

In the 3rd ALT, there were no problems until 74,000 cycles. By parametric ALTs, the HKS with the alternations was ensured to be B1 life of 10 years.

8.2 Increasing the Fatigue Life of a Dispensing System in a Refrigerator

The end user typically uses the refrigerator to dispense the cooled water. The water dispenser in a refrigerator is designed to satisfy these requirements. Dispensing water in a household refrigerator demands: 1) the customer pushes the water cup to the dispenser lever and 2) water dispenses into it. The dispenser therefore might be reliably designed to successfully operate under customer situations (Figure 8.8).

In the field, as dispensers are repetitively exerted, they had been failing due to some design defects, and customers wanted it replaced. To properly function for its expected lifetime, the dispenser system's design defects—a sharp-edged corner, etc. —had to be found and changed by parametric ALT. From the free body diagram of the refrigerator dispenser, the applied loads originated from the cup pressing:

$$\sum M_z = aF_X - bF_Y = 0 \tag{8.5}$$

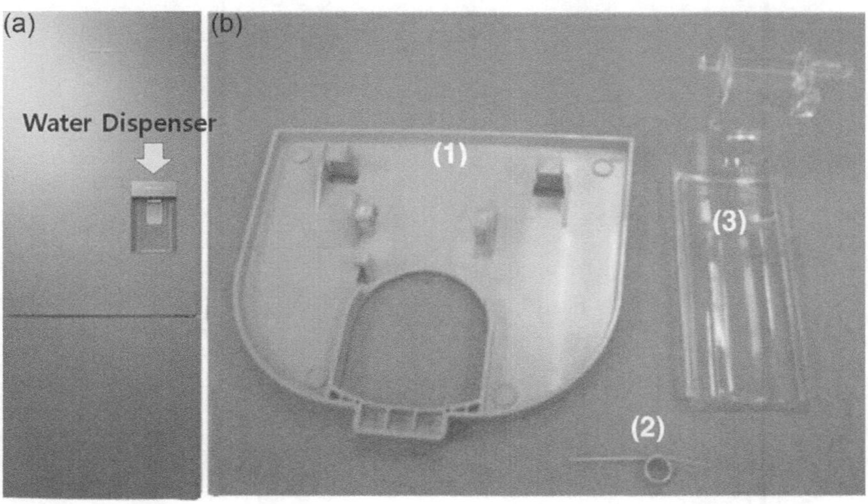

FIGURE 8.8
Refrigerator and dispenser system: (a) refrigerator; (b) dispenser system—(1) dispenser lever, (2) spring, and (3) cover.

As the dispensing stress relies on cup loading, Equation (7.29) may be defined:

$$TF = A(S)^{-n} = A'(F_X)^{-\lambda} = B(F_Y)^{-\lambda} \tag{8.6}$$

where A and B are constants.

So entrenched in Equations (7.30 and 7.31), the AF can be expressed (Figure 8.9):

$$AF = \left(\frac{S_1}{S_0}\right)^n = \left(\frac{F_1}{F_0}\right)^\lambda \tag{8.7}$$

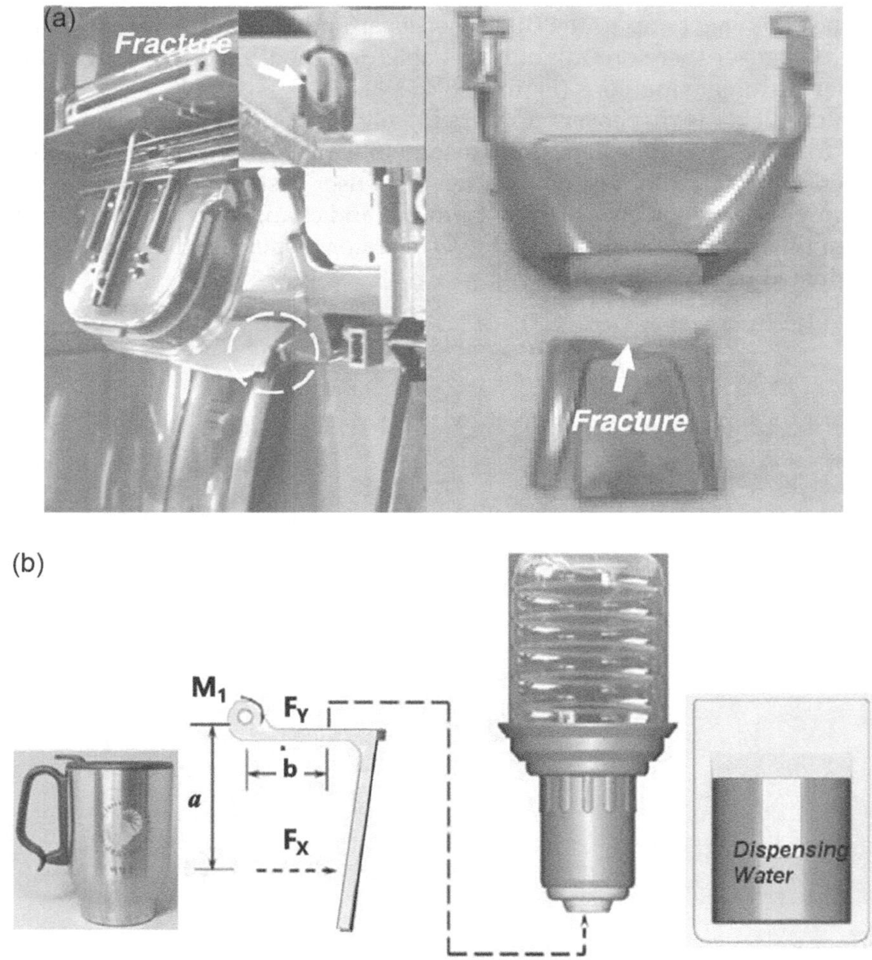

FIGURE 8.9
(a) Failed lever after usage; (b) force equilibrium.

For the dispenser system, the environmental conditions are 0–43°C with 0.2–0.24 g acceleration, and a humidity ranging from 0 to 95%. The working cycles of the daily dispenser went from 4 to 20 cycles. Based on the design life of a product for ten years, the dispenser was applied to 73,000 usage cycles. For the worst case, the greatest load exposed by the customer in dispensing water F_0 was 15–20 N. For ALT, the exerted load F_1 doubled to 35 N. With an accumulative quantity of 2, the AF from Equation (8.7) was 4.0. To retain the life aim—B1 life of 10 years, when the shape parameter was presumed to have 2.0, the assigned mission cycles for ten samples decided in Equation (7.43) were 65,000. The problematic designs of the dispenser therefore may be identified and altered.

8.2.1 Results and Conclusions

In the 1st ALT, the water dispenser was unsuccessful at 25,000 cycles (Figure 8.10). As deliberately identifying the problematic dispenser from the 1st ALT and the field, the failure places were found at the hinge and front corner. Because of stress raisers—no corner rounding and inadequate thickness—the dispenser lever failed.

As the action plan, these designs were revised: (1) increasing the hinge rib rounding, C1, from 0.0 to 1.5 mm (Fillet 1); (2) increasing the front corner rounding, C2, from 0.0 to 1.5 mm (Fillet 2); (3) increasing the hinge rib, C3, from 1.0 to 1.8 mm (Rib 1); and (4) increasing the front side rounding, C4, from 0.0 to 8.0 mm (Fillet 3) (Table 8.1). Stress analysis, united with parametric ALT, was performed. As a consequence, the estimated stress concentrations at the front corner and the shaft hinge in a dispenser lever decreased from 8.37 and 5.66 MPa (Figure 8.11(a)).

In the 2nd ALT, the dispenser lever was cracking and fracturing at roughly 32,000 cycles (Figure 8.4(b)). After a close look at the failure places, the front corner in the dispenser lever had inadequate corner rounding. To increase its product life, it was altered by 1) enlarging the hinge rib rounding (Fillet 1), C1, from 1.5 mm to 2.0 mm and 2) enlarging the front side rounding (Fillet 3), C4, from 8.0 mm to 11.0 mm (Table 8.1).

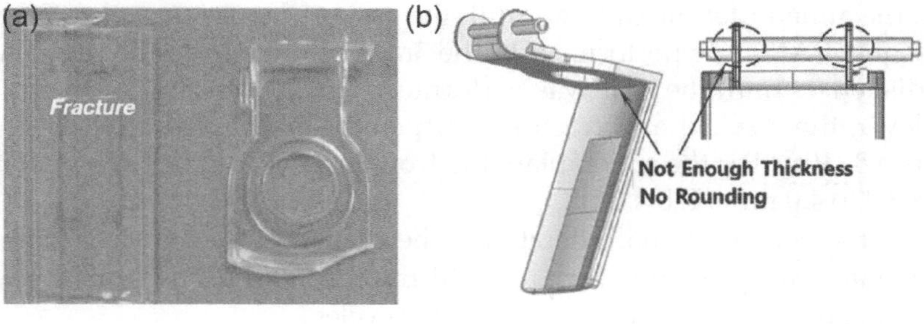

FIGURE 8.10
Troublesome designs of dispenser: (a) 1st ALT failure; (b) basic cause.

TABLE 8.1
Summary for the Redesigned Dispenser Lever

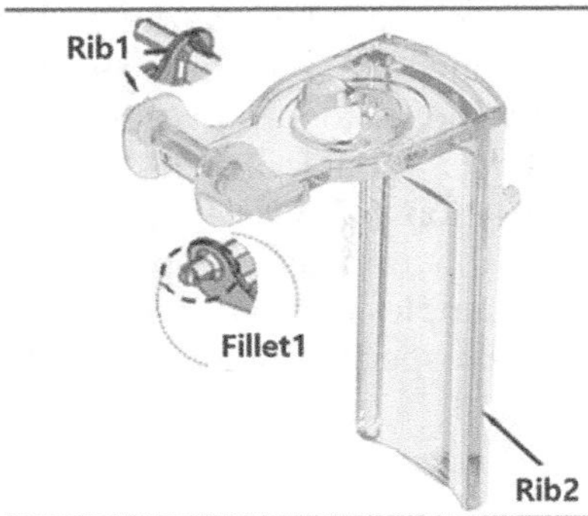	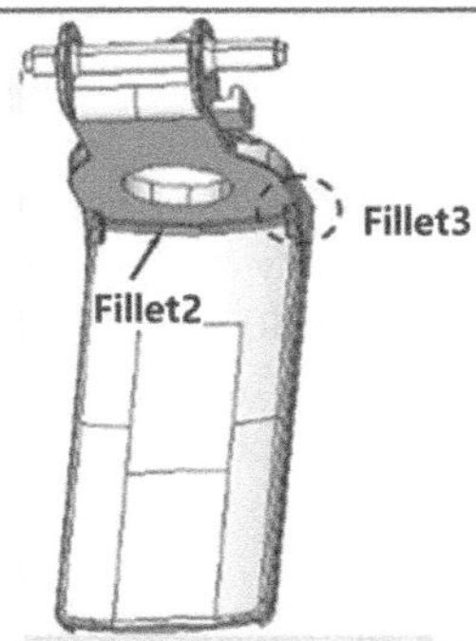
C1: Fillet1 R0 → R1.5 (first ALT) → R2.0 (second ALT) C3: Rib1 T1 → T1.8 (first ALT) C5: Rib2 T3 → T4 (third ALT)	C2: Fillet2 R0 → R1.5 (first ALT) C4: Fillet3 R0 → R8.0 (first ALT) → R11.0 (second ALT)

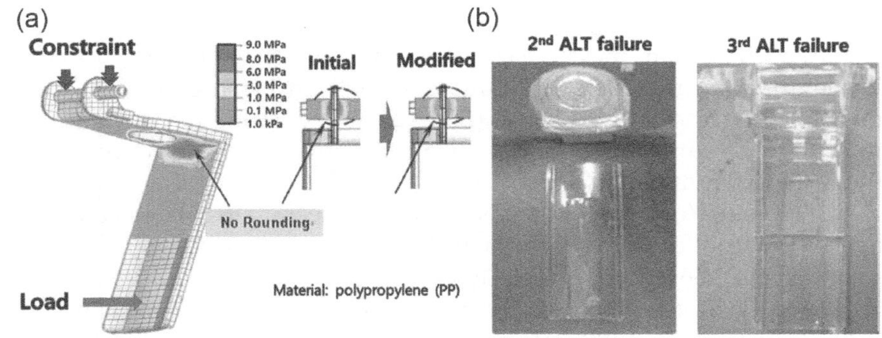

FIGURE 8.11
Problematic designs of dispenser in 2nd and 3rd ALTs: (a) finite element analysis; (b) 2nd and 3rd ALT failures.

Although the design of the water dispenser was enhanced in the 2nd ALT, the aimed lifetime did not fulfill the assigned mission cycles. Thus the subsequent ALT was performed. In the 3rd ALT, there were no problems in the dispenser until the ALT was performed for 68,000 cycles. At the front of the lever, the cracked and fractured dispenser lever failed at 68,000 cycles (Figure 8.11(b)). Per the action plan, the front lever (Rib 2), C5, was increased from 3.0 to 4.0 mm (Table 8.1).

With the constructional alterations, the dispenser system in a domestic refrigerator can properly work to fulfil the life aim—B1 life of 10 years—because there were no problems till 68,000 cycles.

8.3 Enhancing Life of the Ice-Maker Building in Auger Motor with Bearing in a Domestic Refrigerator

Customers want to use the refrigerator to dispense the crushed (or cubed) ice. To fulfill these needs, an ice-maker in the refrigerator is designed. Distributing ice in a household refrigerator requires 1) ice to collect in an ice-maker, 2) the consumer to touch the cup to the lever, and 3) ice to be distributed into it. The main parts are the ice-bucket, blade dispenser, blade, helix support, helix upper dispenser, an auger motor, etc. (Figure 8.12).

In the marketplace, as the lever was repeatedly pressed, the ice-makers, covering the auger motors with a bearing, were cracking and fracturing due to unspecified design flaws, causing customers to insist on replacement. To work correctly for its anticipated lifetime, the design defects had to be found and altered by parametric ALT (Figure 8.13).

From the bond graph, the governing equation may be defined (Figure 8.14):

$$\begin{bmatrix} di_a / dt \\ d\omega / dt \end{bmatrix} = \begin{bmatrix} -R_a / L_a & 0 \\ mk_a & -B / J \end{bmatrix} \begin{bmatrix} i_a \\ \omega \end{bmatrix} + \begin{bmatrix} 1 / L_a \\ 0 \end{bmatrix} e_a + \begin{bmatrix} 1 \\ -1 / J \end{bmatrix} T_L \qquad (8.8)$$

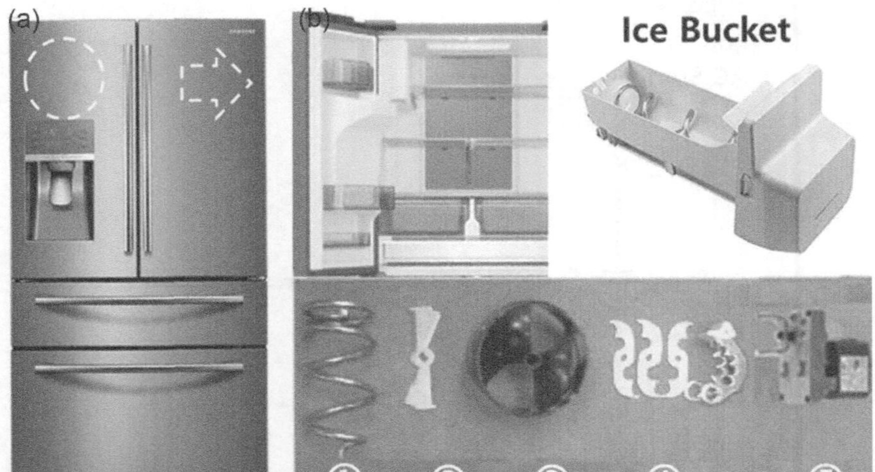

FIGURE 8.12

Refrigerator and Ice-maker system: (a) refrigerator; (b) ice-maker with ① helix support, ② blade dispenser, ③ helix upper dispenser, ④ blade, and ⑤ auger motor.

FIGURE 8.13
Failed auger motor in ice-maker.

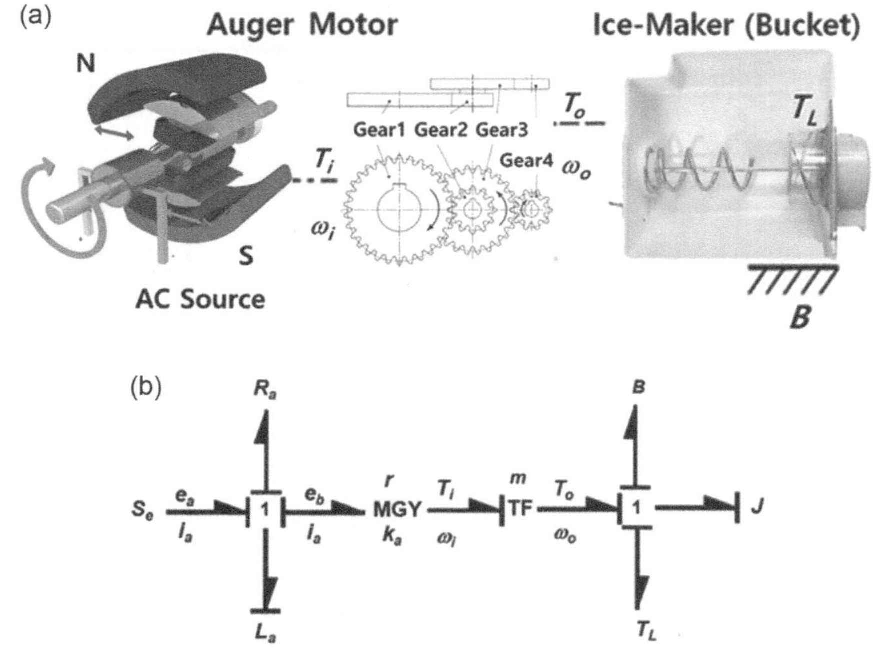

FIGURE 8.14
Design idea of a refrigerator ice-maker: (a) typifying sketch of auger motor, crusher, blade, ice-bucket, etc.; (b) bond graph for ice-maker.

As the stress depends on the exerted torque, Equation (7.29) may be expressed:

$$TF = A(S)^{-n} = AT_L^{-\lambda} = A(F_c \times R)^{-\lambda} = B(F_c)^{-\lambda} \tag{8.9}$$

Thus, based on Equation (7.30), the AF can be restated:

$$AF = \left(\frac{S_1}{S_0}\right)^n = \left(\frac{T_1}{T_0}\right)^\lambda = \left(\frac{F_1 \times R}{F_0 \times R}\right)^\lambda = \left(\frac{F_1}{F_0}\right)^\lambda \tag{8.10}$$

For the ice-maker, the working surrounding conditions are 0–43°C with 0.2–0.24 g acceleration, and a humidity spanning 0–95%. The daily functioning period of the ice-maker was 3–18 cycles. Based on the life of ten years, the ice-maker performed to 65,700 usage cycles. For the worst case, the greatest torque subjected to the customer in dispensing ice T_0 was 0.69 kN-cm. For ALT, the applied torque T_1 increased to 1.47 kN-cm. With a cumulative factor λ of 2, the AF in Equation (8.10) was 4.0. To attain the lifetime goal—B1 lifetime for 10 years, if the shape parameter β was assumed to be 2.0, the allotted mission cycles for ten samples calculated from Equation (7.43) were 42,000 cycles. The troublesome design of the ice-maker thus can be found and modified.

8.3.1 Results and Conclusions

In the 1st ALT, a problematic bearing of the auger motor in the ice-maker replicated 6,500 cycles (Figure 8.15). After a close look at the cracked bearing from the 1st ALT and field, failures were discovered in the outer ring. The troubled outside component on the cross section was an intergranular (IG) crack due to repeated torques under severe conditions (–20°C ↓). Per the action plan, the material of the troublesome bearing in the auger motor was modified from AISI 52100 Alloy Steel to the bearing manufactured with sintered and hardened steel.

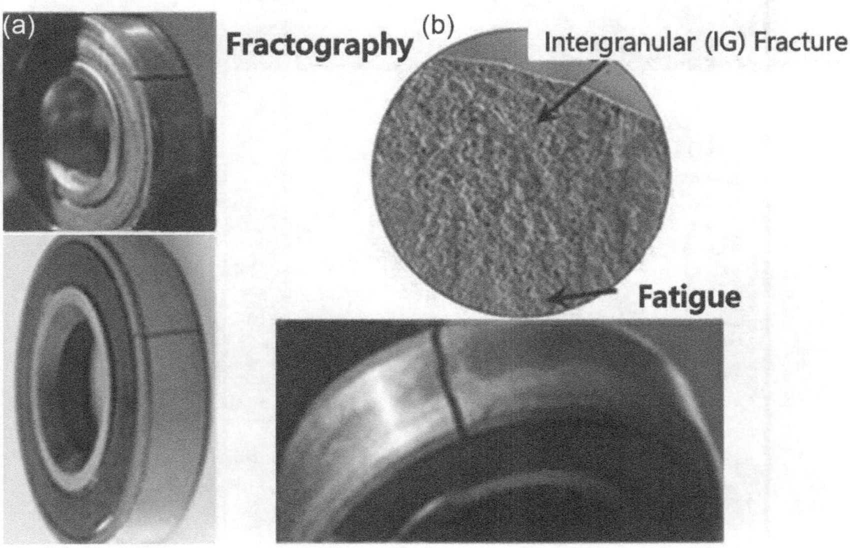

FIGURE 8.15
Problematic designs of the dispenser in the market and 1st ALT: (a) problematic field part; (b) 1st ALT failure.

In 2nd ALT, the helix upper dispenser made of polycarbonates was cracking and fracturing in the exposed region of the blade dispenser. To understand the basic cause of the problematic product, it was found that there was an imperfection in the constructional component—the weld line between the helix blade dispenser and upper dispenser—which had some micro-voids made in the plastic injection process. Per the action plan, a reinforced rib of the helix had been increased after the injection process was altered. So a finite element analysis (FEA), integrated with the ALT result, was carried out. As a result, the component stress by the FEA was lessened from 45.0 to 20.0 kPa (Figure 8.16). As the designs were altered, the ice-maker in a refrigerator may operate adequately way and fulfill the aimed life—B1 life for 10 years—because there were no issues until 60,000 cycles.

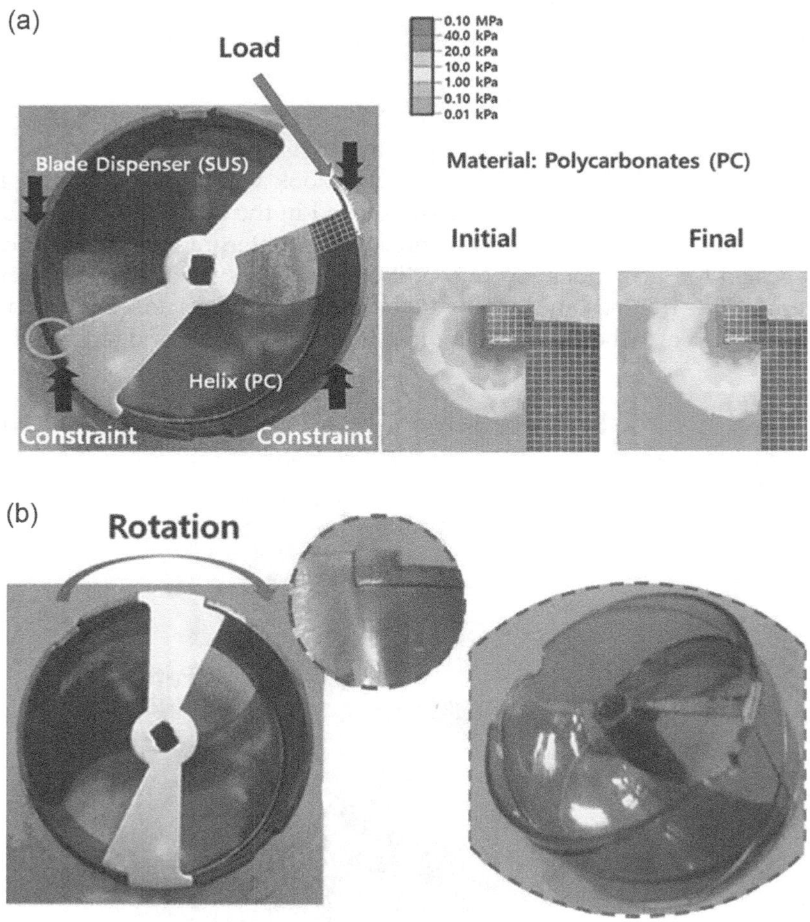

FIGURE 8.16
Problematic designs of dispenser in 2nd ALT: (a) finite element method; (b) 2nd and 3rd ALT failure.

8.4 Pneumatically Enhancing Life of a Cylinder-Operated Machine Tool

The customer utilizes the pneumatic cylinder to easily alter the tools in an automated assembly line. Mechanical components in the cylinder are designed to move a pressure load in a straight line to a designated location. The main parts are a rod cap, piston, head cap, etc. (Figure 8.17).

In the field, when pressure loading is repeatedly applied over its anticipated life, the troublesome piston seal induced the cylinder to leak gas. This led to a crack and stopped operation due to unspecified design flaws, and customers were requesting for replacement. To rightly work the cylinder for its expected life, the design flaws had to be found and modified by parametric ALT (Figure 8.18).

The pressure difference ΔP in the cylinder may be defined:

$$\Delta P = \Delta F_{\text{int}} / A \tag{8.11}$$

where ΔF_{int} = force difference.

As the stress relies on the exerted pressure difference, Equation (7.29) may be defined:

$$TF = A(S)^{-n} = B(\Delta P)^{-\lambda} \tag{8.12}$$

Therefore, based on Equation (7.30) and (7.31), the AF may be defined:

$$AF = \left(\frac{S_1}{S_0}\right)^n = \left(\frac{\Delta P_1}{\Delta P_0}\right)^\lambda \tag{8.13}$$

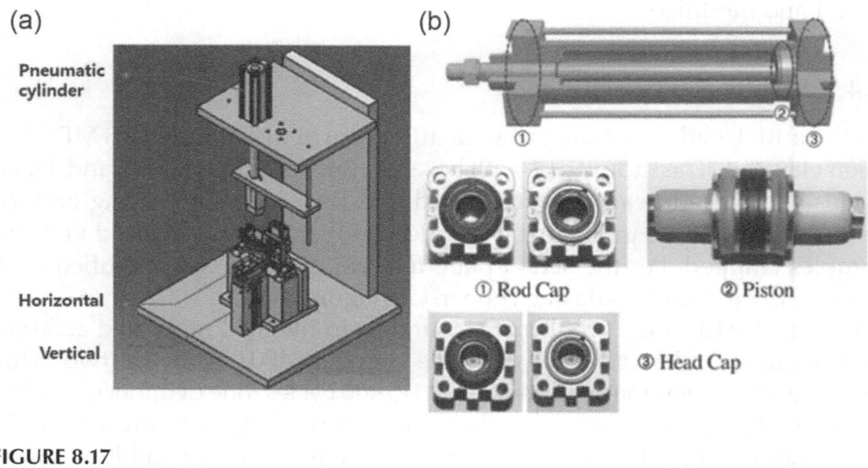

FIGURE 8.17
French door refrigerated and freezer drawer assembly: (a) pneumatic cylinder in an automated assembly; (b) mechanical component: ① rod cap, ② piston, and ③ head cap.

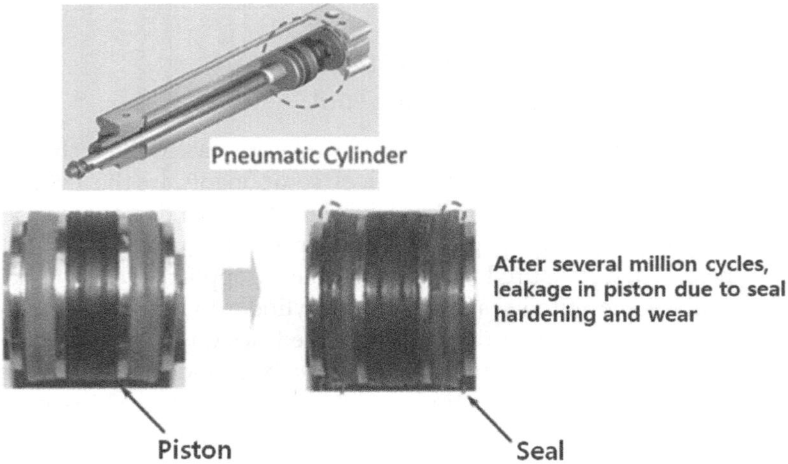

FIGURE 8.18
Failed piston seal in a cylinder.

The environmental conditions for the cylinder in an assembly line are 0–43 °C, 0.2–0.24 g acceleration, and a relative humidity spanning of 0–95%. The daily working duty of the cylinder was 10–30 cycles. Based on the life cycles for ten years, the cylinder performed 109,500 usage cycles. In the worst case, the pressure difference ΔP_0 suggested to the customer in altering tools was 0.69 MPa. For ALT, the applied pressure difference ΔP_1 increased to 0.8 MPa. With a cumulative factor λ of 2, the AF in Equation (8.13) was 1.6. To fulfill the aimed life—B1 life for 10 years—the shape parameter β was anticipated to have to be 2.0, and the assigned mission period for ten samples selected in Equation (7.43) was 200,000 cycles. The design issues of the cylinder had to be found and modified.

8.4.1 Results and Conclusions

At the initial loading, as the pressure difference in the cylinder, 0.8 MPa, had been elevated. The samples (n = 10) became noisy at 6,000, 10,500, and 11,000 cycles, which had peaked at 70 dB and was the source of hearing loss. As disassembling three problematic samples, one sample was cracked and two samples chipped. Per the action plan, the seal material was modified from Fe-36Ni Invar Alloy to silicone rubber, C1 (Figure 8.19).

In the 1st ALT, some problematic cylinders in the field replicated at 50,000 cycles (one cylinder), 65,000 cycles (one cylinder), 100,000 cycles (two cylinders), 110,000 cycles (one cylinder), and 115,000 cycles (one cylinder). The failure mode for cylinders was over the minimum working pressure and stroke time, leading to (1) frequent working stress, (2) friction heat and lube vaporization of the seal, and (3) increasing the hardening and wear of seal. As the

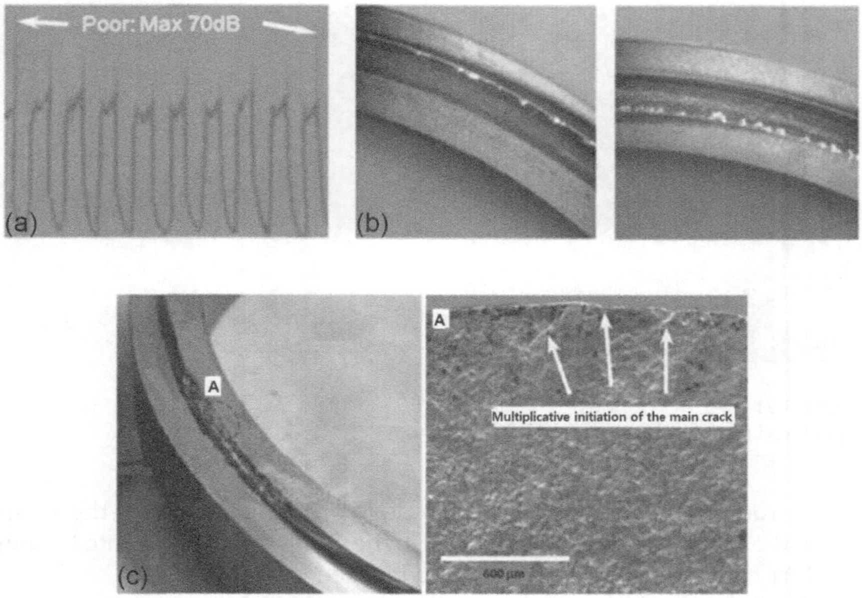

FIGURE 8.19
Design difficulties of samples in initial loading: (a) noised sample; (b) chipped samples; (c) cracked sample.

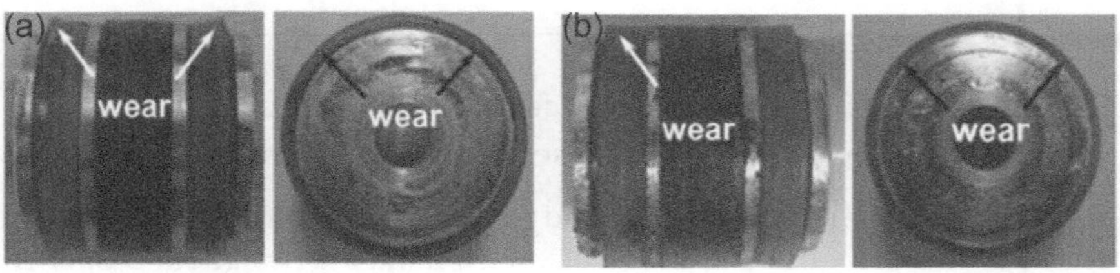

FIGURE 8.20
Design issues of samples in 1st ALT: (a) problematic products at the market; (b) problematic products after parametric ALT.

problematic cylinders were taken apart, the hardening and wear in the piston seal were like cylinders returned from the field. The material of the seal, C2, was altered from silicone rubber to a polyurethane (Figure 8.20).

To discover the cylinder failures under other elevated circumstances of 1.2 MPa and 23 °C, ALT was carried out. As a result, the cylinder was problematic in five out of six samples: 30,000 cycles (one sample), 70,000 cycles (two samples), and 80,000 cycles (two samples). The failure mode was over the minimum operation pressure and piston stroke time. As the problematic samples were disassembled, there were hardening and wear-out in the seal,

FIGURE 8.21
Clogged rod cap found by parametric ALT.

and the rod cap was blocked. The repetitive pressure loading of the sample caused the seal wear, which developed in the slurry and its flow into the port. The slurry at the port accumulated and blocked it (Figure 8.21).

In the 2nd ALT, there were no problems until 220,000 cycles. Because the designs were modified, the cylinder will adequately perform the aimed life—B1 life of 10 years.

8.5 Increasing Lifetime of the Ice-Maker's Auger Motor with Gear System in a Household Refrigerator

As a function of the refrigerator, the end user wants to use the ice-maker to dispense ice. To satisfy this requirement, an ice-maker is designed. Dispensing ice in a domestic refrigerator needs 1) to collect ice in an ice-maker system, 2) the customer presses the cup to the lever, and 3) to dispense ice. The main constructional parts are ① helix support, ② blade dispenser, ③ helix upper dispenser, ④ blade, and ⑤ auger motor with gear system (Figure 8.22).

In the marketplace, as the dispenser lever is frequently exerted, the ice-maker, including the auger motors with gear system, was cracking and fracturing due to unspecified design defects, followed by consumers to demanding replacement. To work the ice-maker for its anticipated life in a manner that is suitable in the circumstances, the design defects in a product may be found and altered by parametric ALT (Figure 8.23).

From the bond graph (Figure 8.24), the governing equation can be derived:

$$\begin{bmatrix} di_a / dt \\ d\omega / dt \end{bmatrix} = \begin{bmatrix} -R_a / L_a & 0 \\ mk_a & -B / J \end{bmatrix} \begin{bmatrix} i_a \\ \omega \end{bmatrix} + \begin{bmatrix} 1 / L_a \\ 0 \end{bmatrix} e_a + \begin{bmatrix} 1 \\ -1 / J \end{bmatrix} T_L \tag{8.14}$$

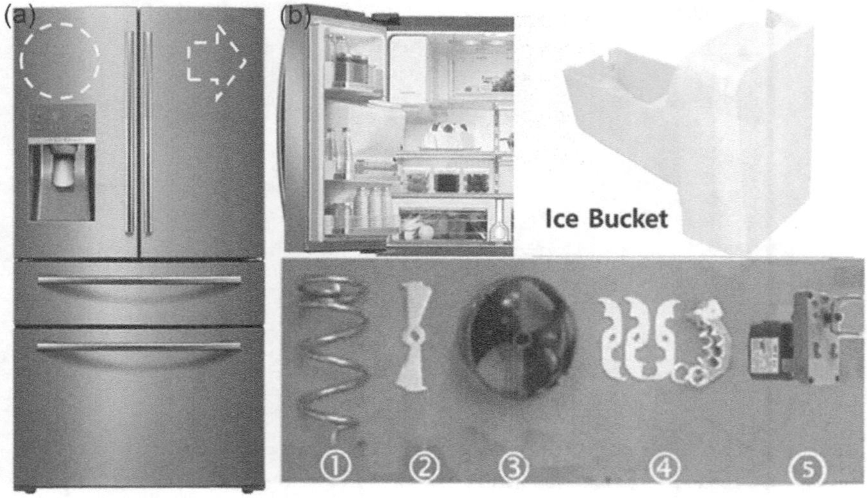

FIGURE 8.22
Refrigerator and ice-maker: (a) refrigerator; (b) ice-maker, involving ① helix support, ② blade dispenser, ③ helix upper dispenser, ④ blade, and ⑤ auger motor with gear.

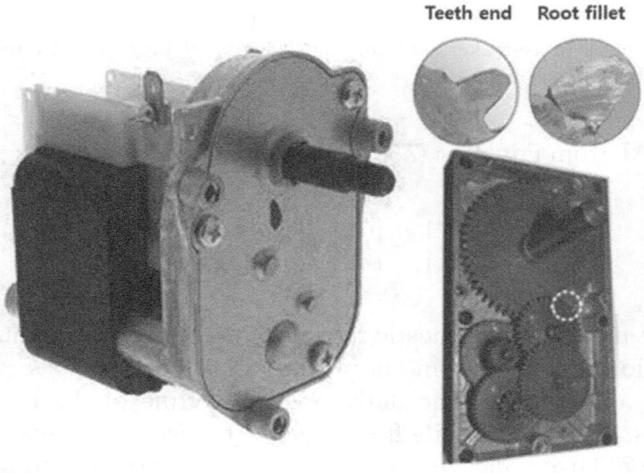

FIGURE 8.23
Failed auger motor in ice-maker.

Because the stress relies on the applied torque, Equation (7.29) can be expressed:

$$TF = A(S)^{-n} = AT_L^{-\lambda} = A(F_c \times R)^{-\lambda} = B(F_c)^{-\lambda} \tag{8.15}$$

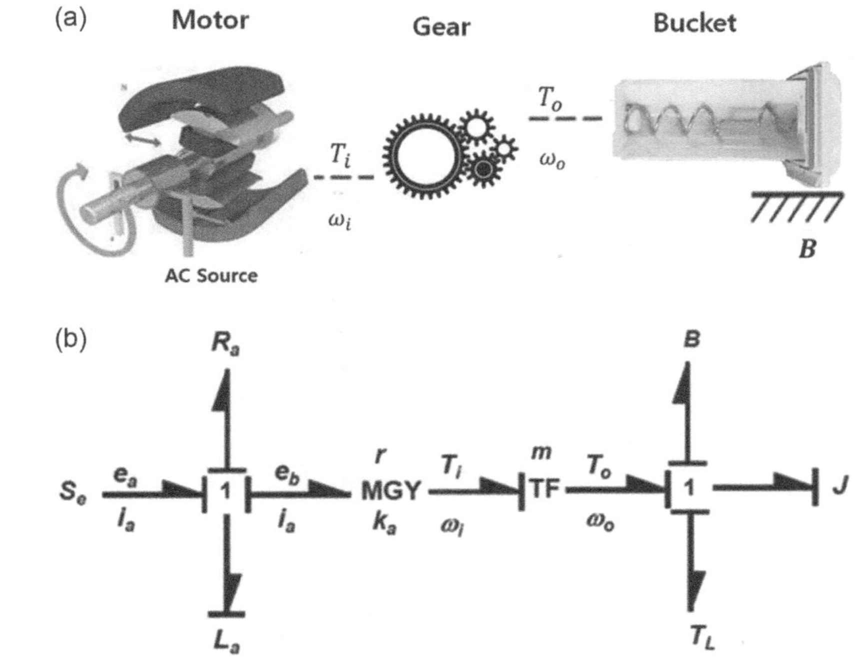

FIGURE 8.24
Failed auger motor in ice-maker: (a) typifying sketching of auger motor, crusher, blade, ice-bucket, etc.; (b) bond graph modeling.

Thus, the AF from Equation (7.30) can be redefined:

$$AF = \left(\frac{S_1}{S_0}\right)^n = \left(\frac{T_1}{T_0}\right)^\lambda = \left(\frac{F_1 \times R}{F_0 \times R}\right)^\lambda = \left(\frac{F_1}{F_0}\right)^\lambda \tag{8.16}$$

For the ice-maker in a domestic refrigerator, the environmental functioning conditions are 0–43°C and humidity ranging from 0 to 95 percent with 0.2–0.24 g acceleration. The daily operating time of the ice-maker was 3–18 cycles. Based on the life for ten years, the ice-maker was expected to achieve 65,700 use cycles. In the worst case, the greatest torque applied by the customer in dispensing ice T_0 was 0.69 kN-cm. For ALT, the applied torque T_1 was increased to 1.47 kN-cm. With a cumulative factor λ of 2, the AF in Equation (8.16) was 4.0. To attain the lifetime target—B1 life for 10 years, if the shape parameter β was assumed to be 2.0, the allotted cycles for ten samples calculated in Equation (7.43) were 42,000 cycles. As an outcome, the troublesome design of an ice-maker had to be found and modified.

8.5.1 Results and Conclusions

In the 1st ALT, the cracked gear in an auger motor of the ice-maker was reproduced at 6,000, 6,900, 8,500, and 8,700 cycles (Figure 8.25). As the cracked and fractured gear system from the 1st ALT and the field were closely inspected by a stereo-microscope, there were design flaws—a cast iron that cannot hold up under severe freezing condition (−20°C↓). As the cast iron gear teeth repeatedly stroked together, they started to crack and finally fracture because the part was brittle under this condition. As action plans, the troublesome material of the gear system utilized in the marketplace was altered from the cast iron (carbon, 3 wt%, and silicon, 2 wt%) to the sinter-hardened powder metallurgy nickel steel.

In the 2nd ALT, the helix upper dispenser possessing the fundamental material of polycarbonates was cracking and fracturing in the exposure area. To perceive the fundamental source of the troublesome system, it was suddenly found that there were design defects in the construction—the weld-line between the helix upper dispenser and blade dispenser—that were numerous voids created in the injection process. Per the action plan, a reinforced rib of the helix had increased after the injection was altered. Finite element analysis (FEA), integrated with parametric ALTs, was carried out. As a result, the component stress after carrying out FEA was lessened from 45.0 to 20.0 kPa (Figure 8.26).

As the designs in the construction were modified, the ice-maker in a refrigerator may perform adequately to fulfill the aimed life—B1 life for 10 years—because there were no issues until 42,000 cycles.

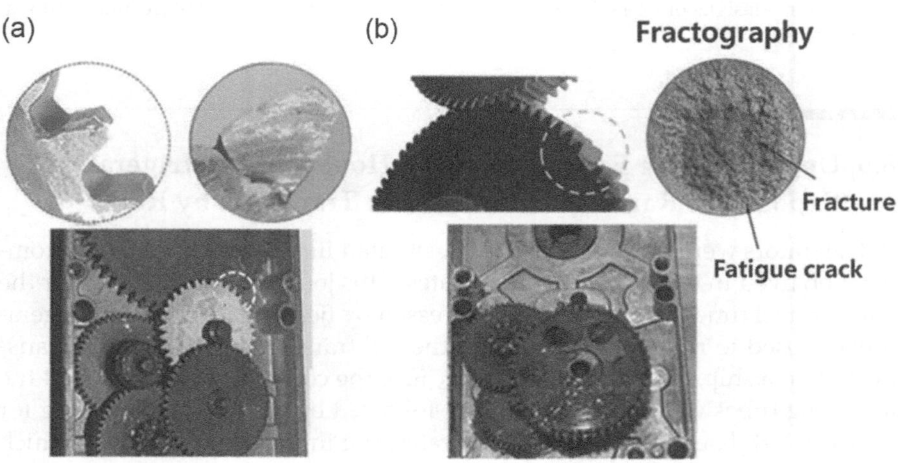

FIGURE 8.25
(a) Problematic designs of a dispenser in the marketplace; (b) 1st ALT failure.

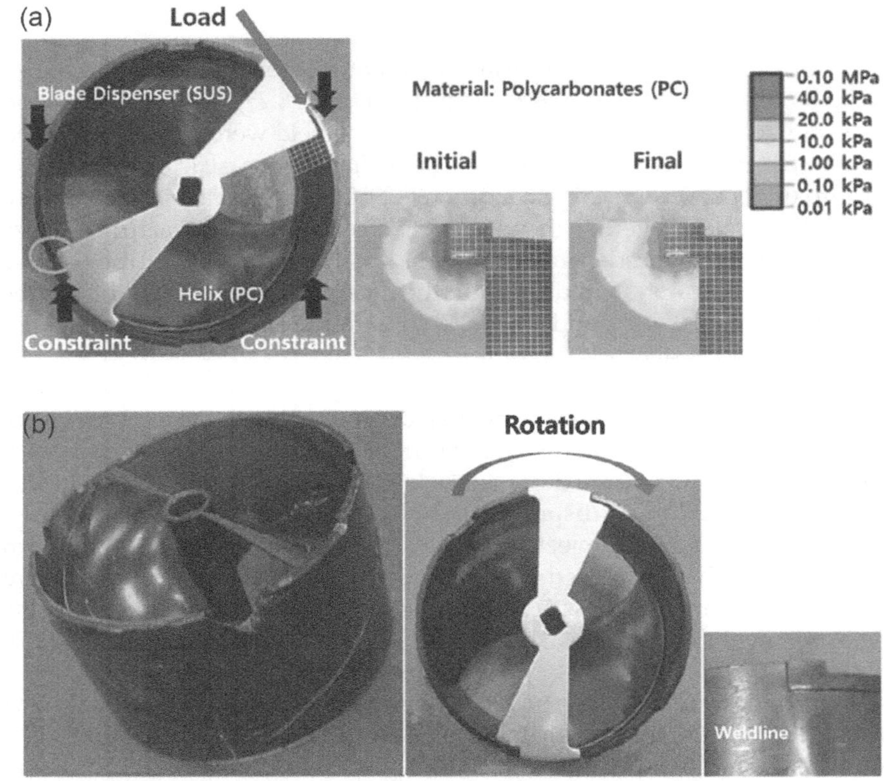

FIGURE 8.26
Troublesome designs of ice-maker in second ALT: (a) FEA; (b) 2nd ALT failure and its fundamental cause.

8.6 Upgrading the Fatigue Life of a Household Refrigerator Exposed to Random Vibrations in Transport by Rail

Refrigerators were moved from the LA situated in the West Coast to customers who lived in the Eastern United States. This journey was 7,200 km for the entire travel time (seven days). As expressed by field data, after the refrigerators exposed to random vibrations in the rail transit route (U.S.) were transported on a ship, they did not work because the compressor rubbers and the attaching tubes were torn and broken, followed by engineers requesting for replacement. Namely, in the United States, the intervening period at which first failure occurred over two days was roughly 2,500 km in rail movement. In Chicago, 27% of the transported products failed. As the refrigerators transported the 7,200-km intervening time from West Coast to Boston over seven days, 67% of the products failed. It was plain that the unsuccessful

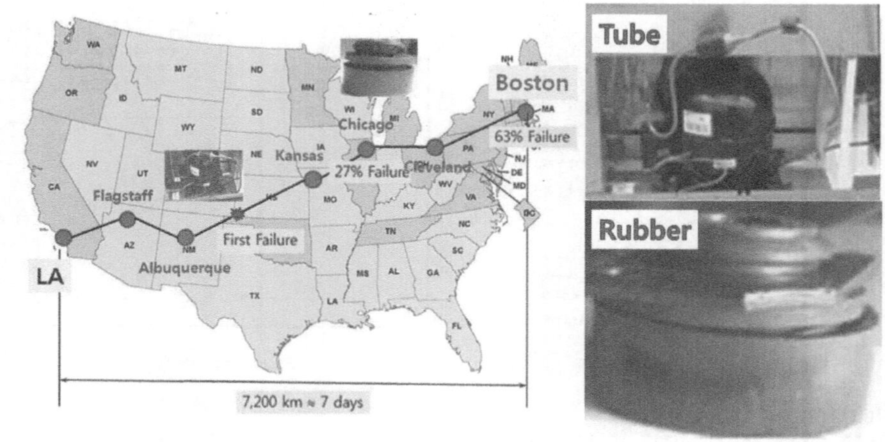

FIGURE 8.27
Unsuccessful locations and product from the field.

refrigerators had design flaws. For the refrigerator to rightly work for its assumed life in transportation, after recognizing them by parametric ALT, the manufacturer had to modify the issues (Figure 8.27).

Due to the foundation load Y the force transmissibility Q can be defined:

$$Q = X(t)/F_T(t) = X(t)/Y(t) = \left[\frac{\left(2\varsigma \dfrac{f_n}{f}\right)^2 + 1}{\left(1 - \left(\dfrac{f_n}{f}\right)^2\right)^2 + \left(2\varsigma \dfrac{f_n}{f}\right)^2} \right]^{1/2} \tag{8.17}$$

where X is the unvarying solution in a physical process, F_T is the induced force, Y is the span of base excitations, r is the frequency proportion $(=\omega_n / \omega = f_n / f)$, ς is the damping proportion ($c/c_c = c/2\omega_n$), k is the spring quantity.

As the stress in transit originates from the transmitted vibration loads F_T, which can be defined as the power spectral density (PSD) level of acceleration for a decided frequency band, Equation (7.29) can be defined:

$$TF = A(S)^{-n} = B(F_T)^{-\lambda} \tag{8.18}$$

So, based on Equation (7.30) and (7.31), the AF can be stated:

$$AF = \left(\frac{S_1}{S_0}\right)^n = \left(\frac{a_1}{a_0}\frac{X}{Y}\right)^\lambda = (R \times Q)^\lambda \tag{8.19}$$

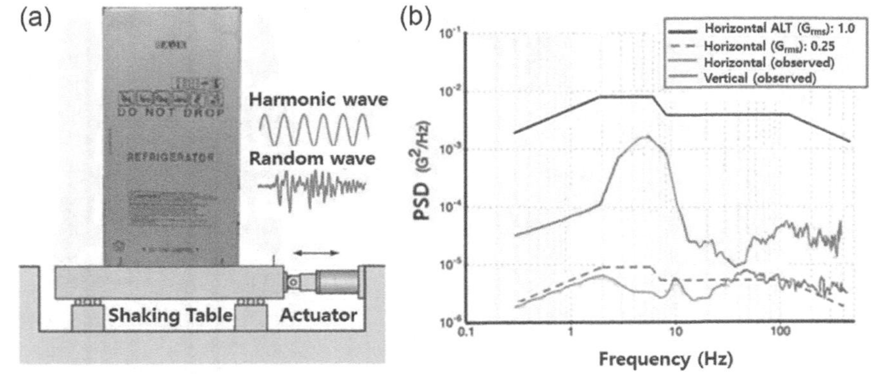

FIGURE 8.28
Programmed PSD loads on shaking table: (a) shaking table; (b) applied PSD loads.

where a_1 is the elevated PSD level for the given frequency band, a_0 is the normal PSD level for the given frequency band, and R the amplitude ratio of gravitational acceleration.

When inspecting the computed vibration spectra, the natural frequencies of vibration (left ↔ right and up ↔ down) were 5 Hz and 9 Hz. The damping proportion was anticipated to be $\zeta = 0.1$ with a settling period of 2 sec and roughly 5% overshoot. The frequency ratio also was anticipated to be $r\ (=\omega/\omega_n) = 1$ at the natural frequency ω_n.

To attain the AF in the vertical/horizontal directions of random vibrations found from the field, elevated PSD loads to the refrigerator were applied on the shaker bench for each orientation (Figure 8.28). The force transmissibility Q had the magnitude of roughly 5.3 from Equation (8.17). On account of acceleration of 1 G_{rms}, the AF was 4.0, roughly calculated to that of worst case of 0.25 G_{rms}. Employing an accumulative damage constant λ, of 2.0, the entire AF in Equation (8.19) was seen to be 450.0. Because the estimated shape parameter in the Weibull plot was 6.41, and the life aim—B1 life for the whole travel interval—the test period attained from Equation (7.31) was approximately 40 min for three samples. Specifically, if the refrigerator failed less than once in the assigned mission period—40 minutes—the refrigerator was allowable for the whole travel interval of 7,200 km to leave the fatigue damaged due to random vibration.

8.6.1 Results and Conclusions

In the 1st ALT, the failures of the stress levels—1.00 G_{rms} at the natural frequency ($r = 1.0, \zeta \approx 0.1$) occurred at 20 min (one sample) and at 40 min (two samples) from the horizontal vibration (left ↔ right). For these circumstances, rubber mounts were torn and tubes were broken because nothing supported the repeated random loads due to horizontal vibrations (left ↔ right). Per the action plan, the refrigerator's compressor rubber mount, C1, was modified (Figure 8.29).

FIGURE 8.29
Fatigued refrigerator from the market and 1st ALT: (a) field; (b) 1st ALT outcomes and design change.

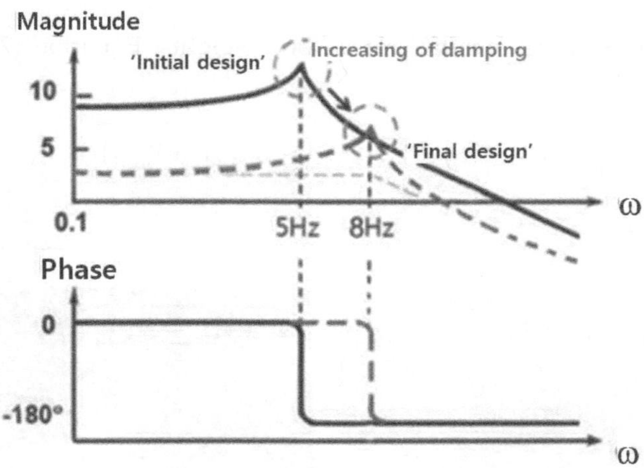

FIGURE 8.30
Changed natural frequency in the horizontal vibration.

In 2nd ALT, the refrigerator was successful for 60 minutes. When the refrigerator achieved an acceleration of 1.00 G_{rms} on the shaking bench, the natural frequency—5 Hz—in the horizonal direction was altered to 8 Hz due to the enlarged damping (Figure 8.30).

As the mounts in a compressor were redesigned, there were no issues for the assigned mission time—40 min. Thus the refrigerator was assured to outlast the induced loads from the bottom and achieve the product life—B1 life for the whole travel interval.

8.7 Drawer in the French-Door Refrigerator Exposed to Repeated Stress

As a function of the refrigerator, customer wants to utilize the drawer that keeps the food fresh. To satisfy these market requirements, the drawer in a refrigerator is designed to keep the essential food over expected customer circumstances within a product life. Keeping food in a drawer thus the repetitive opening and closing: 1) open the drawer to retain food for future use, 2) take the food, and 3) close the drawer. The main constructional components are ① handle, ② drawer box, ③ slide rail, and ④ pocket box (Figure 8.31).

From the marketplace, as load is repetitively exerted, some components in a drawer had been failing because of unrevealed design defects. As a result, end user asks for the troublesome products to be replaced. To function for its expected life, the design flaws of a drawer need to be found and altered by parametric ALT (Figure 8.32).

In Figure 8.33, the free body diagram may be summarized as:

$$F_{\text{draw}} = \mu W_{\text{load}} \tag{8.20}$$

As the stress relies on the stored food weights, Equation (7.29) shall be described:

$$TF = A(S)^{-n} = A\left(F_{\text{draw}}\right)^{-\lambda} = A\left(\mu W_{\text{load}}\right)^{-\lambda} \tag{8.21}$$

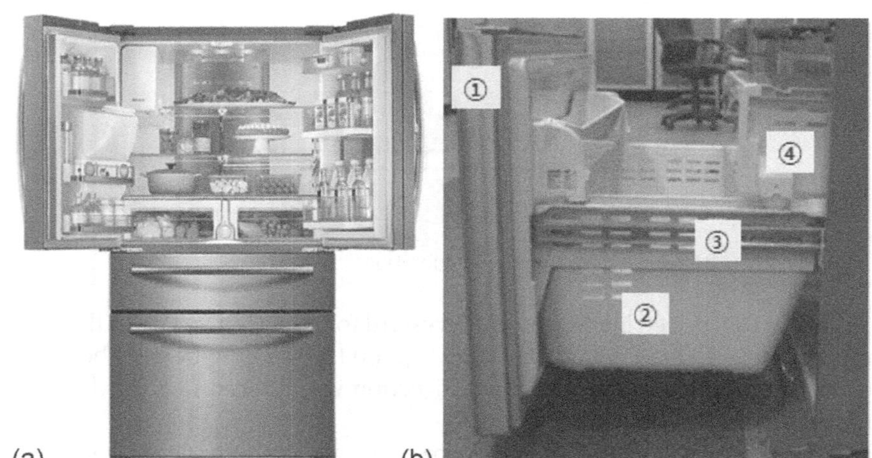

FIGURE 8.31

Freezer drawer in a refrigerator: (a) French-door refrigerator; (b) drawer parts: ① handle, ② drawer box, ③ slide rail, and ④ pocket box.

FIGURE 8.32
Failed product after utilization.

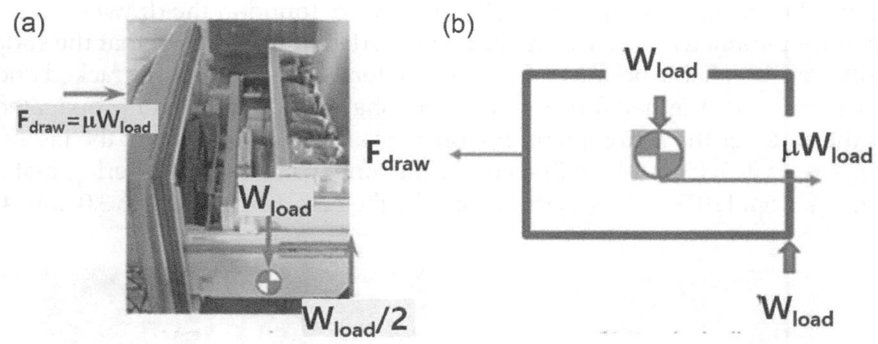

FIGURE 8.33
Practical design idea of the drawer: (a) drawer assembly; (b) free body diagram drawer.

So the AF shaped on the basis of Equations (7.30 and 7.31) can be redefined:

$$AF = \left(\frac{S_1}{S_0}\right)^n = \left(\frac{F_1}{F_0}\right)^\lambda = \left(\frac{\mu W_1}{\mu W_0}\right)^\lambda = \left(\frac{W_1}{W_0}\right)^\lambda \tag{8.22}$$

For the drawer in a refrigerator, the working environment conditions are 0–43°C and humidity ranging from 0 to 95% with 0.2–0.24 g acceleration. In the United States, the functioning of the drawer relies on customer usage. Field data revealed that the end user worked the freezer drawer between five and nine cycles daily. With a design life of ten years, the drawer underwent 36,500 use cycles. For the worst case, the required force due to the food load was 0.34 kN (35 kg_f) on the drawer handle. The food load for the ALT was 0.68 kN (70 kg_f). With a cumulative factor λ of 2, the AF in Equation (8.22) was 4.0. To obtain the life aim—B1 lifetime of 10 years, as the shape parameter β was presumed to be 2.0, the assigned mission cycles for three drawers computed in Equation (7.43) were 67,000 cycles. As a result, the problematic design of the drawer needs to be discovered and altered.

8.7.1 Results and Conclusions

In the 1st ALT, the cracked and fractured drawer handle in a refrigerator failed at 7,000 and 8,000 cycles (Figure 8.34). As the troublesome handles from the 1st ALT and the market were deliberately explored, there were design flaws—insufficiently fastened area of the drawer handle. Subjected to repeatedly raised food loads, these design defects in the drawer handle should be immediately eliminated. To stop the failed freezer handle from the drawer, the handle was altered: (1) increasing the width of the reinforced handle, C1, from 90 to 122 mm and (2) increasing the hooker size in handle, C2, from 8 to 19 mm.

In the 2nd ALT, we found a 10 mm gap between the drawer and freezer section at 15,000 and 16,000 cycles. As the problematic drawers were taken apart, the cracked and fractured slide rails were found in the drawer. As the food for parametric ALT was repetitively loaded, it was known that the slide rails were breakable because of their inadequate strength. These cracked and fractured rails originated from the corner shape of the rails and propagated to the end. Per the action plans, the rail was altered: 1) increasing the fastening screws, C3, from 1 to 2; 2) securing the inner chamber and altering material, C4, from HIPS to ABS; 3) thickening for the boss, C5, from 2.0 to 3.0 mm; 4)

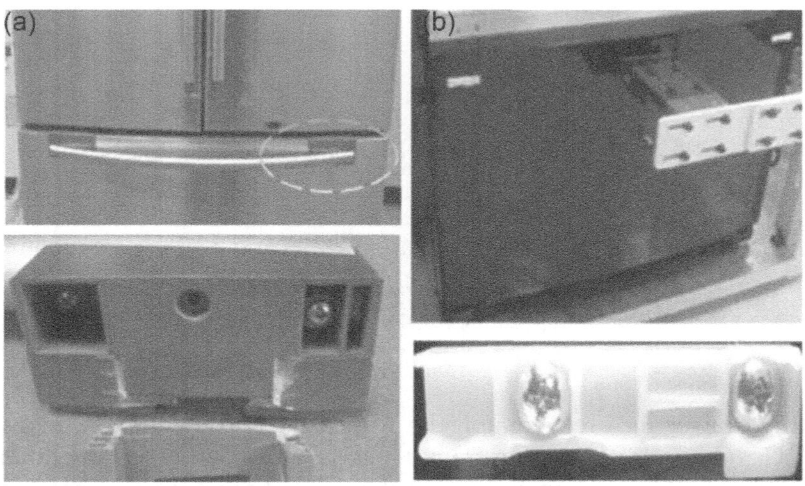

FIGURE 8.34
Troublesome designs of drawer in the market and 1st ALT: (a) problematic field part; (b) 1st ALT failure.

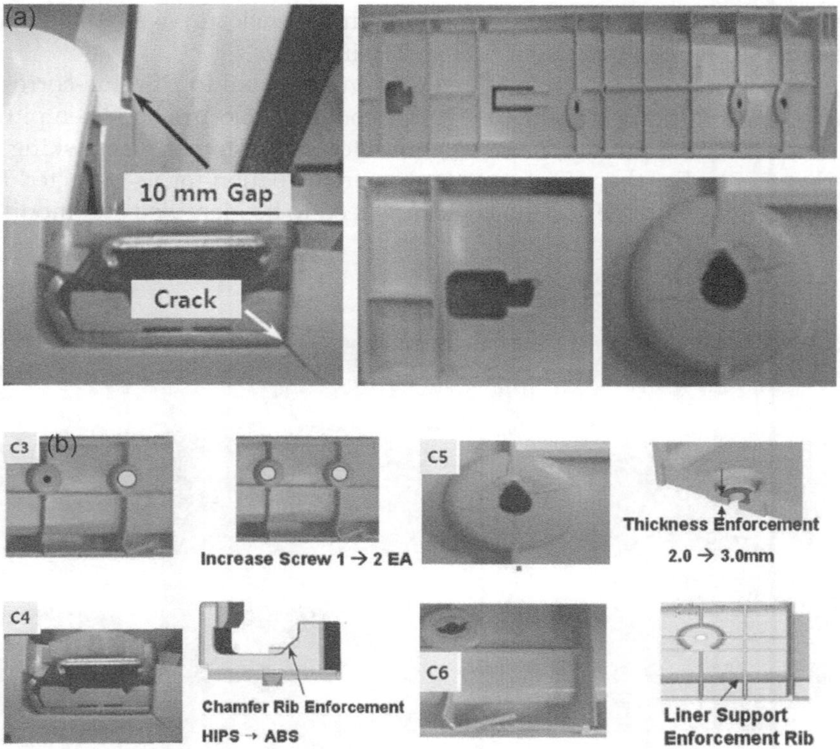

FIGURE 8.35
Troublesome designs of drawer system in 2nd ALT: (a) failed drawer rails; (b) redesigned slide rail.

securing a new support rib, C6. By carrying out the ALT, the design of the problematic slide rail was altered (Figure 8.35).

As the designs in the construction were altered, the refrigerator drawer will stretch out the targeted life—B1 life of 10 years—because of no problems occurred until 67,000 cycles.

8.8 Enhancing the Design Life of the Pitted Cooling Enclosure in the Household Refrigerator

As some Korean traditional food such as kimchi comes in the greatest quantities, customers want to use the refrigerator to keep the food fresh. To fulfill these desires, the cooling enclosure in a refrigerator is designed to stock the totality of the required food for its product life despite predicted consumer conditions. The chip parts in the enclosure consists of: ① inner case, ② evaporator tube, ③ Lokring, and ④ tape. On the other hand, because refrigerators that are down-costed for price competitiveness are necessitated in the field, the new aluminum (1070 Al) tubing in the enclosure was adapted and designed from the first copper tubing (Figure 8.36).

From the marketplace, due to being often subjected to chlorine-corrosive loads, some parts such as the tube in the cooling enclosure had been pitting due to unknown product flaws. As an outcome, customer were asking for replacements. For the refrigerator to properly function for its expected life, the design flaws in the cooling enclosure need to be identified and modified by parametric ALT (Figure 8.37).

FIGURE 8.36
Cost-downed (copper → aluminum) enclosure in a refrigerator: (a) enclosure system in a refrigerator: ① inner case; (b) low-cost pipe ② evaporator tube, ③ Lokring, and ④ tape.

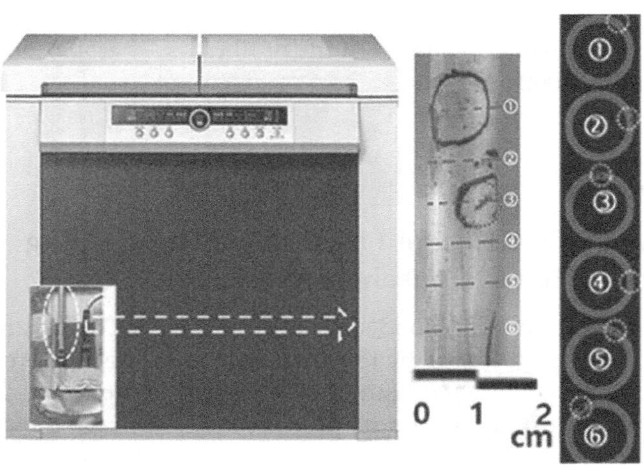

FIGURE 8.37
Failed product after utilization.

The components of the enclosure involve a cathode, anode, water solution (hydrolysis), and conductor (metal) for Aquarius pitting/crevice corrosion. That is, their corrosion processes are as follows: 1) As contacted by the chloride Cl^- of the cotton tape, the passive film of aluminum on the tube breaks down; 2) water condensation on the tube occurs at the evaporator temperature; 3) anode: Al metal oxidation occurs, resulting in pitting, $Al \to Al^{+3} + 3e^-$; in the pit, a hydrolysis reaction takes place: $12H^+ + 12e^+ \to 6H_2 \uparrow (g)$ and $Al^{+3} + 3H_2O \to Al(OH)_3 \downarrow + 3H^-$; 4) frequently repeated electro-migration of Cl^- into the pit through the condensing water; 5) cathode: in the surface of tube, $O_2\,(\text{dissolved oxygen}) + 2H_2O + 4e^- \to 4OH^-$; and 6) the potential drop; that is, the IR drop, as explained by Ohm's law: $V = I \times R$ (Figure 8.38).

The hydrolysis reaction on the pitted crevice of aluminum metal can be defined as:

$$\text{Anode: } 4Al \to 4Al^{3+} + 12e^+ \text{ (oxidation)} \tag{8.23}$$

$$\text{Cathode: } O_2 + 2H_2O + 4e^- \to 4OH^- \text{ (reduction)}$$

$$\text{Hydrolysis reaction: } 4Al^{3+} + 12Cl^{-1} + 12H_2O \to 4Al(OH)_3 \downarrow + 6H_2 \uparrow (g) + 12Cl^{-1}$$

As the stress relies on the corrosive loads, Equation (7.29) can be defined:

$$TF = A(F)^{-\lambda} \exp\left(\frac{E_a}{kT}\right) \approx B(Cl)^{-\lambda} \exp\left(\frac{E_a}{kT}\right) \tag{8.24}$$

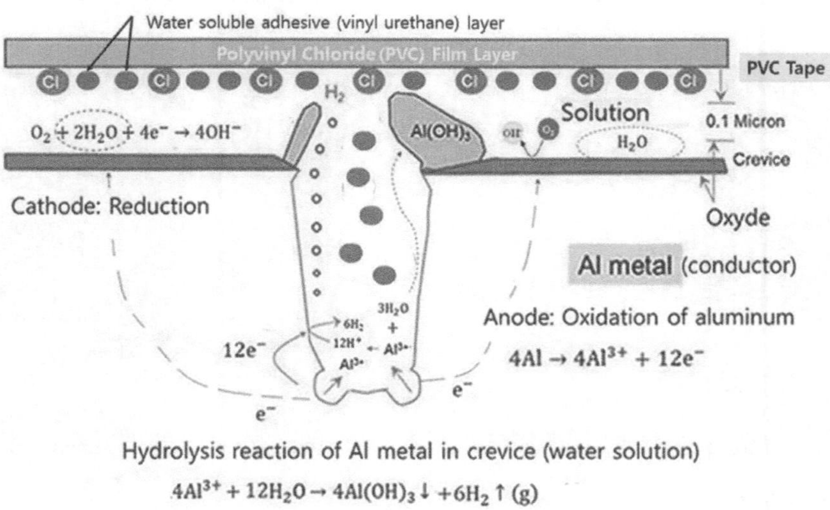

FIGURE 8.38
Mechanism of pitting and crevice corrosion due to low-PH condensing water.

So the AF shaped on Equations (7.30) and (7.31) can be defined:

$$AF = \left(\frac{F_1}{F_0}\right)^{\lambda}\left[\frac{E_a}{k}\left(\frac{1}{T_0}-\frac{1}{T_1}\right)\right] \approx \left(\frac{Cl_1}{Cl_0}\right)^{\lambda}\left[\frac{E_a}{k}\left(\frac{1}{T_0}-\frac{1}{T_1}\right)\right] \tag{8.25}$$

For the refrigerator, the normal environmental circumstances ranging from 0–50°C ambient temperature, 0–85% humidity, and 0.2–0.24 g vibration take place. The lifetime aim for an enclosure in the refrigerator is set to have B1 life for 10 years. The compressor in the enclosure works on daily operating cycles of between 22 and 98. With product life cycles of ten years L_B the compressor is expected to undergo up to 357,700 cycles in use time. To quickly reproduce the pitting corrosion on the tubing, the concentration of chlorine as the accelerated factor was enlarged, compared with 0.35% chlorine concentration equivalent to the saline water solution. The time-to-failure cycles for pitting corrosion were increased at a stress level of 3.5% saline solution (the chlorine concentration for ALT). For a cumulative factor λ of 2, AF might be calculated to be 100.0 from Equation (8.25). To achieve the life aim—B1 life of 10 years, if the shape parameter β was assumed to be 2.0, the assigned mission time for 18 samples calculated from Equation (7.43) was 8,500 cycles. As a result, the problematic design of enclosure may be reproduced and altered.

8.8.1 Results and Conclusions

In the 1st ALT, the refrigerators placed in the 3.5% saline solution were fitted at 1,130 cycles (one), 1,160 cycles (two), 1,680 cycles (four) (Figure 8.39). When the problematic enclosure from the 1st ALT and the field were closely checked, there were structural defects—no contraction tube may be kept safe

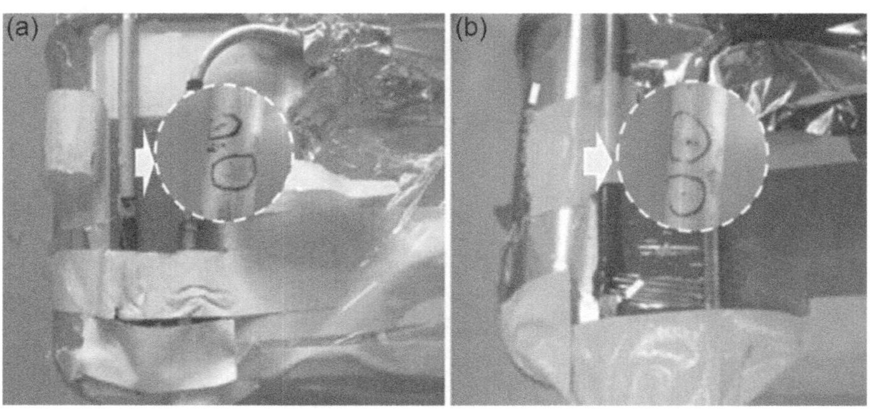

FIGURE 8.39
Troublesome evaporator tube and products in the market and 1st ALT: (a) problematic field part; (b) 1st ALT failure.

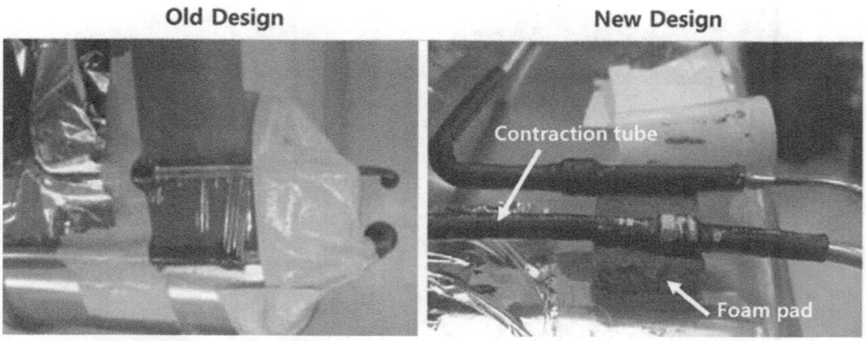

FIGURE 8.40
Troublesome designs of enclosure system in 2nd ALT.

from the chloride attack; the evaporator tubing might directly contact the cotton adhesive tape that held chloride in the polyvinyl chloride (PVC) coating. To protect against the pitting of enclosure, designs were altered: 1) increasing the length of the contraction tube with a nonmetallic coating, which may stop the direct chloride (C1) attack from 50 to 200 mm; (2) modifying the tape C2 from cotton to polyethylene; (3) fastening polyethylene foam pads, which may stop galvanic corrosion, C3 (Figure 8.40).

In the 2nd ALT, there are no issues until 4,700 cycles, the enclosure in a domestic refrigerator may satisfy the target life—B1 lifetime of 10 years.

8.9 Refurbishing the Fatigue Lifetime of a Household Product Subjected to Random Shakes While Transported by Railroad

Refrigerators were transported from the LA located in the western coast of the United States to U.S. resident customers, in some area of the country near the Atlantic Ocean. This journey for the whole travel period—seven days—was 7,200 km. As expressed by the field data, after the products were exposed to random vibrations on the rail transportation, the refrigerators could not function because the mount rubbers and soldering in the compressor were fatigued and the attaching tubes were broken, requiring engineers to ask for replacements. In the United States, the rail transit interval when the first failure happened for two days was roughly 2,500 km. In Chicago, 27% of the transported refrigerators were unsuccessful. When the products were conveyed the 7,200 km interval from Los Angeles to Boston, 67% of the refrigerators were unsuccessful. It was obvious that the failed refrigerators had design weakness. For the units to correctly function for their presumed life

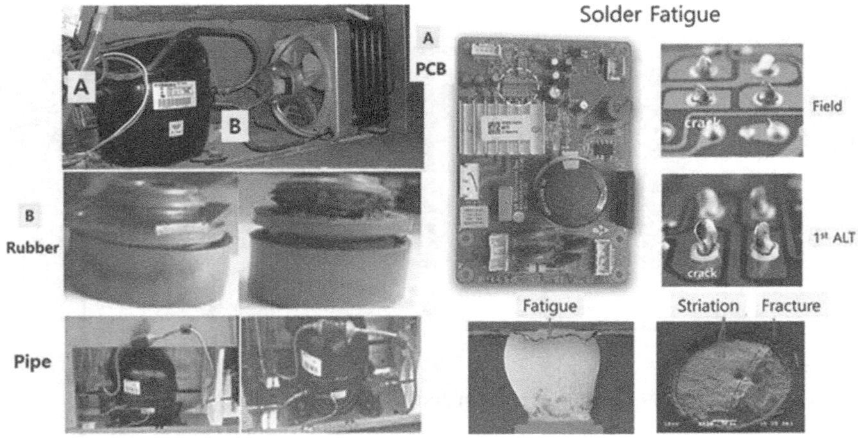

FIGURE 8.41
Unsuccessful product from the market.

after transit, the design defects had to be identified by means of parametric ALT, and the manufacture had to modify them (Figure 8.41).

On account of the base load Y, force transmissibility Q, can be expressed:

$$Q = X(t)/F_T(t) = X(t)/Y(t) = \left[\frac{\left(2\varsigma \frac{f_n}{f}\right)^2 + 1}{\left(1 - \left(\frac{f_n}{f}\right)^2\right)^2 + \left(2\varsigma \frac{f_n}{f}\right)^2} \right]^{1/2} \tag{8.26}$$

where X is the unaltered solution of the mechanical system, F_T is the exerted force, Y is the magnitude of foundation excitations, r is the frequency proportion ($= \omega_n / \omega = f_n / f$), ζ is the damping proportion ($c/c_c = c/2\omega_n$), k is the spring constant.

As the stress in transit starts from the transmitted vibration loads, F_T, that can be defined as the power spectral density (PSD) of acceleration on the adjusted frequency band, Equation (7.29) may be redefined:

$$TF = A(S)^{-n} = B(F_T)^{-\lambda} \tag{8.27}$$

So discovered with Equations (8.30) and (8.31), AF may be indicated:

$$AF = \left(\frac{S_1}{S_0}\right)^n = \left(\frac{a_1}{a_0}\frac{X}{Y}\right)^\lambda = (R \times Q)^\lambda \tag{8.28}$$

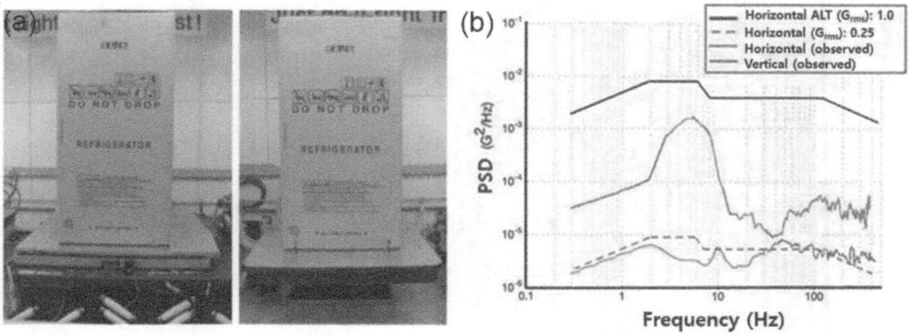

FIGURE 8.42
Scheduled PSD in vibrating bench: (a) shaking table; (b) exerted power spectral density (PSD).

where a_1 is the elevated PSD on the assigned frequency band, a_0 is the common PSD on the assigned frequency band, *R* is the amplitude proportion of gravitational acceleration.

As examining the calculated vibration, the natural frequencies of vibration parallel to the plane of the horizon and vertical to it were 5 and 9 Hz, respectively. The damping proportion was expected to be $\zeta = 0.1$ with a settling period of 2 sec and roughly 5% overshoot. The frequency ratio at the natural frequency ω_n also was expected to be $r\ (=\omega/\omega_n) = 1$ (Figure 8.42).

8.9.1 Results and Conclusions

In the 1st ALT, the failure for the stress—1.00 g_{rms} at the natural frequency ($r = 1.0$, $\zeta \approx 0.1$)—discovered. That is, for these conditions, at 20 min, the mount rubber and soldering were fatigued, and the tubes on the samples also were broken. It occurred due to no support in the region of the stress raiser for the mount rubber to withstand the random loads due to parallel vibrations. Per the action plan, the product was redesigned as the modified rubber mount in compressor, C1 (Figures 8.43 and 8.44).

In the 2nd ALT, the product failed at 60 min. As the product stretched out at an acceleration of 1.00 g_{rms} in the shaking bench, the natural frequency on the parallel orientation was modified from 5 to 8 Hz by the increased damping (Figure 8.45).

To enhance the fatigue life of the refrigerator in transit, parametric ALT was applied. The quantum-conveyed life-stress model and sample size were provided. As a case study, increasing the fatigued product life by random vibrations in rail transit was explored. When the mount rubbers in compressor was redesigned, there were no problems for assigned period—40 min. So the product was ensured to withstand the fatigue load and achieve the refrigerator life—B1 life for the entire travel interval.

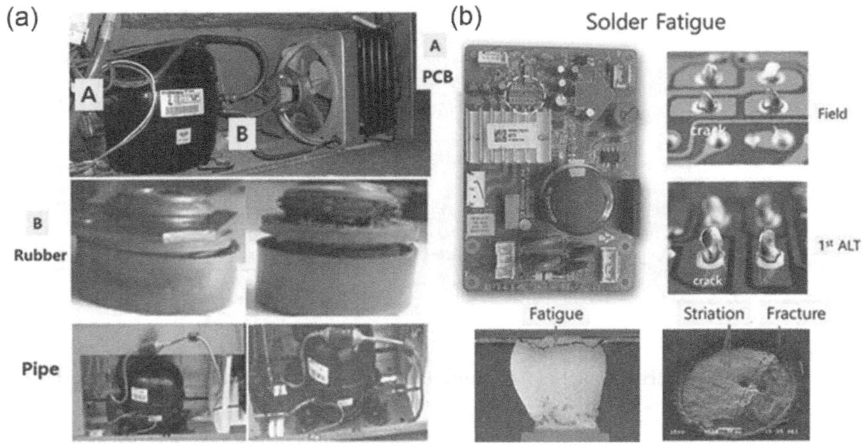

FIGURE 8.43
(a) Damaged refrigerator from the field; (b) 1st ALT outcomes.

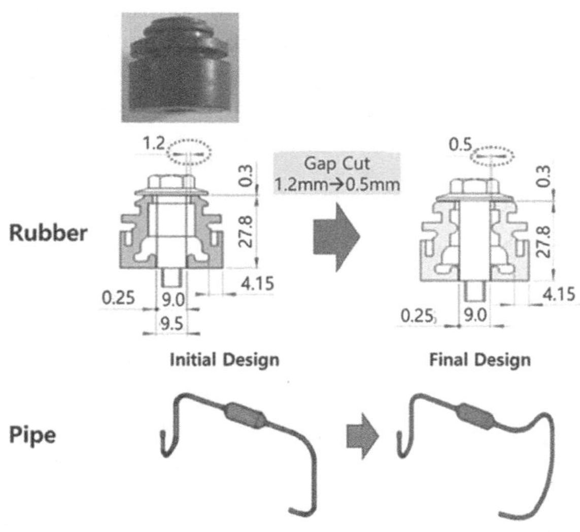

FIGURE 8.44
Design modifications.

8.10 Enhancing the Life of the Compressor with Modified Crankshaft

To keep food fresh during the product life, consumers use the refrigerator, which consists of a condenser, capillary tube, evaporator, and compressor.

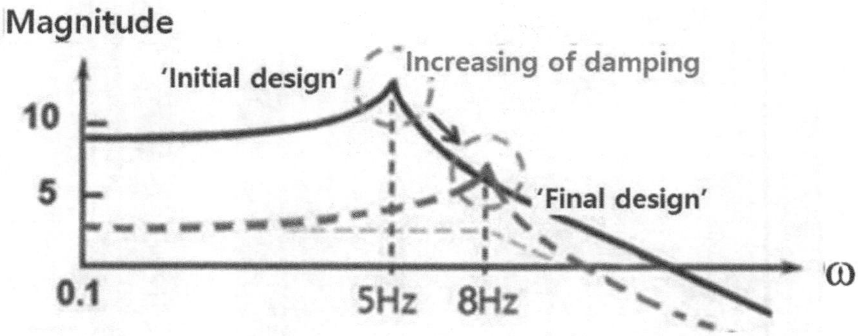

FIGURE 8.45
Changed natural frequency in the horizontal vibration.

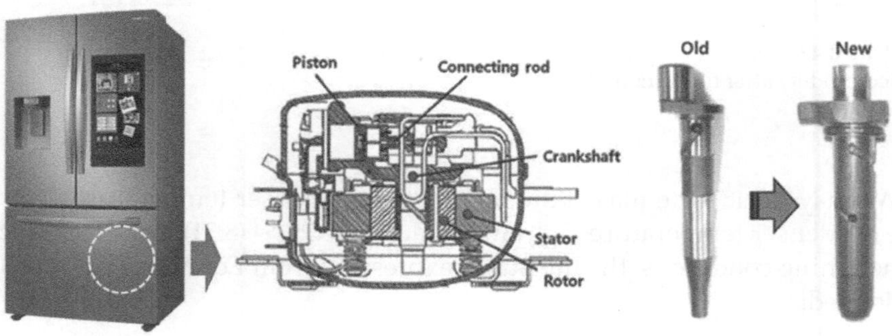

FIGURE 8.46
Newly devised compressor system in a refrigerator.

For energy efficiency, the compressor is a crucial element to enlarge pressure from low status to high status in a vapor-compression refrigerator cycle (Figure 8.46).

From the field, household compressors in the refrigerators were stopping because of defects. Based on the customer functioning circumstances, the compressor was exposed to dissimilar pressure loading in a vapor-compression cycle (Figure 8.47).

The mass conservation in a capillary tube can be defined:

$$\dot{m} = \dot{m}_{cap} \tag{8.29}$$

The energy conservation in the evaporator and condenser can be defined:

$$\dot{Q}_c = \dot{m}(h_2 - h_3) = (T_c - T_o)/R_c \tag{8.30}$$

$$\dot{Q}_e = \dot{m}(h_1 - h_4) = (T_i - T_e)/R_e \tag{8.31}$$

FIGURE 8.47
Failed product after utilization.

We may decide the mass flow rate $\dot{m}$, the condenser temperature T_c, and the condenser temperature T_e, from Equations (8.29) ~ (8.31). Under identical functioning conditions, the life-stress expression from Equation (7.29) can be redefined:

$$TF = A(S)^{-n} \exp\left(\frac{E_a}{kT}\right) = A(\Delta P)^{-\lambda} \exp\left(\frac{E_a}{kT}\right) \tag{8.32}$$

We recognize that system lifetime depends on the applied pressure difference load. Actually, the acceleration factor (AF) in Equations (7.30) and (7.31) may be defined:

$$AF = \left(\frac{S_1}{S_0}\right)^n \left[\frac{E_a}{k}\left(\frac{1}{T_0} - \frac{1}{T_1}\right)\right] = \left(\frac{\Delta P_1}{\Delta P_0}\right)^\lambda \left[\frac{E_a}{k}\left(\frac{1}{T_0} - \frac{1}{T_1}\right)\right] \tag{8.33}$$

For the most severely instance, the compressor temperature in the dome was 90°C, and the pressure difference in the vapor-compression cycle was 1.07 MPa. For the ALTs, we raised the temperature in the dome to 120°C, and the pressure difference was 1.96 MPa. Utilizing an accumulative damage quantity λ of 2.0, AF was computed to be roughly 4.58 from Equation (8.33).

For the life aim—B1 lifetime of 10 years, the assignment cycles for 100 samples calculated from Equation (7.43) were approximately 19,000 cycles if the shape parameter was expected to be 2.0. After recognizing them, the troublesome designs of the compressor may be altered.

8.10.1 Results and Conclusions

In the 1st ALT, the compressor was unsuccessful, locking at 7,500 and 10,000 cycles. In contrasting the compressor that had failed in the field with that from the 1st ALT, it was acknowledged that the failure style in the 1st ALT was similar to those gained from the market (Figure 8.48).

The compressor in both the field and 1st ALT testing failed due to lubrication on the outside of the crankshaft. It was acknowledged that the design flaws of the compressor arose from 1) no lubrication regions, 2) a low starting RPM, and 3) crankshaft material. To reach enough fatigue strength against repeated pressure loading, the troublesome compressor was modified (Figure 8.49).

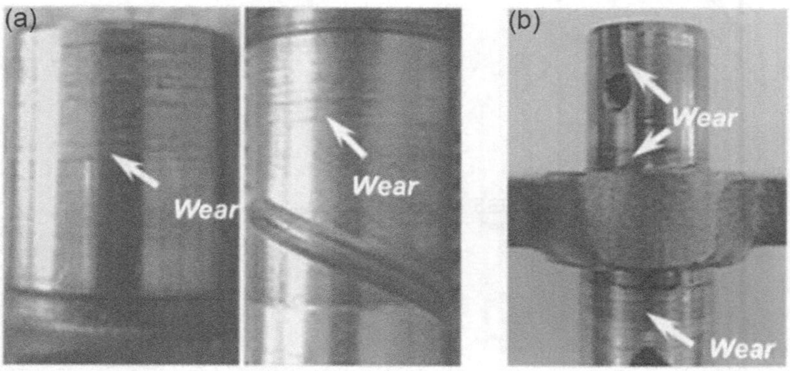

FIGURE 8.48
(a) Problematic field part; (b) 1st ALT failure.

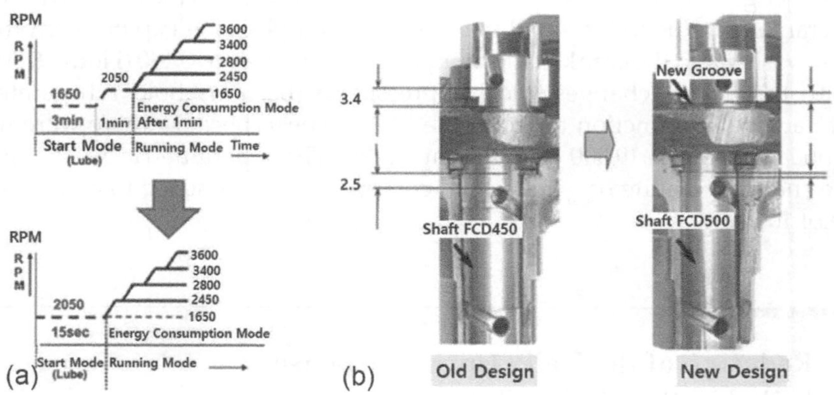

FIGURE 8.49
Modified compressor design: (a) altered starting RPM from 1,650 to 2,050; (b) modified crankshaft.

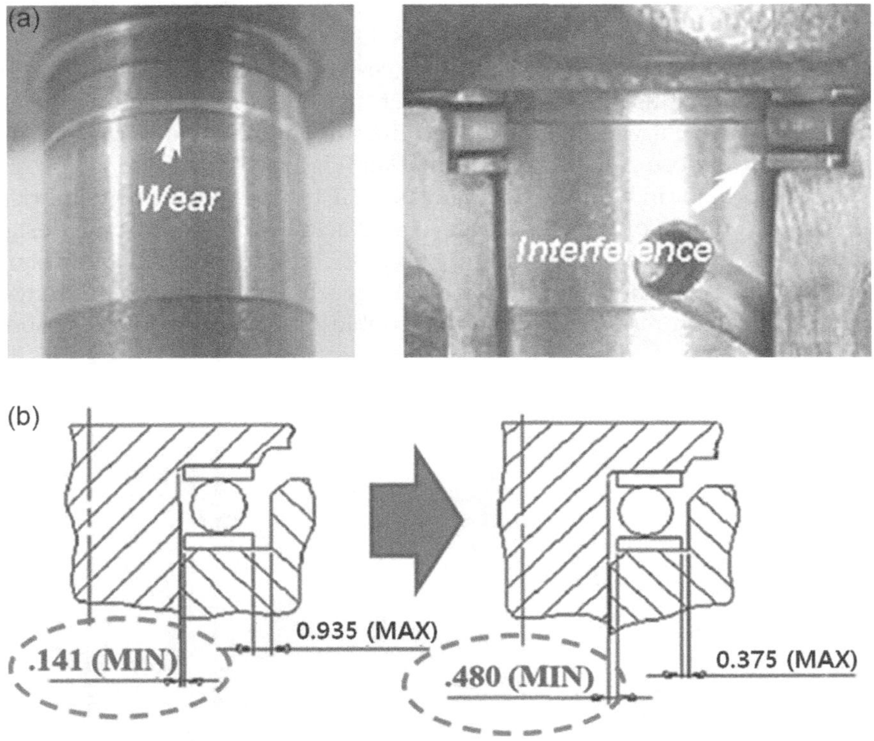

FIGURE 8.50
Redesigned crankshaft: (a) failed crankshaft; (b) modified crankshaft.

In the 2nd ALT, the crankshaft wear was discovered at 19,000 cycles. We acknowledged that this failure arose from intrusion between the thrust washer and crankshaft due to the small gap. As an action plan, the gap between the thrust washer and the crankshaft was enlarged from 0.141 to 0.480 (Figure 8.50).

With the design changes, the compressor in the household refrigerator might accurately function to attain the lifetime need because there were no complications until 19,000 assignment cycles. After parametric ALTs with some changes, the life of a household compressor was assured to be B1 lifetime of 10 years.

8.11 Redesign of the Helix Upper Dispenser Failed in the Marketplace

Because the customer demands ice, the function of an ice-maker is attached to harvest ice. The chief parts are the bucket case, auger motor involving a gear system, helix blade dispenser, helix support, helix upper dispenser, blade, and dispenser clamp (Figure 8.51).

In the marketplace, the helix upper dispenser in an ice-maker was unsuccessful due to design defects under unknown end user situations. Field data demonstrated that the unsuccessful component might have had design defects: 1) a 2 mm gap and 2) a weld line. The blade dispenser made of stainless-steel strokes the helix upper dispenser made of plastic under –20°C. The helix dispenser cracked (Figure 8.52).

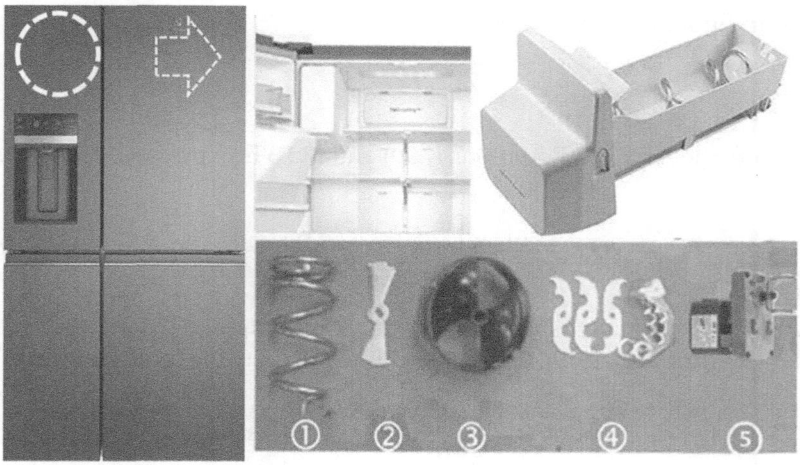

FIGURE 8.51
Ice-maker in refrigerator.

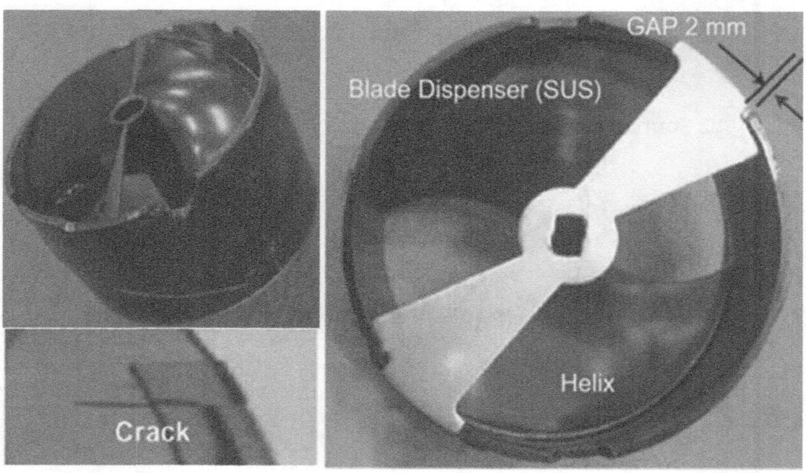

FIGURE 8.52
Unsuccessful helix upper dispenser in the market.

The model of the mechanical load transfer in an ice-maker may be established by utilizing a bond graph. From its derivation, the governing equations may be abridged:

$$\begin{bmatrix} di_a / dt \\ d\omega / dt \end{bmatrix} = \begin{bmatrix} -R_a / L_a & 0 \\ mk_a & -B / J \end{bmatrix} \begin{bmatrix} i_a \\ \omega \end{bmatrix} + \begin{bmatrix} 1 / L_a \\ 0 \end{bmatrix} e_a + \begin{bmatrix} 1 \\ -1 / J \end{bmatrix} T_L \tag{8.34}$$

From Equation (8.34), it was recognized that the lifetime of the ice-maker depends on the exerted torque needed to crush the ice. The life-stress (LS) model in Equation (7.29) is defined:

$$TF = A(S)^{-n} = AT_L^{-\lambda} = A(F_c \times R)^{-\lambda} = B(F_c)^{-\lambda} \tag{8.35}$$

Therefore, the AF from Equations (7.30) and (7.31) is expressed (Figure 8.53):

$$AF = \left(\frac{S_1}{S_0}\right)^n = \left(\frac{T_1}{T_0}\right)^\lambda = \left(\frac{F_1 \times R}{F_0 \times R}\right)^\lambda = \left(\frac{F_1}{F_0}\right)^\lambda \tag{8.36}$$

Relying on customer usage, an ice-maker is employed daily 3–18 times. Under maximum use for ten years, the ice dispensing undergoes roughly

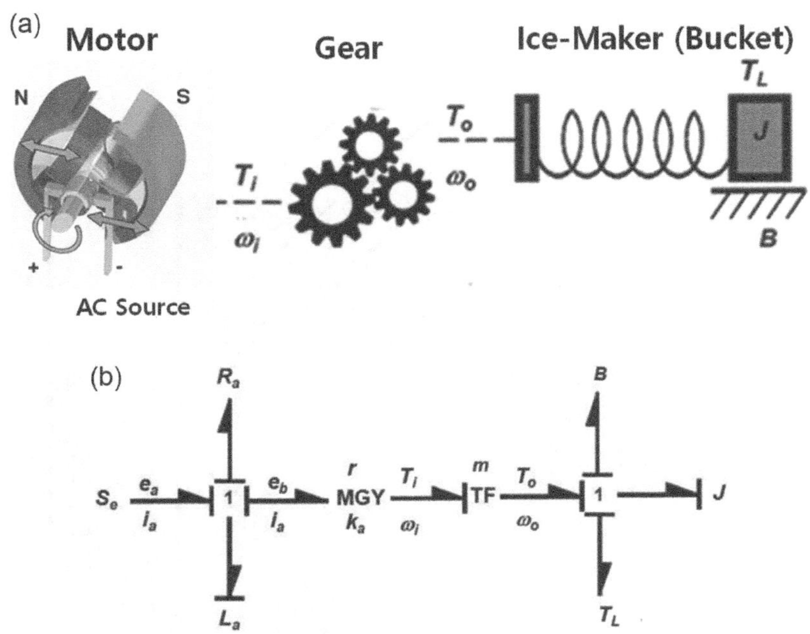

FIGURE 8.53
Design idea of an ice-maker: (a) schematic diagram of an ice-making system; (b) bond graph representation,

65,700 usage cycles. Manufacturer data shows that the normal torque is 0.69 kN-cm and the maximum torque is 1.47 kN-cm. Assuming the damage factor λ is 2, AF is approximately 5 in Equation (8.36).

For B1 life of 10 years, if the shape parameter was assumed to be 2.0, the allotted mission cycles for ten samples calculated from Equation (7.43) were approximately 42,000 cycles. This ALT is designed to ensure a B1 life of ten years if the helix upper dispenser is unsuccessful less than once in 42,000 cycles.

8.11.1 Results and Conclusions

In the 1st ALT, the cracked helix upper dispenser found at 170, 5,200, 7,880, 8,800, and 11,600 cycles. Figure 8.54 indicates a picture comparing the fractured components from the market and that from the 1st ALT. As they are alike, the helix upper dispenser failed from the market might be reproduced by ALT.

There was a design defect—a 2 mm gap. As the blade dispenser made of stainless-steel stroke the helix upper dispenser made of plastic, it finally might be cracking and fracturing. To withstand repetitive impact loads, the helix upper dispenser failed from the field was altered: the 2 mm gap was eliminated (Figure 8.55).

In the 2nd ALT, the helix upper dispensers were unsuccessful at 17,000, 25,000, 28,200, and 38,000 cycles. In the 1st ALT, though the gap was removed and the life of the helix upper dispenser increased, the helix upper dispenser could not have adequate strength because 42,000 mission cycles into the 2nd ALT still wasn't achieved. As a plan of action, a rib on the outside of the helix was strengthened. In the 3rd ALT, there were no issues until 75,000 cycles. During three rounds of ALTs with the modifications, the ice-maker involving the helix upper dispenser was ensured to have B1 life of 10 years.

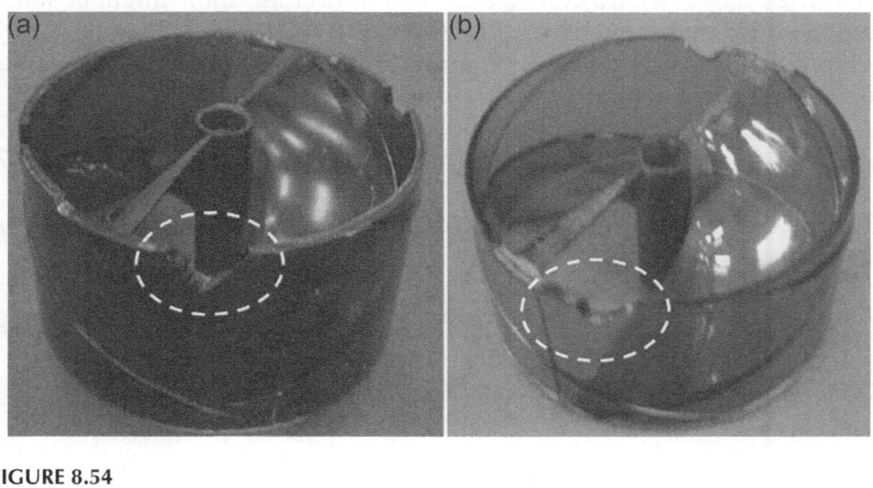

FIGURE 8.54
(a) Fractured parts from the field; (b) products fractured after the 1st ALT.

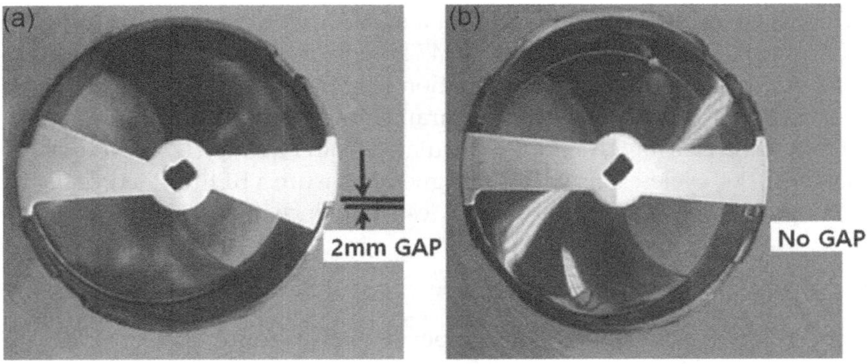

FIGURE 8.55
Modified helix upper dispenser: (a) original design; (b) modified design.

8.12 Increasing the Fatigue Lifetime of a Refrigerator Drawer

End users utilize the refrigerator to keep food fresh during its life. To fulfill this requirement, the drawer in the refrigerator is designed. Storing food in a household refrigerator entails: 1) opening the drawer in the refrigerator, 2) putting stocked food into it, and 3) closing the drawer. The drawer must be designed to withstand the functioning situations exerted on it by the end user (Figure 8.56).

From the field, as repeated loading of drawer was exerted, drawer had been failing due to design defects, compelling consumers to request replacement. For the drawer system to work properly for expected life, design defects such as a stress raiser—sharp corner, etc.—in a drawer assembly might be found and modified by parametric ALT (Figure 8.57).

From the free body diagram (Figure 8.57 (b)), it is known that the applied load on the refrigerator drawer originated from the food weight:

$$F_{\text{box}} = \mu W_{\text{load}} \tag{8.37}$$

Because the drawer stress depends on the food loads, Equation (7.29) can be defined:

$$TF = A(S)^{-n} = A(F_{\text{box}})^{-\lambda} = A(\mu W_{\text{load}})^{-\lambda} = B(W)^{-\lambda} \tag{8.38}$$

Thus, based on Equation (7.30 and 7.31), the AF can be formulated:

$$AF = \left(\frac{S_1}{S_0}\right)^n = \left(\frac{W_1}{W_0}\right)^\lambda \tag{8.39}$$

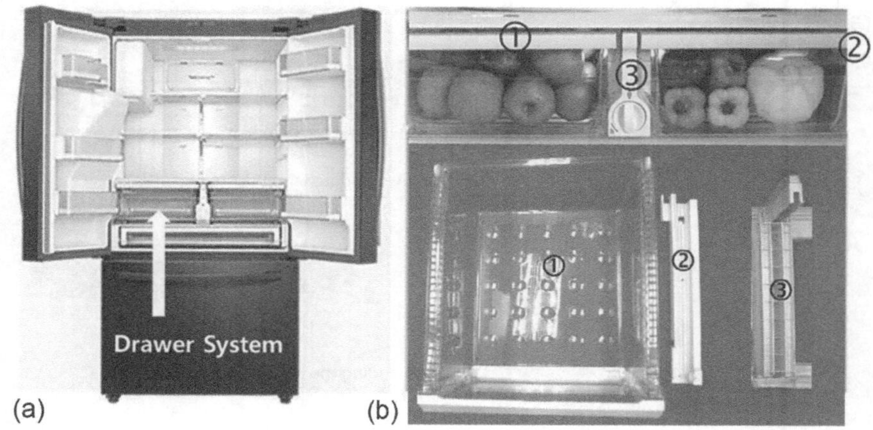

FIGURE 8.56
Refrigerator and drawer system: (a) refrigerator; (b) drawer system: ① box, ② rail, and ③ support.

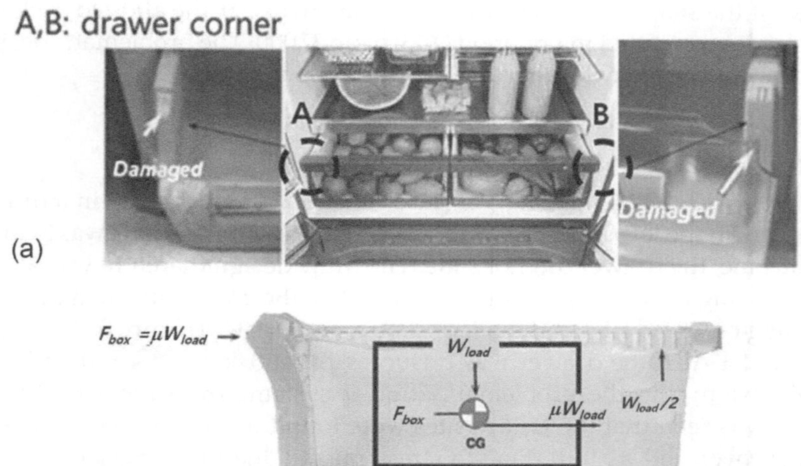

FIGURE 8.57
Fractured drawer after usage: (a) failed drawer in the field; (b) free body diagram.

For the refrigerator drawer, the functional environmental conditions are 0–43°C with 0.2–0.24 g acceleration, and a humidity ranging from 0 to 95%. The daily functioning cycles of the drawer were from 5 to 10 cycles. Established on the design of a product life for ten years, the drawer was applied to 36,500 usage cycles. For the most severe condition, the greatest force exposed to the customer in storing food W_1 was 0.059 kN (6 kg$_f$). For ALT, the applied force W_2 was doubled to 0.117 kN (12 kg$_f$). With an accumulative exponent λ

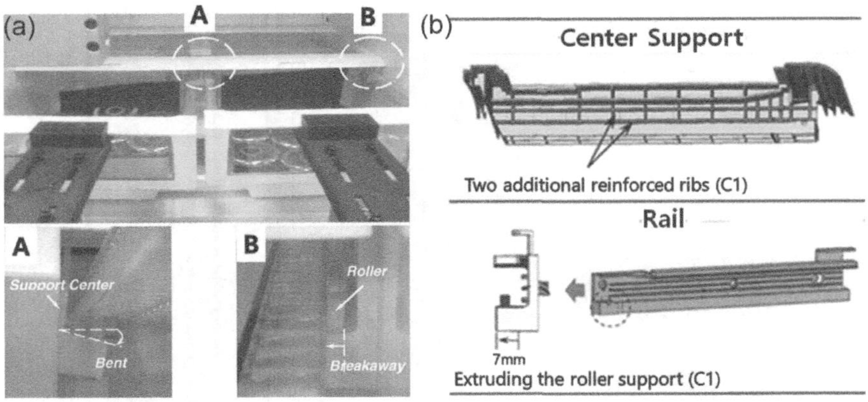

FIGURE 8.58
(a) Design alternations of drawer in initial loading; (b) modified designs in drawer.

of 2, the AF from Equation (8.38) was 4.0. To fulfill the life aim—B1 life of 10 years—if the shape parameter was presumed to be 2.0, the allotted cycles for six samples computed in Equation (7.43) were 37,000. The problematic designs of —drawer thus may be altered after recognizing them.

8.12.1 Results and Conclusions

At the 1st ALT, 1.17 kN (12 kg_f) of raised food load was exerted in a drawer, the rollers in the left/right rail were taken away, and the support was bent. As an outcome, the drawer did not slide. The draw design therefore was altered by increasing the roller support to 7 mm, C2, in the guide rail and adding the fastening enforced ribs in the rail of center support, C1 (Figure 8.58).

In the 1st ALT, the drawer was unsuccessful at 3,800 and 4,800 cycles. By closely examining the problematic drawer systems from the 1st ALT and field, it was seen that the failure sites were found in the intersections of the drawer cover and its hull. Due to stress raisers due to no cornering in the intersection regions (a), the drawer system had been failing the drawer cover. These designs were altered by (1) increasing strengthened ribs in terms of thickness, Rib 1, C3, from T2.0 to T3.0 mm; (2) fastening the fillets, Fillet 1, C4, from R0.0 to R1.0 mm (Figure 8.59).

In the 2nd ALT, it was known that the roller was sunken and the guide rail in center support was cracked at 6,000 cycles. Cautiously examining the failure places made it clear that the guide rail in the construction of a drawer had no strengthened rib and inadequate corner rounding to hold out against the loading. To improve the design of the guide rail, it was modified by (1) fastening strengthened ribs, C5; (2) increasing the rounding, C6, from R3 to R4 mm. The center support in the structure of a drawer also was altered by (3) increasing the roller rib, C7, from L0.0 to L2.0 mm (Figure 8.60).

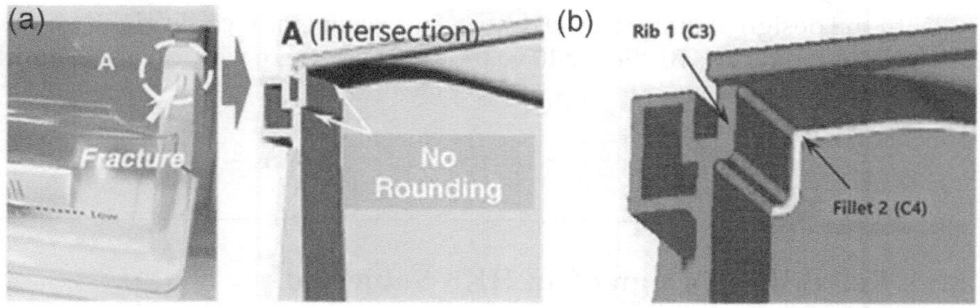

FIGURE 8.59
Problematic designs of drawer in 1st ALT: (a) original design; (b) modified design.

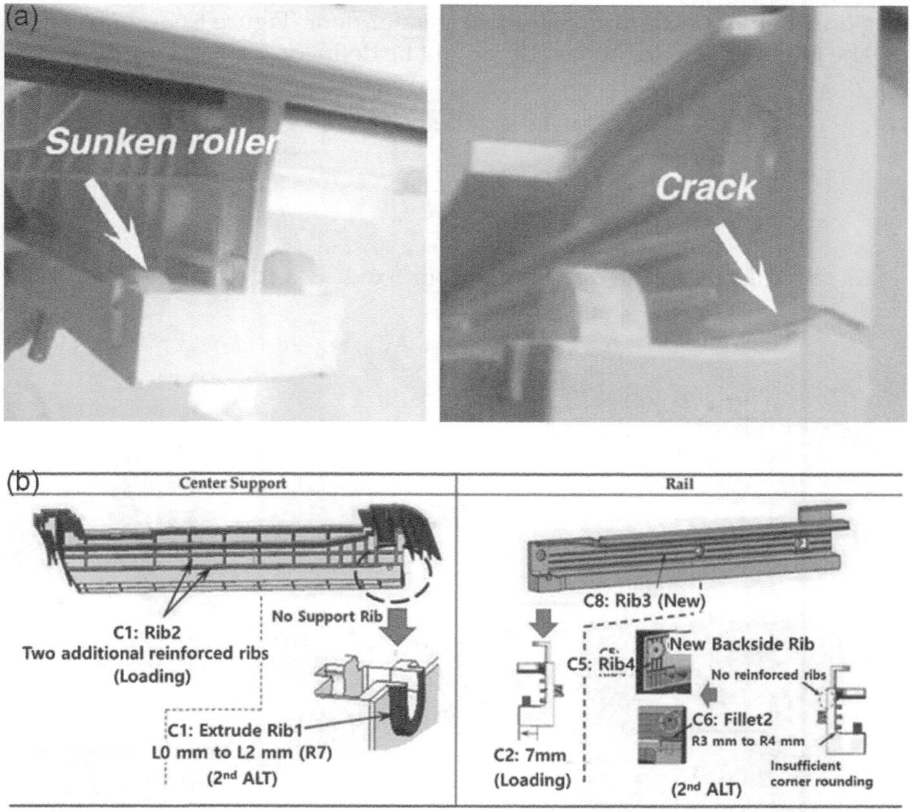

FIGURE 8.60
Design issues of refrigerator drawer in second ALT: (a) failed roller of the support rail and cracked guide rail; (b) altered center support and rails.

With the design alterations, the drawer in a refrigerator could rightly work to reach the life aim—B1 life for 10 years—because there were no issues until 19,000 mission cycles.

8.13 Reliability Design of an HKS Subjected to Repeated Stresses

As a customer uses the refrigerator, one wants to comfortably close the door reinforce. For this aim, the hinge kit system (HKS) is newly designed in a household refrigerator (Figure 8.61).

From the marketplace, HKS components in a household refrigerator were unsuccessful in fulfilling the lifetime aim due to design defects. Established on the end user functioning conditions, HKS were exposed to different loading during the functioning of the refrigerator door (Figure 8.62).

The moment balance at the HKS might be defined:

$$M_0 = W_{\text{door}} \times b \tag{8.40}$$

$$(8.40) = T_0 = F_0 \times R \tag{8.41}$$

As raised loads on the refrigerator door fixed:

$$\sum M = M_1 = M_0 + M_A = W_{\text{door}} \times b + W_A \times a = T_1 = F_1 \times R \tag{8.42}$$

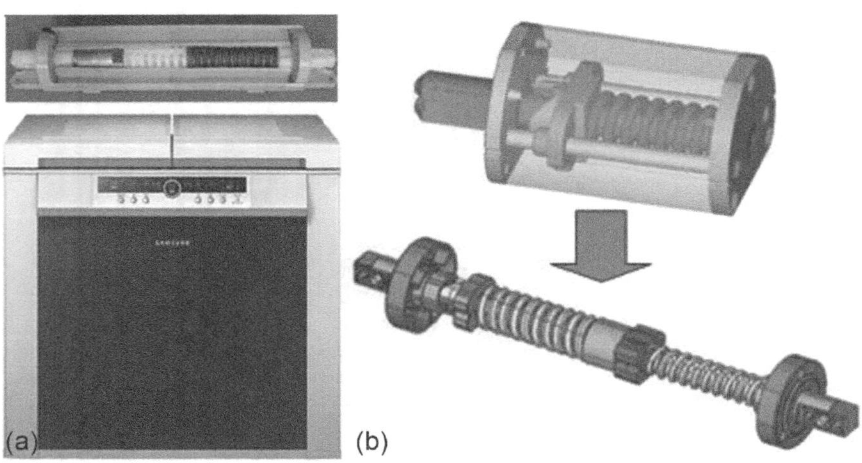

FIGURE 8.61
(a) Domestic refrigerator; (b) HKS.

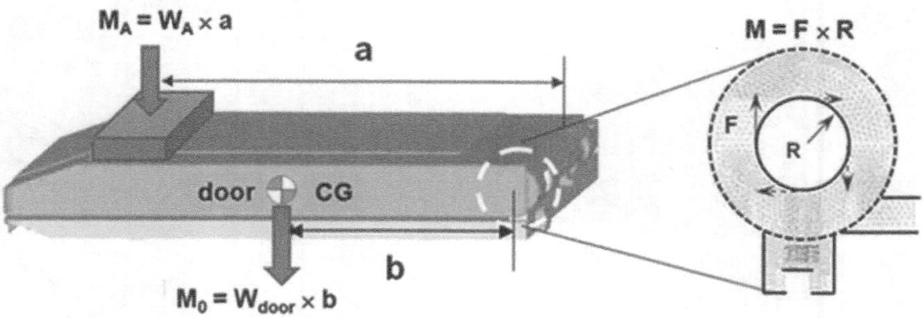

FIGURE 8.62
Design ideas of HKS.

$$(8.42) = T_1 = F_1 \times R \tag{8.43}$$

Under the functioning circumstances, the life-stress model from Equation (7.29) can be expressed:

$$TF = A(S)^{-n} = AT^{-\lambda} = A(F \times R)^{-\lambda} = B(F)^{-\lambda} \tag{8.44}$$

It is recognized that the product life relies on exerted impact load. Thus the acceleration factor (AF) in Equations (7.30) and (7.31) can be expressed:

$$AF = \left(\frac{S_1}{S_0}\right)^n = \left(\frac{T_1}{T_0}\right)^\lambda = \left(\frac{F_1 \times R}{F_0 \times R}\right)^\lambda = \left(\frac{F_1}{F_0}\right)^\lambda \tag{8.45}$$

As the life of the new HKS was aimed to be B1 life of 10 years, the number of mission test cycles attained for allotted samples in Equation (7.43) is 24,000. In ALTs, the troublesome designs of HKS had to be identified to achieve the life aim.

8.13.1 Results and Conclusions

In the 1st ALT, the housing of the HKS was unsuccessful at 3,000 cycles. When comparing the fractured products from the field and the 1st ALT, it was known that the look and arrangement of the failure in the 1st ALT were alike to those attained from the market (Figure 8.63).

The failed HKS in both the market and the 1st ALT test happened at its housing. It was known that the design defects of the HKS originated from the construction in its housing where there are no supporting ribs. To obtain the adequate strength against repetitive loading, the troublesome HKS was altered by fastening inside supporting ribs (Figure 8.64).

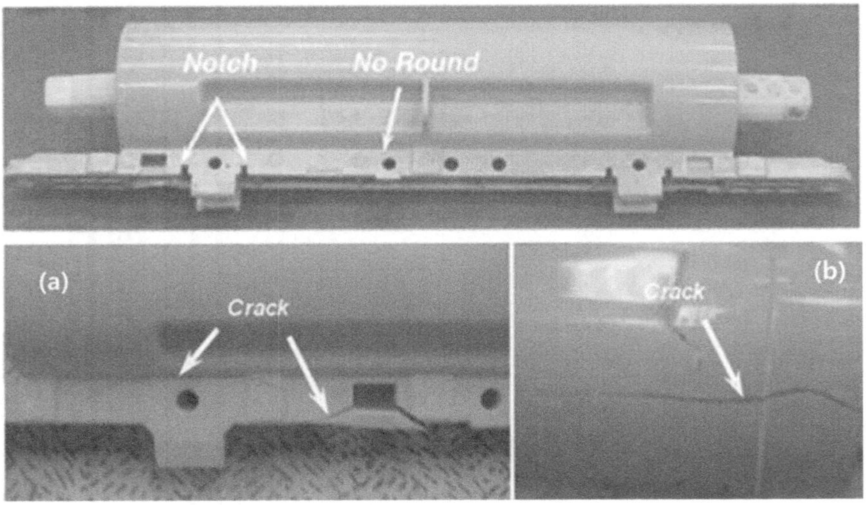

FIGURE 8.63
(a) Failed products from the marketplace; (b) fracture1st ALT.

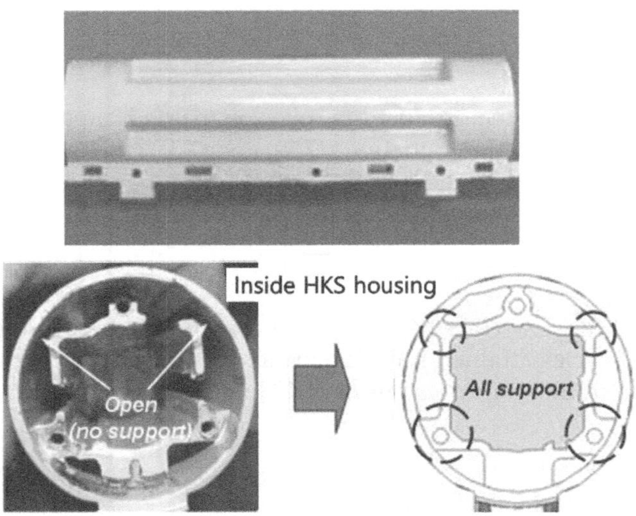

FIGURE 8.64
Redesigned HKS housing structure.

In the 2nd ALT, the cracked torsional shaft in HKS failed at 12,000 cycles. It was known that the basic cause of the failed torsional shaft started from its adequate rounding due to repetitive stress. As an action plan to give the adequate strength for repeated impact loads, its torsional shaft was redesigned to give it more rounding from R0.5 to R2.0 mm (Figure 8.65).

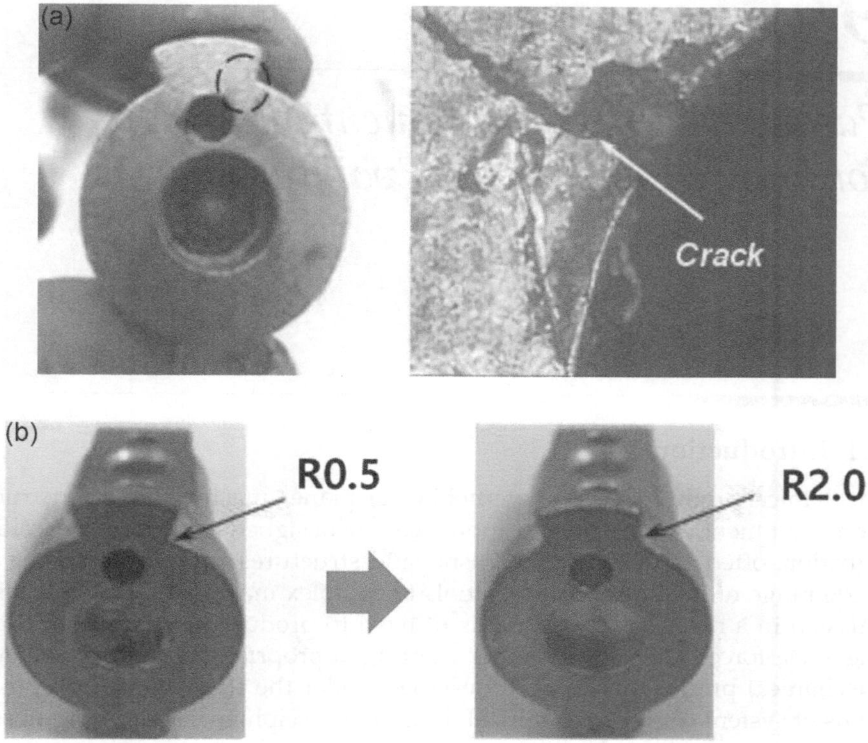

FIGURE 8.65
Redesigned torsional shaft: (a) failed torsional shaft; (b) modified torsional shaft.

With the design alterations, the HKS in an household refrigerator could satisfactorily work to achieve the life aim—B1 life of 10 years—during its life because there were no problems until 24,000 assigned mission cycles. Over the round of three ALTs, the life of the HKS was guaranteed to have B1 life of 10 years.

9

Parametric ALT: An Influential Device for Future Engineering Evolution

9.1 Introduction

Mechanical products such as automobiles, airplanes, machine tools, construction equipment, and household appliances are designed to perform a specific function, often involving mechanisms and structures and encompassing a wide range of items from simple tools to complex machinery. Power transmission in a mechanical product is utilized to produce mechanical advantages via forces and movement by exerting appropriate mechanisms. New mechanical products have been designed under the fundamental propositions of system engineering (SE). SE is an interdisciplinary area of engineering on how complicated engineering projects might be designed to run over product life cycles. Based on user requirements, SE will develop numerous functions and architecture. Issues—reliability, logistics, coordination of dissimilar teams, evaluation measurements, and dissimilar disciplines—become harder when carrying out big and complicated projects. In SE, all features of a project are contemplated and combined into an entire product system (Figure 9.1).

9.2 Parametric ALT as a Quantitative Assessment of Product

Because of hostile forces in the global marketplace, manufacturers have to regularly develop new. After being combined into the product's design specifications, they are quickly developed, combined into the system, and shipped to the marketplace. Under these unknown environmental/working circumstances in the field, there may not be enough time to test them in the space provided in scientific facilities. If a significant amount of technology failures occur in the marketplace, the manufacturer's brand and business can be impacted negatively.

DOI: 10.1201/9781003451471-9

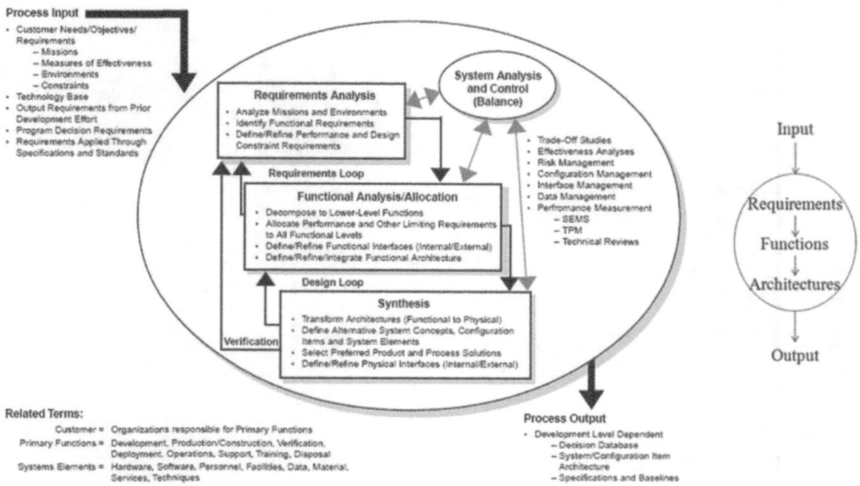

FIGURE 9.1
Systems engineering process [1].

Engineers should think about reliability as a critical factor in designing the product. The fundamental question is how to think about the reliability idea in the design process. The organized development process of product in a company may be mostly arranged in categories as research and development (R&D) and quality assurance (QA). R&D activities is a core portion of the modern company because crucial manufacturer design decisions are carried out by departments with personnel specializing in relevant techniques based on their technical levels. On the other hand, QA assesses and verifies a (detailed) design of product by each manufacturer's specifications that can be explained as perception quantity of the customer's expectations.

Two typical design tools can be distinguished as qualitative and quantitative methods in product development. The traditional qualitative methods—design review, FMEA, FTA, and capability maturity model integration (CMMI)—are judged by looking for the design issues on the distributed documents whether the design qualitatively accomplishes the product specifications. In the planning, design, and production of product, they only perform to collect the past experience or design plans by the representatives. However, they often do not find a critical data such as design flaws in the design phase. From a viewpoint of quantitative quality, there is no established methodology in the design process to achieve the reliability targets. These organized procedures consist of 1) allocation of the reliability target, 2) reliability testing and Weibull analysis, 3) design feedback, and 4) the examination of the market failure data (Figure 9.2).

As a form of reliability testing, the parametric ALT might be a quantitative way to find out the design defects in the developing process because it

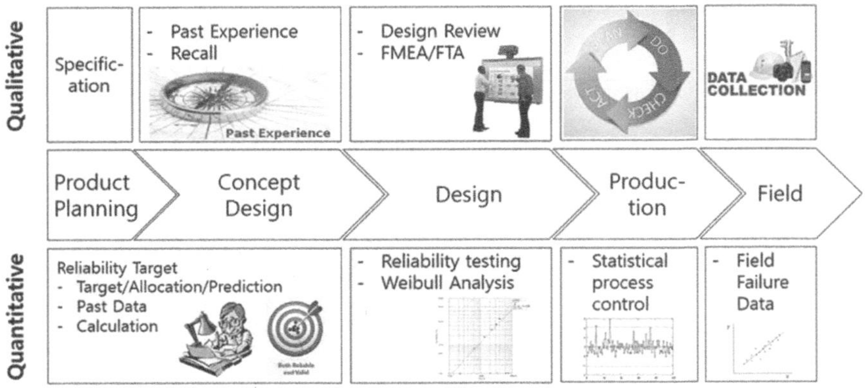

FIGURE 9.2
New product development process including reliability concept.

utilizes the ALT scheme, load examination, and elevated testing. Under system engineering, all mechanical products that are created from a multiple of structure will develop the customer-oriented intended functions, which may tend to deteriorate or stop working suddenly by (random) loads in the marketplace. As mechanical products are exposed to random loads, they might originate from the design defects, spread, and fracture it. If failure such as fatigue or fracture happens, the product cannot formally satisfy its required operation. To avoid failure, the mechanical product might be designed to identify the failure mode through testing over time and robustly endure a number of different loads in a life based on reliability testing, such as parametric ALT. The fundamental ideas of parametric ALT are as follows: 1) placing the whole parametric ALT scheme for product; 2) failure mechanics, design, and load analysis; 3) parametric ALT with the design modifications; and 4) judgment whether the designs achieve the reliability target (see Chapters 7 and 8) (Figure 9.3).

Utilizing this organized reliability method can resolve design problems such as fatigue, vibration and noise, etc. This methodology may be extended to other mechanical systems such as automobiles, airplanes, construction equipment, etc. These fundamental propositions of parametric ALT are also relevant to the region of civil engineering to design structures such as bridges. To enhance the reliability of these systems, the design defects need to be found and altered. That is, if there are problems in the structure, the product may be unsuccessful during its anticipated lifetime. By performing parametric ALT, engineers may replicate the failure and alter it. These organized reliability methods also supply the reliability quantitative (RQ) specifications. As an outcome, the reliability-embedded design procedure will put aside the cost of design alteration because the number of design problem decreases in the field (Figure 9.4).

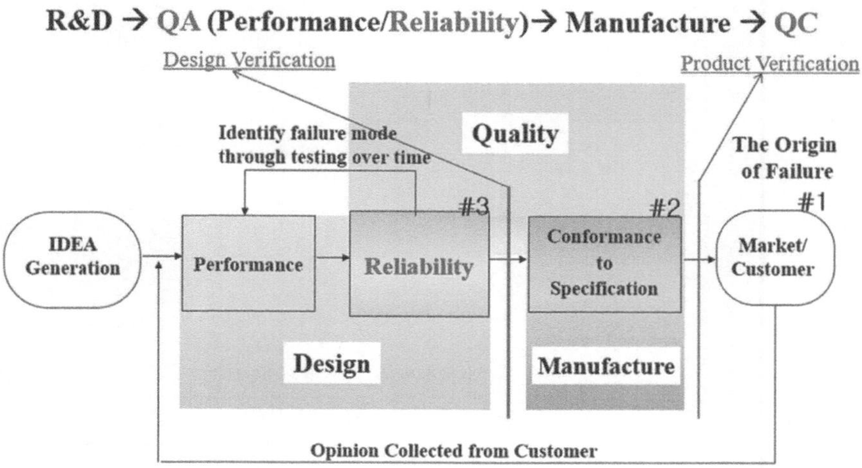

FIGURE 9.3
Basic quality for creating a first-class product.

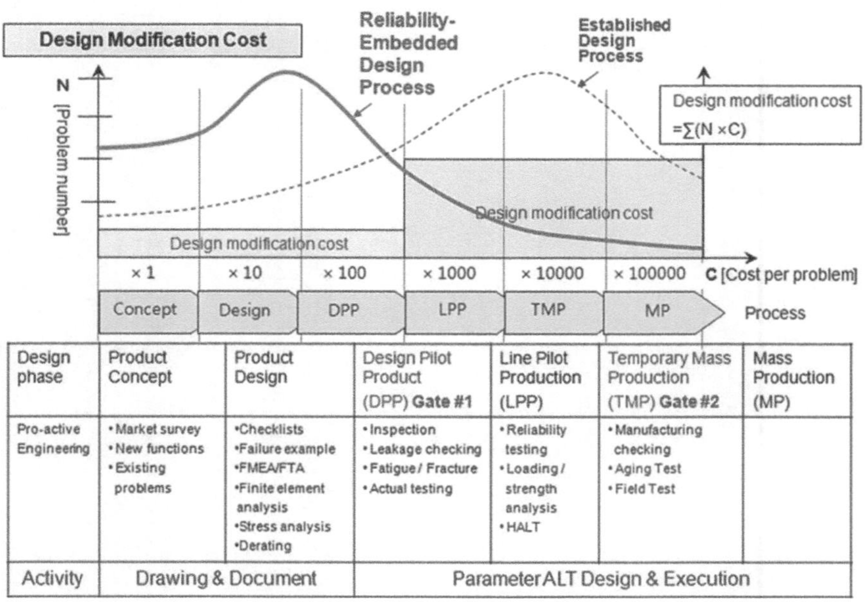

FIGURE 9.4
Reliability-embedded design process.

References

1. Defense of Department (2001) *Systems Engineering Fundamentals*. Defense Acquisition University Press: Fort Belvoir, VA, p. 31.

Index

R

S

T

V

W